SUGARCANE CULTIVATION AND MANAGEMENT

Challenges and Opportunities

SUGARCANE CULTIVATION AND MANAGEMENT

Challenges and Opportunities

Krishan K. Verma, PhD
Xiu-Peng Song, PhD
Munna Singh, PhD
Renato de Mello Prado, PhD
Jian-Ming Wu, PhD
Yang-Rui Li, PhD
Editors

AAP | APPLE ACADEMIC PRESS

First edition published 2025

Apple Academic Press Inc.
1265 Goldenrod Circle, NE,
Palm Bay, FL 32905 USA

760 Laurentian Drive, Unit 19,
Burlington, ON L7N 0A4, CANADA

CRC Press
2385 NW Executive Center Drive,
Suite 320, Boca Raton FL 33431

4 Park Square, Milton Park,
Abingdon, Oxon, OX14 4RN UK

Apple Academic Press exclusively co-publishes with CRC Press, an imprint of Taylor & Francis Group, LLC

Library and Archives Canada Cataloguing in Publication

Title: Sugarcane cultivation and management : challenges and opportunities / Krishan K. Verma, PhD, Xiu-Peng Song, PhD, Munna Singh, PhD, Renato de Mello Prado, PhD, Jian-Ming Wu, PhD, Yang-Rui Li, PhD, editors.
Names: Verma, Krishan K., editor.
Description: First edition. | Includes bibliographical references and index.
Identifiers: Canadiana (print) 20240466489 | Canadiana (ebook) 20240466497 | ISBN 9781774917091 (hardcover) | ISBN 9781774917176 (softcover) | ISBN 9781003504122 (ebook)
Subjects: LCSH: Sugarcane. | LCSH: Sugarcane industry—Management.
Classification: LCC SB231 .S84 2025 | DDC 633.6/1—dc23

Library of Congress Cataloging-in-Publication Data

CIP data on file with US Library of Congress

ISBN: 978-1-77491-709-1 (hbk)
ISBN: 978-1-77491-717-6 (pbk)
ISBN: 978-1-00350-412-2 (ebk)

About the Editors

Krishan K. Verma, PhD
Foreign Expert, Sugarcane Research Institute, Guangxi Academy of Agricultural Sciences, Nanning, Guangxi, China

Krishan K. Verma, PhD, is working as a Foreign Expert at the Sugarcane Research Institute, Guangxi Academy of Agricultural Sciences/Sugarcane Research Center, Chinese Academy of Agricultural Sciences/Guangxi Key Laboratory of Sugarcane Genetic Improvement/Key Laboratory of Sugarcane Biotechnology and Genetic Improvement (Guangxi) of the Ministry of Agriculture and Rural Affairs in Nanning, Guangxi, China. He has 14 years of research experience in the field of biotic and abiotic stresses, such as environmental toxicology, plant physiology, bioengineering, and nanobioscience. His research focuses on environmental toxicology, bioengineering modeling, plant physiology, and molecular biology, and their impacts on growth, physiological, and molecular adaptation strategies of plants to stressful environments and how they affect plant structure. He is working with Prof. Yang-Rui Li, a world-renowned Plant Scientist. He was also conferred the prestigious Award of Excellence for outstanding research in agricultural sciences (2021). Prof. Bruce Spittle, Editor-in-Chief of *Fluoride* in New Zealand, rated our paper (46:83-89) as excellent in 2013. He has published more than 130 scientific research articles, including books and book chapters. He is an internationally recognized reviewer and has been serving as an editorial board member of various peer-reviewed journals.

Xiu-Peng Song, PhD
Associate Professor, Sugarcane Research Institute, Guangxi Academy of Agricultural Sciences, Nanning, Guangxi, China

Xiu-Peng Song, PhD, is working as an Associate Professor at Sugarcane Research Institute, Guangxi Academy of Agricultural Sciences/Sugarcane

Research Center, Chinese Academy of Agricultural Sciences/Guangxi Key Laboratory of Sugarcane Genetic Improvement/Key Laboratory of Sugarcane Biotechnology and Genetic Improvement (Guangxi) of Ministry of Agriculture and Rural Affairs, Nanning, Guangxi, China. His ongoing research focuses on disease resistance breeding and molecular basics, and comprehensive control of diseases and insect pests in sugarcane. He received his Master's and Ph.D. (Biochemistry and molecular biology) from Guangxi University, Guangxi, China. He presided over 6 projects, such as the National Natural Science Foundation China, and participated in 16 projects, such as national and international cooperation. He won 8 awards, such as the Guangxi Science and Technology Progress Award. He has published 66 scientific publications, 3 book chapters, and 1 book from Springer.

Munna Singh, PhD
Professor and Head, Botany Department,
Chairman Board of Studies, Agriculture,
University of Lucknow, Lucknow, India

Munna Singh, PhD, served as Vice-Chancellor at Chandra Shekher Azad University of Agriculture and Technology, Kanpur, India (2013-2016). He is currently working as Professor and Head, Botany Department, and Chairman of the Board of Studies, Agriculture, University of Lucknow, Lucknow, India, after serving at G.B. Pant University of Agriculture and Technology, Pantnagar, for nearly two decades, as Professor and Head, Plant Physiology. He is a Fellow of the Japan Society for the Promotion of Science. The Department of Science and Technology, Ministry of Science & Technology, Govt. of India, awarded him with a Better Opportunities for Young Scientists in Chosen Areas of Science and Technology (BOYSCAST) Fellowship to pursue postdoctoral research at the University of California, USA (1998). He was also conferred with other various prestigious awards, including a Jawahar Lal Nehru National Award of ICAR for outstanding research in agricultural sciences (1991), Honorary Fellow of the Indian Society of Biochemist (2013), Saraswati Sammaan (2014) for outstanding contribution in education, etc. He has completed nearly a dozen research projects and published over 100 research papers in national and international journals of high repute. He has published seven books based on the doctoral research work of his students. He has guided over 50 PhD and postgraduate students. He completed his MSc in Botany (1985) and his doctoral research on "Photosynthetic characteristics of Populus deltoides under light and temperature stress including chloroplast genome (cpDNA) isolation and characterization" at the University of Lucknow.

Renato de Mello Prado, PhD
Professor of Plant Nutrition, São Paulo State University, São Paulo, Brazil

Renato de Mello Prado, PhD, is currently working as a Professor of Plant Nutrition at São Paulo State University, São Paulo, Brazil. He also coordinates a research group (Genplant) focused on the nutrition of tropical plants. He has published over 500 articles in peer-reviewed journals and has authored/co-authored about 30 academic books. He holds a PhD in Agronomy from São Paulo State University, Brazil, and completed his postdoctoral program at the University of Cordoba and the Public University of Navarra, Spain. He coordinated dozens of national and international research projects. These projects provided guidance/co-supervision for more than one hundred students (Masters/Doctorate) and thirty postdocs from Brazil and other countries.

Jian-Ming Wu, PhD
Professor, Sugarcane Research Institute, Guangxi Academy of Agricultural Sciences, Nanning, Guangxi, China

Jian-Ming Wu, PhD, is working as a Professor at the Sugarcane Research Institute, Guangxi Academy of Agricultural Science/Sugarcane Research Center, Chinese Academy of Agricultural Sciences/Guangxi Key Laboratory of Sugarcane Genetic Improvement/Key Laboratory of Sugarcane Biotechnology and Genetic Improvement (Guangxi) of Ministry of Agriculture and Rural Affairs, Nanning, Guangxi, China. His ongoing research focuses on comprehensive control of physiology, biochemistry, and molecular biology in sugarcane crop. He has presided over 10 projects, such as the National Natural Science Foundation of China, and participated in 20 projects, such as a national international cooperation project, and won four awards, such as a Guangxi Science and Technology Progress Award. He has published more than 120 scientific publications with five granted invention patents. He received his master's and doctoral degrees in Agronomy from Guangxi University, Nanning, Guangxi, China.

Yang-Rui Li, PhD
Director, Sugarcane Research Center, Chinese Academy of Agricultural Sciences, Nanning, Guangxi, China

Yang-Rui Li, PhD, is a lifetime Professor and has been serving as Director of the Sugarcane Research Center, Chinese Academy of Agricultural Sciences, since September 2007. He is Chief Expert of the National Joint Research Program for Elite Sugarcane Variety Development in China (since 2018). He worked in various capacities, including as Professor in the Department of Agronomy at Guangxi Agricultural University; Vice-President of Guangxi University; President of Guangxi Academy of Agricultural Sciences, etc. He is President of the International Association for Professionals of Sugar and Integrated Technologies (IAPSIT); Vice-President of the Society for Sugar Research and Promotion (SSRP); and President of the Chinese Sugarcane Industry Association for Technological Innovation (CSIATI). His research interests include sugarcane physiology, biochemistry, and molecular biology, sugarcane genetics, breeding, and cultivation, and chemical regulation of growth and development of sugarcane. He has published 16 books and more than 1000 papers and received 23 scientific research achievement awards from the Chinese government to date as well as 14 awards from international organizations, including a Lifetime Achievement Award—Society for Sugar Research and Promotion (2011), Lifetime Achievement Award—International Association of Professionals in Sugar and Integrated Technologies (2014), and Leadership Excellence Award from Thailand Society of Sugarcane and Sugar Technologists (2018). He earned his BA degree at the Department of Agronomy, Guangxi Agricultural University, China, and MSc and PhD degrees from the Department of Agronomy, Fujian Agricultural University, China, respectively.

Contents

Contributors

Shahid Afghan
Sugarcane Research and Development Board, Ayub Agricultural Research Institute, Faisalabad, Punjab, Pakistan

Musa Opeyemi Ahmed
Nigeria Sugar Institute, Ilorin, Kwara State, Nigeria; Department of Agricultural and Environmental Engineering, University of Ibadan, Oyo, Nigeria

V. Anjaly
Department of Agronomy, Punjab Agricultural University, Ludhiana, Punjab, India

Tamilselvan Arumugam
ICAR–National Institute of Abiotic Stress Management, Baramati, Pune, Maharashtra, India

Marissa Corinna Asen
School of Law and Social Sciences, University of the South Pacific, Suva, Fiji

Karishma Behera
Department of Plant Breeding and Genetics, Jawaharlal Nehru Krishi Vishwavidyalaya, Jabalpur, Madhya Pradesh, India

Gayani K. M. Chanchala
Sugarcane Research Institute, Udawalawe, Sri Lanka

Milton Garcia Costa
São Paulo State University (UNESP), School of Agricultural and Veterinarian Sciences, Jaboticabal, Brazil

Munmun Dash
Tamil Nadu Agricultural University, Coimbatore, Tamil Nadu, India

Mohseen Riaz Ud Dean
Anthropology Program, Faculty of Arts and Social Sciences, University of Waikato, Hamilton, New Zealand

Sunita Devi
ICAR–Central Soil Salinity Research Institute, Karnal, Haryana, India

Pooja Dhansu
ICAR–Sugarcane Breeding Institute, Regional Center, Karnal, Haryana, India

Praveen Kumar Divvela
Contec Global Agro Limited, Abuja, Nigeria

Faith Eweluegim Enahoro-Ofagbe
Nigeria Sugar Institute, Ilorin, Kwara State, Nigeria; Department of Agronomy, University of Ilorin, P.M.B., Ilorin, Kwara, Nigeria

Vinod Kumar Garg
College of Agriculture, Jawaharlal Nehru Krishi Vishwavidyalaya, Ganj Basoda, Vidisha, Madhya Pradesh, India

Vinay Hegde
ICAR–National Institute of Abiotic Stress Management, Baramati, Pune, Maharashtra, India

Adepoju Moronkola Idris
Nigeria Sugar Institute, Ilorin, Kwara State, Nigeria; Department of Agronomy, University of Ilorin, P.M.B., Ilorin, Kwara, Nigeria

Kingsley O. Iwuozor
Nigeria Sugar Institute, Ilorin, Kwara State, Nigeria; Department of Pure and Industrial Chemistry, Nnamdi Azikiwe University, P.M.B., Akwa, Nigeria

M. Jagadesh
Tamil Nadu Agricultural University, Coimbatore, Tamil Nadu, India

Jonas Pereira de Souza Júnior
São Paulo State University (UNESP), School of Agricultural and Veterinarian Sciences, Jaboticabal, Brazil

Muhammad Ehsan Khan
Sugarcane Research and Development Board, Ayub Agricultural Research Institute, Faisalabad, Punjab, Pakistan

A. Krishnaveni
Tamil Nadu Agricultural University, Coimbatore, Tamil Nadu, India

Ashwani Kumar
ICAR–Central Soil Salinity Research Institute, Karnal, Haryana, India

Ravinder Kumar
ICAR–Sugarcane Breeding Institute, Regional Center, Karnal, Haryana, India

Aradhana Kumari
Department of Plant Physiology, College of Agriculture, Jawaharlal Nehru Krishi Vishwavidyalaya, Ganj Basoda, Vidisha, Madhya Pradesh, India

Mukesh Kumar Malviya
Institute of Biological Science, SAGE University, Indore, Madhya Pradesh, India

Eid. M. Mehareb
Sugar Crops Research Inst., ARC, Giza, Egypt

P. K. Mishra
Department of Agronomy, College of Agriculture, Jawaharlal Nehru Krishi Vishwavidyalaya, Ganj Basoda, Madhya Pradesh, India

Amin Nikpay
Department of Plant Protection, Sugarcane and By-Products Development Company, Salman Farsi Agro-Industry, Ahwaz, Iran

Kamilla Silva Oliveira
São Paulo State University (UNESP), School of Agricultural and Veterinarian Sciences, Jaboticabal, Brazil

Satpal S. Paalli
Department of Agronomy, Punjab Agricultural University, Ludhiana, Punjab, India

S. K. Pandey
ICAR–Sugarcane Breeding Institute, Regional Center, Karnal, Haryana, India

Adhini S. Pazhani
ICAR–Sugarcane Breeding Institute, Coimbatore, Tamil Nadu, India

Renato de Mello Prado
São Paulo State University (UNESP), School of Agricultural and Veterinarian Sciences, Jaboticabal, Brazil

Rahul Ravneel Prasad
Pacific Center for Environment and Sustainable Development, University of the South Pacific, Suva, Fiji

Kumar Preet
ICAR–Sugarcane Breeding Institute, Regional Center, Karnal, Haryana, India; Maharishi Markandeshwar (Deemed to be) University, Mullana, Haryana, India

Jagadish Rane
ICAR–National Institute of Abiotic Stress Management, Baramati, Pune, Maharashtra, India

R. K. Samaiya
Department of Plant Physiology, Jawaharlal Nehru Krishi Vishwavidyalaya, Jabalpur, Madhya Pradesh, India

Sudip Sarkar
Department of Soil Science, ICAR-RCER, Patna, Bihar, India

Prachi Sharma
Institute of Biological Science, SAGE University, Indore, Madhya Pradesh, India

Rajni Sharma
Department of Agronomy, Punjab Agricultural University, Ludhiana, Punjab, India

Gilmara Pereira da Silva
São Paulo State University (UNESP), School of Agricultural and Veterinarian Sciences, Jaboticabal, Brazil

Bhaskar Pratap Singh
Agricultural Engineering, ANDUAT–Krishi Vigyan Kendra, Amethi, Uttar Pradesh, India

Santosh Kumar Singh
Dr. Rajendra Prasad Central Agricultural University, Pusa, Samastipur, Bihar, India

Trisha Sinha
Narayan Institute of Agricultural Sciences, Gopal Narayan Singh University, Jamuhar, Sasaram, Rohtas, Bihar, India

Anjali Chandrol Solanki
Department of Agriculture Science, Mansarovar Global University, Bhopal, Madhya Pradesh, India

Manoj Kumar Solanki
Plant Cytogenetics and Molecular Biology Group, Institute of Biology, Biotechnology, and Environmental Protection, Faculty of Natural Sciences, University of Silesia in Katowice, Katowice, Poland

Cherukumalli Srinivasarao
ICAR–National Academy of Agricultural Research Management (NAARM), Hyderabad, Telangana, India

Gelza Carliane Marques Teixeira
São Paulo State University (UNESP), School of Agricultural and Veterinarian Sciences, Jaboticabal, Brazil

Krishan K. Verma
Sugarcane Research Institute, Guangxi Academy of Agricultural Sciences/Key Laboratory of Sugarcane Biotechnology and Genetic Improvement (Guangxi), Ministry of Agriculture and Rural Affairs/Guangxi Key Laboratory of Sugarcane Genetic Improvement, Nanning 530007, Guangxi, China

R. G. Vyshnavi
Department of Plant Physiology, Jawaharlal Nehru Krishi Vishwavidyalaya, Jabalpur, Madhya Pradesh, India

VKASM Wanasinghe
Sugarcane Research Institute, Udawalawe, Sri Lanka

Sapna Yadav
ICAR–Sugarcane Breeding Institute, Regional Center, Karnal, Haryana, India

Preface

Sugar is one of the most important agriproducts that is exported globally. The sugar sector plays a vital role for the economy of sugarcane growing countries, with over 120 countries depending upon sugar production. Sugarcane is the source of over three quarters of the world's sugar and is cultivated widely in the arid and semi-arid areas. Sugar production faces challenges in terms of stagnating production, price regulations, unstable sugarcane market, and water scarcity/drought. Average yields have not enhanced significantly, partly because of its vulnerability to environmental stresses, such as biotic and abiotic stresses.

Worldwide, 80% of sugar produced from the sugarcane crop, while the remaining 20% is from sugar beets, classified as C_4 and C_3 crop plants, respectively, based on their carbon fixation stratigies, regulated by Kranz anatomy and operation of specific enzymatic reactions, associated with C_4 and C_3 carbon assimilation cycles, located in the chloroplats of the green leaves. Interestingly, sugarcane, being a C_4 plant, may prefer an elevated growing temperature (more than 30°C) throughout, while sugar beet always maximizes its growth and acquisition biomass optimally within the range of 20–25°C.

The largest top sugar-producing countries of the world are Brazil, India, Thailand, China, and the United States. Indeed, in totality over 120 countries are involved for sugar production, as it attracts towards better agri-economics.

Sugarcane (*Saccharum* spp.) represents a crop of great economic importance and is remarkably relevant for agro-industries to produce bioenergy from renewable sources, based on transformation of light energy into biochemical energy through the unique biological process of nature called photosynthesis. The process operates enzymatically, regulated by enzymes and agro-variables: light, temperature, humidity, water with intensive use of fertilizers, pesticides, and agrochemicals.

Organic farming has been accepted as an environmentally friendly way of crop production. However, sugarcane cropping strategies linked with microbiota on plant growth, development, and its performance is yet to be explored for plant/biomass yield to enhance sugarcane productivity, as sugarcane rhizosphere found to be a potential ecosystem, characterized as an environment of interaction for events between plants and plant growth promoting microorganisms (PGPMs), which may dynamically contribute for improved absorption of water-minerals and their mobilization for conferring physiological fitness in sugarcane crop plants with loss in pathogenic activities.

This book summarizes the cultivation and management of sugarcane crop prodution strategies and its related sustainability issues. This book is intended as a reference for sugarcane growers and policymakers with one aim to enhance the income of farmers and sugarcane-associated industry people. The various possibilities have been incorporated from the eye of improved and more effective cultivation, management, and production strategies of sugarcane crop linked with adequate literature to ensure significant information for further exploring sugarcane plant productivity and economic returns to sustain an agri-socio-economy in years to come.

With distinguished editors and a learned team of expert authors, we have tried our level best to present the most importhant agri-techniques that may be useful for sugarcane crop productivity to sustain the production of sugar and bioenergy, globally.

—Krishan K. Verma, PhD
Nanning, Guangxi, China
Xiu-Peng Song, PhD
Nanning, Guangxi, China
Munna Singh, PhD
Lucknow, Uttar Pradesh, India
Renato de Mello Prado, PhD
SAO, Paulo, Brazil
Jian-Ming Wu, PhD
Nanning, Guangxi, China
Yang-Rui Li, PhD
Nanning, Guangxi, China

CHAPTER 1

Prospects in Sugarcane Cultivation: An Overview

BHASKAR PRATAP SINGH,[1] TRISHA SINHA,[2] SUDIP SARKAR,[3] and ARADHANA KUMARI[4]

[1]*Agricultural Engineering, ANDUAT–Krishi Vigyan Kendra, Amethi, Uttar Pradesh, India*

[2]*Narayan Institute of Agricultural Sciences, Gopal Narayan Singh University, Jamuhar, Sasaram, Rohtas, Bihar, India*

[3]*Department of Soil Science, ICAR-RCER, Patna, Bihar, India*

[4]*Department of Plant Physiology, College of Agriculture, Jawaharlal Nehru Krishi Vishwavidyalaya, Ganj Basoda, Vidisha, Madhya Pradesh, India*

ABSTRACT

Man has a long-term relationship with sugar. So, the practice of sugarcane cultivation started long ago with the sole purpose of obtaining sugar. The word 'sugar' originated from the Sanskrit word 'Sakkara,' meaning sweet. Sugarcane (*Saccharum officinarum* L.) has a long-rooted history in cultivation, as it was first cultivated as a garden crop in 2500 BC. It is a tall-growing monocotyledonous crop belonging to the family Poaceae. Sugarcane crop is dominantly cultivated in tropical and subtropical regions all over the world. It is mainly cultivated for its ability to store high concentrations of sucrose or sugar in the stem. Agronomic management of this crop requires a huge labor force, which becomes expensive throughout its cultivation process. Irrigation is crucial for its growth and sugar content, particularly in the formative

Sugarcane Cultivation and Management: Challenges and Opportunities. Krishan K. Verma, Xiu-Peng Song, Munna Singh, Renato de Mello Prado, Jian-Ming Wu, and Yang-Rui Li (Eds.)

stage. Yield security is also unsatisfactory as water availability is uncertain in countries like India, and frequent fertilizer scarcities are due to the lack of production and exports from other countries. The shortage of groundwater for irrigation purposes has been a concern for a long time. Thus, a sustainable sugarcane initiative has been proposed and started. Potassium builds the transporting sugar chain from its source, i.e., leaf, to the sink or ultimate destination, i.e., stem. Other nutrients like B, N, and P also contribute to sugarcane growth. Thus, sugarcane, one of the largest growing crops in the world, is categorized as a specific crop due to its over-dependence on water and nutrients for growth and yield. Sugarcane has shifted from its sole purpose of cultivation, i.e., sugar obtainment, to receiving by-products and ethanol in return for the same amounts of inputs used. The increasing threats due to ongoing climatic changes have been standing as huge barriers to the production and productivity of crops. Being highly dependent on specific climates, sugarcane is at the top of the list of dangerous crops. Not limited to that, sugarcane has to go through a lot of uncomfortable and stressful situations throughout its life cycle. Diseases like red rot and pests like shoot borer have deadly effects on this crop, leading to a huge loss of productivity. Abiotic factors like nutrient scarcity, moisture scarcity, and temperature cost a lot for the growing crop. Thus, alternative eco-friendly approaches like smart cultivation methods and sustainable management practices must be adopted to let this long-duration crop stand strong and yield more with the least possible use of resources. Post-production management and applications also have to be ensured for sugar quality maintenance following the obtainment of by-products.

1.1 INTRODUCTION

As the name suggests, sugarcane is a familiar crop with sugar content. Man has a long-term association with sugarcane for its sugar content. The Sanskrit word "Sakkara," which meaning sweet, is where the term "sugar" first appeared. The cultivation of sugarcane (*Saccharum officinarum* L.) dates back to 2500 BC, when it was initially grown as a garden crop. Some people have known it since the Vedic period as sugarcane is mentioned in 'Atharvaveda,' which is 5,000 years old (Gujja et al., 2009). People who used to chew sugarcane to enjoy sweetness initially thought of sugar crystallization as a permanent source of sweetness available on the table. The first-ever preparation of crystallizing sugar was reported from the Indus Valley around 3,000 years ago (Bhatt, 2020).

The purpose of sugarcane cultivation was not limited to consuming sugar for people. Not only sugar in itself, but sugarcane is also important for producing jaggery, which is a very common food for many people in countries like India, Bangladesh, Pakistan, etc. It is a crop of the Poaceae family, thus being monocotyledonous, like other family members. However, it is morphologically different from other cereals in terms of its tall stature and heavier stems, which are distinct. The crop is also unique in its growth habit as it is grown in two types, namely *eksali* (complete cycle ends in one year) and *adsali* (entire cycle ends in one and a half years). The same crop demands different practices when grown in these two different cycles.

Sugarcane crop is predominantly cultivated in tropical and subtropical regions worldwide (Singh et al., 2018; Sanghera et al., 2018; Verma et al., 2022a). The countries include Brazil, India, Thailand, and others (USDA, 2020) as the primary contributors to its production. As India has favorable climatic conditions for sugarcane, this country provides a good environment for this crop. All it needs is a tropical environment with sufficient rainfall, which is prevalent in India (Shanthy & Ramanjaneyulu, 2014). In India, Uttar Pradesh secures the first position in terms of production, while Tamil Nadu ranks first in productivity. The love for sugar among people has kept this crop in trend, leading to increased production. The last two decades have witnessed a significant increase in global sugar production from 56 to 80%, which is very impressive (Singh et al., 2018).

When talking about India, it has improved productivity levels in the past. Still, the rise in productivity has stagnated over the last two decades, as Sundara (2011) reported. Over the years, the reduction in the area and production of sugarcane can be attributed to frequent drought, the declining trend of soil health, high input costs, and insufficient sugarcane prices fixed by the government (Suguna & Jayanthi, 2021; Verma et al., 2022b, c). Efforts must be made to meet the demands of the ever-increasing population pressure all over the world. A lot of techniques have now been known to the farmers for coping with all the drawbacks coming in the path of extracting its full potential. As the crop requires huge amounts of water, water management strategies must be adopted for sustainable cultivation. Weeds are also a big problem for this crop. Thus, weed management is of utmost need. Throughout its crop growth, sugarcane has also faced a number of serious diseases, red rot being one of them. Some pests are deadly for this crop, making no productive output (Song et al., 2021, 2022). Thus, disease and pest management in the most appropriate way is a must to do. Sugarcane leaves a lot of residues in the soil that, in turn, cause big trouble for the soil. Most of the time, those residues are simply burnt, making the soil degraded. This practice not only adds pollution to the environment but also washes away valuable properties of soil, reducing soil fertility and productivity. Thus, soil and nutrient management is also necessary as they play a big role in deciding the future of soil health after sugarcane cultivation. The urge for sugarcane cultivation lies in its cane juice which is processed to produce sugar and sugar products in the industry (Anas et al., 2021a, b; Deng et al., 2022; Xu et al., 2022). But this costs a lot of things, most importantly water at the rate of even 2,500 kg water to have 1 kg of cane. This creates a big question mark in the utility of its production as sustainability becomes uncertain. Thus, the uncertainty in sustainability must be nullified by adopting proper management practices in sugarcane cultivation. It should be remembered that our target is not only limited to improved productivity but also to ensure the sustainability of natural resources.

1.2 PROSPECTIVE OF SUGARCANE IN THE GLOBE AND INDIA

For commercial purposes, sugarcane was first grown from the cross between *Saccharum officinarum* and *Saccharum spontaneum* (Huang et al., 2021). The wild cultivation of the natural hybrid has been reported in India and China (Bull, 2000). These two species vary in sugar content and sensitivity toward various stresses. According to Cox et al. (2000), *S. officinarum* is a rich source of sugars, but it is also more susceptible to diseases, contrary to *S. spontaneum*, which accumulates little sucrose and shows improved resistance to several diseases and pests (Jackson, 2005). Sugarcane has a huge prospect globally, especially in India. India is blessed with a good amount of annual rainfall and a tropical climate, the major requisites for sugarcane cultivation. India solely depends on sugarcane for sugar, making it a dominant crop of the nation. Harender et al. (2021) reported that India shelters around 50 million farmers dependent on sugarcane cultivation, and the Indian sugar industry is worth 550 billion rupees, making it the second largest industry in India, just after the cotton and textile industry. India contributes about 6% of the agricultural GDP through sugarcane (https://eagri.tnau.ac.in/eagri50/AGRO301). FAO (2020) reported that India cultivates sugarcane in a total area of 47.3 lakh ha, from which an annual production of 3,769 lakh tons is a good amount. Suguna & Jayanthi (2021) supported this fact and concluded that India is one of the largest sugarcane producers worldwide.

1.2.1 CURRENT STATUS

Sugarcane is cultivated in more than 100 countries; the production potential is substantially higher in tropical and sub-tropical regions than in other global regions (Prasanna & Pattar, 2014; Verma et al., 2021a, b, c). Brazil ranks first among the world's highest sugar-producing countries, followed by India and China (Li & Yang, 2015a). Other major contributing countries include Thailand, Pakistan, Australia, Cuba, and Indonesia (Sarwar et al., 2008; Singh et al., 2018). FAO (2020) reported that Brazil tops in the export of sugar globally, followed by Thailand and Australia. India shares 16.6% of the total sugarcane production in the world (Harender et al., 2021). In terms of consumption also, India stands 2nd in the world with the 14th position in exporting. Surprisingly, India imports around 0.67 million tons of sugar, ranking it 5th among the sugar-importing countries of the world (Parajuli et al., 2019). India seems to have to produce nearly 33 million tons of white sugar by 2030 for the sole purpose of domestic consumption. This challenging task looks more difficult, predicting the national population pressure will increase to 1.65 billion by 2050 (Parajuli et al., 2019). India adds 376.91 million tons in sugarcane production from an area of 4.73 million hectares with a productivity of 80.10 tons per hectare (Bhatia & Nimbrayan, 2020; FAO, 2020).

1.2.2 PRODUCTION TECHNOLOGIES AND PRACTICES IN SUGARCANE CULTIVATION IN INDIA

Based on sugarcane cultivation, India is divided into two agro-climatic zones, *viz.* tropical and sub-tropical zones. The tropical location includes the western and southern regions like Maharashtra, Karnataka, Gujarat, and Tamil Nadu, whereas the latter covers the northern areas like Punjab, Haryana, Uttar Pradesh, and Bihar (Gujja et al., 2009). In terms of area and production, Uttar Pradesh is the leading Indian state, followed by Maharashtra. Still, Tamil Nadu outshines all the states employing productivity (Anonymous, 2019) because of its prevalent black soil that is extremely suitable for sugarcane cultivation (Suguna & Jayanthi, 2021). The other south Indian states *viz.* Karnataka and Andhra Pradesh stand second and third, respectively, in productivity. Two dominant cultivation practices are there, *viz.* adsali and eksali planting. All the cultivation practices depend on the zone and climate. According to Shukla et al. (2017), sowing in *adsali* planting, practiced in

Maharashtra and Karnataka, is performed from July-August. The crop takes 16–18 months to mature while sowing is completed in spring, February-March in the case of eksali planting (crop of one year), mainly practiced in Gujarat and Andhra Pradesh. North India shares around 60% of the total national production (Kamboj et al., 2008), while the rest 40% comes from South India (Shukla et al., 2017). Another technique in practice is ratooning, in which the ratoon of the previous season's cane is planted. This practice is mostly prevalent in Maharashtra.

To grow well, sugarcane requires an annual rainfall of 1,100–1,500 mm with adequate distribution. The optimum temperature for sugarcane germination (sprouting) ranges from 32 to 38°C, with a gradual reduction in germination below 25C and above 35°C. The ripening stage requires 12–14°C of temperature above which sucrose (soluble sugar) is converted to fructose and glucose, deterring the juice quality (Bhatt, 2020). Sugarcane fits well within a large range of soils with heavy textures. However, well-drained, deep, and medium-textured soil with good water-holding capacity and neutral pH is suitable for sugarcane (Singh et al., 2018; Verma et al., 2021d, e, f).

1.2.3 SCOPE OF SUGARCANE CULTIVATION

The sea of the world's population depends on several crops for food, fiber, fodder, and fuel (4Fs). Sugarcane is India's 3rd most important cereal crop, next to wheat and rice (Harender et al., 2021). The provision of sugar, molasses, ethanol, bagasse, and press mud altogether makes sugarcane an excellent cash crop and able to sustain the livelihood of a number of people around the world. According to Parajuli et al. (2019), sugarcane has a great future as it has all the 4Fs of agriculture. The 4Fs of agriculture and other important products can be obtained from different plant parts or components of sugarcane (Table 1.1). Sugarcane also plays a great role in industrial purposes, comprising an industry of Rs. 30,000 crores in the agro-processing sector, representing the livelihood of 35 million farmers in India. According to the Indian Sugar Mills Association (2019), each farmer effectively contributes to producing 2.9 million tons of sugar every year. Sugarcane grows with complete faith as the revenue generation from sugar and other by-products has been estimated to grow from Rs. 11.20 lakhs in 2019 to Rs. 14.60 lakhs in 2024 with an increment of 30% (Chakraborty & Sahu, 2020).

TABLE 1.1 List of Products Obtained from Sugarcane and Their Uses (Dotaniya et al., 2016)

Products	Parts or Components of Sugarcane Used	Uses/Properties
Sucrose, syrup, jaggery, fructose	Cane juice	Food
Cellulose	Cane stalk	Fiber
Residue	Bagasse or juice-extracted wastes	Fuel
Alcohol	Molasses	Chemicals
Press mud	Sugar	Chemicals

1.2.3.1 PROCESS OF SUGAR MAKING

According to Bhatt (2020), sugarcane and sugarbeet are used to make white sugar, *khandsari*, and jaggery, contributing their share of 80% and 20%, respectively. Sugarcane juice, containing high sucrose content, is required to make sugar and molasses. Sugarcane mills receive canes from transportation and extract the juice. The process of evaporation concentrates the juice. By the end, some parts are crystallized, and some remain, known as sugar and molasses (Montiel-Rosales et al., 2022). To make it saleable, sugar, after crystallization, is followed by drying, and the molasses is stored for disposal by a food chain. An estimated production of 4.46 tons of molasses can be obtained from 100 tons of sugarcane (Chakraborty & Sahu, 2020). Molasses contains about 50% sugar and is being used as a sub-stratum of procedures to produce nourishment yeast or cattle feed (Chakraborty & Sahu, 2020). The molasses is also collected in large volumes to prepare ethanol in the alcohol industries further.

1.2.3.2 PROCESS OF OBTAINING BY-PRODUCTS

Alcohol generation takes place during the treatment of molasses. An alcohol yield of 4,000 liters from one hectare of sugarcane can be obtained, as reported by Chakraborty & Sahu (2020). The making of alcohol involves fermentation and distillation, from which a separate by-product, is known as vinasse (da Silva et al., 2022). One liter of alcohol can even produce 10–18 liters of vinasse, as reported by Coelho et al. (2018); de Jesus et al. (2019); and Tamashiro et al. (2022). This range depends on a number of factors, such as the sweetness of molasses and the distillery process (Montiel-Rosales et al., 2022). Vinasse is a rich source of nutrients like nitrogen, phosphorus, and potassium (Cortes-Rodríguez et al., 2018), and organic residues help improve soil fertility. Vinasse also leads to the production of other bioproducts that, in turn, help in various purposes such as the bioenergy industry (43.51%), agriculture (13.74%), biotechnology (13.74%), and waste and wastewater management (22.14%), as reported by Montiel-Rosales et al. (2022). Applications of vinasse in various fields have been discussed in Table 1.2.

TABLE 1.2 Applications of Sugarcane Vinasse in Agricultural, Industrial, and Other Sectors

Areas	Applications	References
Agriculture	Nitrogen supply to soil	Ventorino et al. (2019)
Pisciculture	Fish feed	Rulli et al. (2021)
Agriculture	Fertilizer for sugarcane and other crops	Bettani et al. (2019); Fregolente et al. (2021)
Industrial	Bioenergy production	Sousa et al. (2019); Magrini et al. (2021)
Bioplastic	Production of biodegradable plastics	Satos et al. (2020)

Press mud, also known as filter cake, is another byproduct obtained during the purification of sugar, mainly by the process of carbonation or sulfitation (Ghulam et al., 2012), leading to the separation of clear juice on the top and letting the mud precipitated at the bottom (Dotaniya et al., 2016). According to Gupta et al. (2011), about 3 tons of press mud is obtained from 100 tons of sugarcane. The raw spent wash is produced following the process of fermentation and distillation of molasses. It has a lower pH, unpleasant odor, higher COD, and BOD at 1,00,000 mg L^{-1} and 45,000 mg L^{-1}, respectively. Treating raw spent wash with a process known as biomethanation makes it rich in plant nutrients and suitable for use as liquid manure for crops (Dotaniya et al., 2016). Bagasse is another by-product of sugarcane, and it is the leftover fibrous material following juice extraction in the mills. An estimated 135 kg and 240–280 kg of bagasse can be obtained from each ton of clean cane (Macedo et al., 2008) and processed cane, respectively (Cheavegatti-Gianotto et al., 2011).

1.3 CROP MANAGEMENT

1.3.1 SOIL AND NUTRIENT MANAGEMENT

Soil management includes a set of practices to minimize soil degradation by wise application of fertilizers and their post-application control measurements. This is complicated, especially for sugarcane, as it is a long-duration crop. However, soil management in sugarcane is also possible if the right steps are taken at the right time and magnitude. Sugarcane is an exhaustive crop for which most nutrient reserves in the soil get worn out after harvesting. Around 95% of the fresh weight of a sugarcane plant is composed of the structural elements, i.e., carbon, hydrogen, and oxygen, while the rest 5% is comprised of some essential mineral nutrients (de Souza et al., 2015). According to Prasanna & Pattar (2014), nutrient absorption in ratoon cane is disturbed due to the altered soil physio-chemical properties to a severe extent after sugarcane cultivation. Sugarcane is similar to almost all crops in terms of requiring major and minor nutrients like N, P, K, Mg, Zn, B, etc. However, the fondness of sugarcane for silicon has been found as the crop has been seen to gather a large amount of silicon that, in turn, helps to yield better (Berthelsen et al., 2001). The requirement of various nutrients by sugarcane depends on several factors such as soil texture, agro-climatic conditions, indigenous technical knowledge (ITK), as well as cultivars of sugarcane also (Bhatt, 2020). Reports from PAU (2018–2019) state that sugarcane crops, whether plant or ratoon, have different nitrogenous fertilizer requirements, which are 150 and 225 kg N ha^{-1}, respectively. For ratoon crops, three major nutrients – nitrogen, phosphorus, and potash – are needed at rates of 125, 30, and 75 kg $acre^{-1}$, respectively (Prasanna & Pattar, 2014). According to Vuyyuru et al. (2019), ratoon sugarcane is more sensitive to N-fertilization compared to the seed crop. ITK-based fertilizer application sometimes fails to maintain the recommended doses, leading to soil toxicity. A significant portion of nitrogen in the soil becomes immobilized in the trash, making it unavailable to the crop. Therefore, proper resource management is crucial to preserve the soil and ensure plant sustainability. Timely and accurate fertilizer application is vital as it determines nutrient availability for sugarcane. The correct placement of fertilizers is also a key factor in nutrient use efficiency. Splitting the fertilizer dose near the row enhances nutrient use efficiency and reduces off-farm impacts (PAU, 2018–2019).

Similarly, Sanghera et al. (2018) reported that biofertilizer treatment of *Azotobacter/Consortium* at 4 kg per acre in the furrows during planting would be beneficial in enhancing the yield of sugarcane. Failing to follow any of the correct steps might lead to loss of soil and crop. Soil management can be successful by applying biofertilizers with properties of fixing nutrients like nitrogen and phosphorus, and others like the synthesis of phytohormones, increasing the length of internodes and enhancing yield as the most important ones. Shankaraiah & Nagaraju (2002) reported that soil inoculated with *Azotobacter chrococcum* and *Azospirillum brasiliense* at 2.5 kg per hectare each resulted in improved jaggery yield in both the main and ratoon crops. The functions of different biofertilizers have been depicted (Table 1.3).

TABLE 1.3 Different Biofertilizers with Their Potential Functions in Sugarcane (Verma, 2019)

Availability of Nutrients	Name of Biofertilizer	Characteristics	Other Functions
N	*Gluconacetobacter diazotrophicus*	Endophytic	Increases both the size and length of internodes. It helps in increasing sugar content. Promotes IAA biosynthesis. Yield increment of about 5–15 t ha^{-1}.
N	*Azospirillum lipoferum*	Rhizospheric	Promotes synthesis of auxin and cytokinin. Enhances traits like germination and vigor. Yield increment up to 20%.
P	*Aspergillus awamori, Pseudomonas spp.*, and *Bacillus* spp.	Soil bound	Enhances vigor and increases yield by up to 15%.
N	*Azotobacter chrococcum*	Free-living	Excretes important antibiotics. It improves seed vigor and also increases yield by up to 25%.

1.3.2 RESIDUE MANAGEMENT

Sugarcane produces a significant amount of residues in and out of the field, raising concerns about environmental threats. In ratoon crops, trash is usually burned to prevent waste accumulation, which can make it difficult for the next crops to grow (Prasanna & Pattar, 2014). However, open disposal of various residues in soil and water can lead to contamination. If this continues, the original properties of natural resources will be altered, resulting in suppressed crop growth and yield. Different types of residues can be effectively managed by taking appropriate actions. According to Dotaniya & Kushwah (2013), crop residues can be an excellent option for recycling various macro- and micronutrients. Moreover, the sugar contents in organic residues can enhance soil organic acids, potentially improving phosphorus use efficiency (PUE), crop quality, and yield (Shukla et al., 2013). Additionally, trash can serve as excellent mulch, aiding in moisture conservation, weed management, and enriching soil organic matter. Therefore, the future potential of sugarcane crops largely depends on the optimal utilization of the generated by-products. The potential application and management of residues are discussed in Table 1.4.

TABLE 1.4 Applications of Various By-Products Generated from Sugarcane

Product Name	Application/Management	References
Press mud	As soil ameliorate	Razzaq (2001)
	As organic manure	Bokhtiar et al. (2001)
	As biofertilizer	Patil et al. (2013)
	As an enhancer of cation exchange capacity	Rodella et al. (1990)
	As an enhancer in P fixation in crops	Dotaniya & Datta (2014)
Bagasse	Helps in the mobilization of P after application	Dotaniya & Datta (2014)
	Substitute for cement to be used in concrete	Kawade et al. (2013); Malyadri & Supriya (2015)
	Production of resin and fibers	Hoareau et al. (2006)
Spentwash	Role in integrated nutrient management	Dotaniya et al. (2016)
Vinasse	Yield improver in rice	Raza et al. (2021)
	Waste and wastewater management	Ferreiro & Kronemberger (2021)

1.3.3 WATER MANAGEMENT

Water is an essential component for growth and development, as well as other physiological processes in the plant, and the absence of water alters these activities (Sawhney & Singh, 2002). Water used to be available for its natural abundance, but now it has become an expensive commodity due to random misuse and wastage by people. According to Qureshi et al. (2001), a system's ability to apply water consistently and effectively is a key aspect that directly impacts the agronomic and financial feasibility of production for many crops, including sugarcane (Verma et al., 2022).

Irrigation is vital for crops, for which water management becomes necessary. Water management refers to using water resources by sensible means per established regulations and standards. Sugarcane plays a significant role in the economy of sugarcane-producing countries, although the water consumption scenario is unfavorable (Dingre & Gorantiwar, 2020). In terms of water management, sugarcane appears to be under a peculiar condition. Sugarcane's love for water is known to all. This is so high that sugarcane might take even 10,000 m^3 of water in order to yield 100 tons per hectare

(Choudhary et al., 2013). Nevertheless, this is not the strict figure, as the requirement of water for sugarcane varies with the agro-climatic zone, such as dry winds associated with moisture scarcity and higher temperatures increasing the requirement of water for sugarcane. It also goes with soil type, use of manures and fertilizers, and method of planting. Drought greatly obstructs the growth and development of sugarcane, as reported by Lakshmanan & Robinson (2014). Ramesh (2000); Basnayake et al. (2012); and Gentile et al. (2015) also reported that areas of uneven rainfall are completely impossible for commercial cultivation of sugarcane. Thus, the requirement of water for sugarcane cannot be reduced without compromise in its yield, as a yield loss of even 60% can be faced depending on the duration and magnitude of the water scarcity (Verma et al., 2022a, c). Wise and efficient irrigation application in sugarcane is essential as various crop phases face multiple situations of water availability. It has been seen that drought circumstances prevail during the summer months when the crop goes through its formative phase, while the waterlogging situation prevails during the grand growth period of the crop that appears during monsoon.

Apart from being an exhaustive water crop, sugarcane has primarily been grown using surface irrigation, where water consumption efficiency can be limited to 35–40%, which is quite low. The reason is attributed to significant losses during distribution and evaporation (Sivanappan, 1994; Rosegrant & Meinzen-Dick, 1996; Rosegrant, 1997). Under these circumstances, water should be applied wisely to avoid waste and maximize possible benefits. In order to improve water consumption efficiency in sugarcane, it is crucial to understand crop evapotranspiration (ET) and periodic or stage-based crop coefficients (Inman-Bamber & McGlinchey, 2003). According to Shrivastava et al. (2011), a sugarcane crop consumes 88–118 kg of water per kg of cane and 884–1,157 kg of water per kg of sugar produced. Sugarcane requires less than 2,000 mm of water, but evapotranspiration loss makes the crop receive an average water application of 3,000–4,000 mm based on the irrigation technique (Shrivastava et al., 2011). Easy adoption of flood irrigation to irrigate sugarcane, delivering 2–3 times more water than needed owing to farmers' perceptions of improved yield from more applications, is a big problem. Furrow irrigation, the most popular way of water application, can be run effectively but is often poorly managed as the crop utilizes only about 50% of the water provided.

There are several water-saving techniques, but optimum utilization of those techniques needs accurate estimations of crop water requirements or other water availability indicators. Both the quality and quantity of water for irrigation are important. According to Prasanna & Pattar (2014), using poor quality water, viz. saline and alkaline water, continuously impedes soil health, while even good quality water, when used excessively, renders soils unproductive. Precise application and consistent distribution of water for sugarcane can be made possible with the adoption of water-saving approaches such as overhead sprinklers, surface, or subsurface drip irrigation, etc. Sugarcane is suitable for drip irrigation (Narayanamoorthy, 2005; Prasanna & Pattar, 2014). Although sugarcane has been grown using drip irrigation by farmers across various states, its coverage is extremely limited. This is large because of a knowledge gap regarding the significance of drip irrigation for sugarcane cultivation as well as the higher installation cost of drip irrigation. Understanding the water conditions in sugarcane could be used more successfully in irrigation management. Underestimating crop evapotranspiration in irrigated agriculture can lead to suboptimal yields due to water scarcity while overestimating can lead to excessive water use, less available water for other purposes or additional land, and a reduction in economic competitiveness. Thus, with a full understanding of crop evapotranspiration, the execution of irrigation planning would result in improved irrigation efficiency and water savings.

1.3.4 WEED MANAGEMENT

Sugarcane is infested with a number of weeds that damage the crop. Some of the deadly weeds, *viz. Trianthema portulacastrum, Digera arvensis, Amaranthus viridis, Parthenium hysterophorus, Cleome gynandra,* and *Datura fastuosa* have been reported to affect the crop (Suganthi & Muthukrishnan, 2017). Grassy weeds like *Dactyloctenium aegyptium, Echinochloa colona, Eleusine indica*, and *Digitaria sanguinalis;* and among sedges, *Cyperus rotundus* damages the crop to a great extent (Singh et al., 2017). A crop loss ranging from 10% to complete failure in sugarcane has been reported in India due to weed interference (Chauhan & Srivastava, 2002). In India, *Aeginetia pedunculata* leads to a crop loss of up to 37% due to suppressed stalk growth and poor juice quality (Ray & Dasgupta, 2006). Upon considering these factors, adequate weed control measures must be adopted to eliminate the weed problem in sugarcane.

The mechanical method of weed control includes earthing up and blind hoeing, which is helpful in the management of annual and perennial weeds (Harender et al., 2021). Culturally, a number of possible management strategies could be adopted to control weeds in sugarcane. Intercropping with crops like winter maize, pulses, oilseeds, and wheat; and vegetables such as potato, garlic, pea, radish, etc., can benefit autumn-planted sugarcane. For spring-planted sugarcane, intercrops like summer mung, menthe, etc., are recommended (Shanghera et al., 2018). Intercropping not only helps restore soil fertility but also aids in weed control and boosts yield in sugarcane (Singh, 1993). Mulching with trash is also a good cultural method of weed control. According to Harender et al. (2021), 7.5 t/ha of trash is generally used between rows of sugarcane 45 days after planting, with an average thickness of 10–12 cm for mulching. Chemical control of weeds like *Cyperus* spp. is done with Sulfentrazone and Halosulfuron (Anonymous, 2009). Atrazine and 2,4-D are best suited to control the *Striga* population (Prasanna & Pattar, 2014; Harender et al., 2021).

1.3.5 INSECT AND PEST MANAGEMENT

A number of insects and pests attack sugarcane and damage the crop significantly. Insect pests like early shoot borer, scales, root grub, mealy bugs, and termites are important ones (Prasanna & Pattar, 2014). Managing the pests at the economic threshold level (ETL) involves the best approach, which includes integrated pest management. Some important measures have been discussed in Table 1.5.

TABLE 1.5 Some Useful Practices for Insect Pest Management in Sugarcane

Insect and Pest	Control Measures	References
Top borer	Applying 10 kg granules of Ferterra 0.4 GR or Thimet encapsulated 10 G (Phorate) at the shoot base from June end to the first week of July after damage crosses 5% level. Slight earthing up to check the granules from flowing with the irrigation water following immediate irrigation is helpful to control the third brood of the top border, causing the maximum damage.	Sanghera et al. (2018)
Early shoot borer	Applying 10 kg granules of Regent 0.3 G/Fipronil in the furrows along the cane setts before soil covering or spraying 45 ml Imidagold 17.8 SL (Imidacloprid) or 2 liters of Chlorpyriphos 20 EC in 400 liters of water acre^{-1} along the rows 45 DAP (post-germination stage).	Sanghera et al. (2018)
	By natural enemy *Trichogramma* sp.	Verma (2019)
Termites	Removal of dead wood from fields of cane.	OECD (2016)
Aphids	Natural enemies	OECD (2016)
Pyrilla	*Epiricania melanoleuca* @ 4,000–5,000 cocoons and 4–5 lakh eggs ha^{-1}.	Verma (2019)

1.3.6 DISEASE MANAGEMENT

Sugarcane crops struggle with some deadly diseases like smut, grassy shoot disease, ratoon stunting disease (RSD), red rot, and others. Bacterial diseases like RSD, caused by *Leifsonia xyli* subsp. *xyli*, and leaf scald, caused by *Xanthomonas albilineans*, lead to maximum damage to the crop. Fatal fungal diseases include brown and orange rust, caused by *Puccinia melanocephala* and *Puccinia kuehnii*, respectively (Braithwaite et al., 2009), and red rot, caused by *Colletotrichum falcatum* (Sundar et al., 2015). Smut (*Ustilago scitaminea*) causes slender stalks, deteriorating juice quality, and poor yield. Viral diseases such as sugarcane mosaic virus disease and yellow leaf disease are severely infectious (Gilbert et al., 2009). Management of diseases is, therefore, a must-needed approach for sugarcane. Planting resistant varieties developed by breeding approaches is the most appropriate method for controlling most of the fungal, bacterial, and viral diseases in sugarcane (Sundar et al., 2015). Physical disease control methods include heat treatment (Almeida, 2008) and water management (Izaguirre et al., 2020). Mechanically, diseases in sugarcane can be controlled by roguing infected plants (Izaguirre et al., 2020) and selecting healthy cane setts (Sundar et al., 2015). Red rot-tolerant varieties like CoJ88, CoPb91, and Co0238 have been recommended to grow (Sanghera et al., 2018).

1.4 DRAWBACKS AND THEIR PROBABLE SOLUTIONS

All the sweet facts about sugarcane are known to many people. Sadly, but it's real, people are hardly aware of the bitterness hiding inside sugarcane. Over requirements of water and nutrients make sugarcane categorized as an exhaustive crop. The capacity of soil to supply nutrients to the plants declines steadily under continuous and intensive cropping systems like sugarcane (Shukla et al., 2016). In order to produce 100 tons of sugarcane per hectare of land, costs the removal of 140 kg, N, 34 kg, P, and 332 kg, K, respectively, as reported by Bokhtiar et al. (2001). Sometimes, limited sugarcane production can result from a deficiency in fertilizer supply (Gholve et al., 2001) due to unaffordable prices (Khandagave, 2003). Besides, its long duration sometimes creates tension among the farmers as the cultivation costs are also high for sugarcane. From sowing to harvesting, a number of agronomic practices are carried out, which often bankrupt the sugarcane farmers. Insecurity in rainfall and shortage of irrigation water will lead to the complete failure of this crop.

The post-cultivation measures are also difficult as those include leftover residues and reduced soil fertility. Sugarcane attracts a lot of pests for its sugar content and others. Diseases are another big problem for sugarcane; it becomes very difficult to manage the disease completely even after harvesting as it is a ratoon crop. Xiao et al. (2017) reported that several factors, such as proper harvest timing, diseases and pests, and other cultivation practices, determine sugarcane juice quality. Any alteration in the optimum level for the mentioned factors has been reported to deteriorate juice quality. As per recent reports, sugar industries are facing challenges due to the lower productivity of sugarcane. The reasons for lower productivity in sugarcane can be owed to the limited release of resistant varieties and sluggish growth in genetic improvement, unpredictable water availability (Chand et al., 2016), degrading soil health (Suguna & Jayanthi, 2021), imbalanced crop management, and increasing input and labor costs (Loganandhan et al., 2012). The recent changing pattern in climate has been adding fuel to it (ICRISAT, 2009). The depleting water table, global warming, and pollution are damaging the crop on a serious note.

Moreover, unwise disposal of residues and byproducts can alter the soil and water properties leading to environmental contamination. Silva et al. (2021) reported that vinasse could change soil properties like pH and conductivity. Therefore, a serious concern is being raised to produce sugarcane sustainably by overcoming the fundamental obstacle in its production. Thus, proper agronomic management, water management, and other sustainable practices should be conducted to break all the barriers to obtaining the optimum potential of sugarcane. To make this possible, sugarcane farmers must access all the latest resource management techniques such as soil test reports, drip irrigation for better water use efficiency, seed treatment, new hybrids, proper ratoon management techniques, and the use of the latest machines (Bhatt, 2020). Some recommended varieties of sugarcane have been depicted in Table 1.6.

1.5 SUSTAINABILITY THROUGH SUGARCANE

A lot of possibilities for attaining sustainability could be made through sugarcane. Although a separate section of this chapter on residue management has been discussed, some future also lies in the leftover residues in the form of bagasse. Bagasse is the leftover material after juice extraction from sugarcane (Frías et al., 2011). The preparation of ethyl alcohol for

TABLE 1.6 List of Recommended Varieties of Sugarcane for Different States of India (Shukla et al., 2017; Sanghera et al., 2018)

State	Recommended Varieties	Nature of the Varieties
Uttar Pradesh	Co0238, Co88230	Early
	CoS767, CoS96275, BO91	Mid-late
Maharashtra	Co94012, CoC671	Early
	Co86032, Co92005, CoM0265	Mid-late
Andhra Pradesh	Viswamithra (87A298) (CoA92081), Bharani (2003V46) (CoV09356), Krishna (91V83) (CoV95101)	Early
	Kanaka Durga (83V15) (CoV92102), 93V297	Mid-late
Karnataka	Co86032, CoC671	Early
	CoSnk03632, CoM0265	Mid-late
Tamil Nadu	CoC90063, CoC671	Early
	Co86032, CoSi96071	Mid-late
Bihar	CoP9301, CoS8436, BO102	Early
	BO91, CoP9206	Mid-late
Punjab	Co118	Early
	CoPb91	Mid-late

commercial purposes, followed by power generation from sugarcane bagasse, has been adopted in most countries as a sustainable approach (Bhatt, 2020). Nowadays, bagasse from stalks or any source is collected and stored in a place. After that, they are burned to produce heat energy used in operating turbines and others. It has even been possible to expect 70% of the efficiency of energy production from bagasse burning. Brazil generates 19.3% of its energy from bagasse (Hofsetz & Silva, 2012). Ash produced from sugarcane bagasse can be used in concrete as a replacement for cement that produces CO_2, contributing to 5% of greenhouse emissions (Kumar et al., 2017). The benefit of utilizing bagasse ash includes a reduction in the emission of greenhouse gases without compromising the strength and ductility of engineered cementitious composite (ECC), generally used in concrete material (Shafiq et al., 2018). Sugarcane trash can contribute essential nutrients to the soil, control weeds, and conserve soil moisture when used as mulch. Prasanna & Pattar (2014) further reported that this trash could be a source of phosphorus, nitrogen, and potassium at the rate of 50, 100, and 140 kg ha^{-1}, respectively.

Similarly, farmyard manure at 20 t ha^{-1} is recommended to be very beneficial for the crop (Bhatt, 2020). To improve soil-plant-atmosphere interaction and also cane growth, *Azotobacter* at 10 t ha^{-1} is recommended. Sugarcane trash serves as an excellent substrate for soil microorganisms and conserves moisture, reinvigorating soil health. Additionally, trash is an important source of secondary nutrients like Ca, S, and Mg, and micronutrients like Fe, Mn, Cu, Zn, Mo, and B (Prasanna & Pattar, 2014).

Potential sustainability through sugarcane was confirmed, and thus a NABARD-based sustainable approach, i.e., sustainable sugarcane initiative (SSI), was started to change the conventional practices in sugarcane cultivation. Moreover, it is costlier than the conventional method of sugarcane cultivation. The multipurpose SSI aims to promote sustainability by adopting viable practices that can reduce the ecological footprint among the farmers engaged in sugarcane cultivation and create market linkages with sugar factories (Shanthy & Ramanjaneyulu, 2014). A brief comparative discussion between the conventional method and SSI in sugarcane cultivation has been provided (Table 1.7). The SSI operates with lower seed costs, higher income-generating ability, soil enrichment, reduced crop duration, and increased crop yield (Shanthy & Ramanjaneyulu, 2014). Therefore, the project is a significant benefit for small and marginal farmers as it can provide improved yield with less fertilizer and water use, and an improved cropping system.

TABLE 1.7 Comparative Discussion Between Conventional and SSI Methods of Sugarcane Cultivation (Parajuli et al., 2019)

Components	Conventional Method	SSI Approach
Number of setts required	Single-budded setts: 60,000 or two-budded setts: 30,000.	Single-budded setts: 5000
Requirement of nursery preparation	Not required	Required
Method of planting	Direct	Transplanting with nursery-raised seedlings aged 20–25 days.
Requirement of water	Higher (flooding condition)	Lesser (moisture maintenance in the furrows is required).
Row-row spacing	1.5–2.5 ft	At least 5 ft
Number of tillers	10–15 per plant	20–25 per plant
Number of millable canes	4–5 per clump	9–10 per clump
Intercropping scope	Lesser than SSI	Greater than the conventional method

1.6 CONCLUSION

The cultivation of sugarcane has offered a lot of benefits to humankind. Imagining a world free of sugar and sugar products is difficult. Though people are becoming health conscious daily, a complete rejection of sugar worldwide is impossible. Keeping aside health issues, sugarcane also leaves some drastic negative impacts like residues, soil erosion, and groundwater depletion, which are the biggest finger-pointing whether to go for sugarcane or not. Degraded soils and poor water sources used for irrigation are other problems that directly affect juice quality. But with the rise of every problem in the agricultural sector, scientists have tried their best to find the best possible solutions to overcome the problem as we know that no big target is easily fulfilled. Therefore, to achieve the big target of improved production along with quality maintenance and sustainability, a proper and concrete idea of sustainable management practices should be adopted immediately. Like other problems, the complications in sugarcane cultivation can also be surpassed with the help of remedies like the sustainable sugarcane initiative. The SSI covers all three pillars of agriculture, viz. land, labor, and water. The proper implementation of SSI will surely lead to improved crop productivity, water management, and soil health. This way, we could outshine all the barriers to turn those into opened blossoms. Then it can be said that sugarcane is not a curse, rather, it's a boon for humankind.

KEYWORDS

- **by-products**
- **diseases-pests**
- **moisture scarcity**
- **nutrients**
- **post-production management**
- **residue management**
- **sugarcane cultivation**
- **sustainable sugarcane initiative**

REFERENCES

Almeida, I. M. G. (2008). Doenças causadas por bactérias. In L. L. Dinardo-Miranda, A. C. M. Vasconcelos, & M. G. A. Landell (Eds.), *Cana-de-açúcar* (pp. 437–450). Instituto Agronômico.

Anas, M., Verma, K. K., Haq, I. U., Naeem, M., Li, Q., Liao, F., Xiang, Y. X., & Li, Y. R. (2021a). Characterization of exotic, native and wild-type genotypes of sugarcane (Saccharum spp. Hybrids) for internal nitrogen use efficiency under different nitrogen levels. *Sugar Tech.*, *23,* 1258–1267. https://doi.org/10.1007/s12355-021-01014-1.

Anas, M., Verma, K. K., Riaz, M., Qiang, L., Liao, F., Liu, Y., & Li, Y. R. (2021b). Physio-morphological and biochemical mechanism of nitrogen use efficiency in sugarcane (Saccharum spp.) genotypes under different growth stages and nitrogen levels. *Journal of Plant Interactions*, *16*, 332–343.

Anonymous. (2009). Louisiana chemical weed management guide. Louisiana State University Agricultural Center, Louisiana Cooperative Extension Service, *1565*, 57–76.

Anonymous. (2019). Pocketbook of Agricultural Statistics. Directorate of Economics & Statistics, Ministry of Agriculture & Farmers Welfare, Government of India.

Basnayake, J., Jackson, P. A., Inman-Bamber, N. G., Lakshmanan, P. (2012). Sugarcane for water-limited environments. Genetic variation in cane yield and sugar content in response to water stress. *Journal of Experimental Botany*, *63*(16), 6023–6033.

Berthelsen, S., Hurney, A. P., Kingston, G., et al. (2001). Plant cane response to silicated products in the Mossman, Innisfail and Bundaberg districts. *Proceedings of the Australian Society of Sugar Cane Technology*, *23*, 297–303.

Bettani, S. R., de Oliveira, R. G., Santos, L. N., Kieckbusch, G. T., Bastos, R. G., Soares, M. R., & da Silva, M. A. (2019). Sugarcane vinasse and microalgal biomass in the production of pectin particles as an alternative soil fertilizer. *Carbohydrate Polymers*, *203*, 322–330.

Bhatia, J., & Nimbrayan, P. K. (2020). Pocketbook of Haryana Agriculture Statistics. Directorate of Human Resource Management, CCSHAU, Hisar.

Bhatt, R. (2020). Resources Management for Sustainable Sugarcane Production. In S. Kumar et al. (Eds.), Resources Use Efficiency in Agriculture. Springer Nature Singapore Pte Ltd. https://doi.org/10.1007/978-981-15-6953-1_18.

Bokhtiar, S. M., Paul, G. C., & Rashid, M. A. (2001). Effect of press mud and organic nitrogen on soil fertility and yield of sugarcane grown in high Ganges River flood plain soils of Bangladesh. *Indian Sugar*, *1*, 235–240.

Brackin, R., Robinson, N., Lakshmanan, P., & Schmidt, S. (2014). Soil microbial responses to labile carbon input differ in adjacent sugarcane and forest soils. *Soil Research*, *52*(3), 307–316.

Braithwaite, K. S., Croft, B. J., Magarey, R. C., & Scharaschkin, T. (2009). Phylogenetic placement of the sugarcane orange rust pathogen Puccinia kuehnii in a historical and regional context. *Australasian Plant Pathology*, *38*(4), 380–388.

Bull, T. (2000). The sugarcane plant. In M. Hogarth & P. Allospp (Eds.), *Manual of cane growing* (pp. 91–108). Bureau of Sugar Experimental Stations.

Chakraborty, M., & Sahu, P. (2020). Scope of sugarcane by-products in reviving sugar industry. *Indian Journal of Research*, *10*(1), 3–16.

Chand, R. A., Parwar, B. R., & Krishna, M. (2016). Constraints faced in production and marketing of sugarcane in Parbhani district of Maharashtra. *Agriculture Update*, *11*(3), 209–213.

Chauhan, R. S., & Srivastava, T. K. (2002). Influence of weed management practices on weed growth and yield of sugarcane. *Indian Journal of Weed Science*, *34*, 318–319.

Cheavegatti-Gianotto, A., de Abreu, H. M. C., Arruda, P., Filho, J. C. B., Burnquist, W. L., Creste, S., di Ciero, L., Ferro, J. A., de Oliveira Figueira, A. V., de Sousa Filgueiras, T., de Fátima Grossi-de-Sá, M., Guzzo, E. C., Hoffmann, H. P., de Andrade Landell, M. G., Macedo, N., Matsuoka, S., de Castro Reinach, F., Romano, E., da Silva, W. J., de Castro Silva Filho, M., & Ulian E. C. (2011). Sugarcane (Saccharum × officinarum): A reference study for the regulation of genetically modified cultivars in Brazil. *Tropical Plant Biology*, *4*, 62–89.

Choudhary, H. R., Singh, R. K., Prajapat, K., et al. (2013). Water management in sugarcane. Paper presented at the National Seminar on Enhancing Water Productivity in Agriculture.

Coelho, M. P. M., Correia, J. E., Vasques, L. I., de Castro Marcato, A. C., de Andrade Guedes, T., Soto, M. A., Basso, J. B., Kiang, C., & Fontanetti, C. S. (2018). Toxicity evaluation of leached of sugarcane vinasse: Histopathology and immunostaining of cellular stress protein. *Ecotoxicology and Environmental Safety*, *165*, 367–375.

Cortes-Rodríguez, E. F., Fukushima, N. A., Palacios-Bereche, R., Ensinas, A. V., & Nebra, S. A. (2018). Vinasse concentration and juice evaporation system integrated to the conventional ethanol production process from sugarcane – heat integration and impacts in cogeneration system. *Renewable Energy*, *115*, 474–488.

Cox, M., Hogarth, M., & Smith, G. (2000). Cane breeding and improvement. In M. Hogarth & P. Allsopp (Eds.), *Manual of Cane Growing* (pp. 91–108). Bureau of Sugar Experimental Stations. Retrieved from www.sugarresearch.com.au/icms_docs/166961_Chapter_5_Cane_Breeding_and_Improvement.pdf

da Silva, J. J., da Silva, B. F., Zanoni, M. V. B., & Ramos Stradiotto, N. (2022). Sample preparation and antibiotic quantification in vinasse generated from sugarcane ethanol fuel production. *Journal of Chromatography A*, *1666*, 462833.

de Jesus, G. C., Gaspar, B. R., & da Silva, A. M. (2019). Production and characterization of alginate beads for growth of immobilized desmodesmus subspicatus and its potential to remove potassium, carbon and nitrogen from sugarcane vinasse. *Biocatalysis and Agricultural Biotechnology*, *22*, 101438.

de Souza, R. B., de Menezes, J. A., de Souza, R. D., Dutra, E. D., & de Morais Jr., M. A. (2015). Mineral composition of the sugarcane juice and its influence on the ethanol fermentation. *Applied Biochemistry and Biotechnology*, *175*(1), 209–222.

Deng, Z. N., Huang, H. R., Huang, C. M., Verma, K. K., Luo, H. B., Jiang, S. L., Li, Y. J., Cao, H. Q., Wu, K. C., & Xu, L. (2022). Short-term limited water irrigation influences the photosynthetic pigments, enzymatic and non-enzymatic activities in *Saccharum spontaneum* L. at vegetative stage. *Sugar Tech.*, *25*, 110–118. https://doi.org/10.1007/s12355-022-01175-7.

Dingre, S. K., & Gorantiwar, S. D. (2020). Determination of the water requirement and crop coefficient values of sugarcane by field water balance method in semiarid region. *Agricultural Water Management*, *232*, 106042.

Dotaniya, M. L., & Datta, S. C. (2014). Impact of bagasse and press mud on availability and fixation capacity of phosphorus in an Inceptisol of north India. *Sugar Tech.*, *16*(1), 109–112.

Dotaniya, M. L., & Kushwah, S. K. (2013). Nutrients uptake ability of various rainy season crops grown in a Vertisol of central India. *African Journal of Agricultural Research*, *8*(44), 5592–5598.

Dotaniya, M. L., Datta, S. C., Biswas, D. R., Dotaniya, C. K., Meena, B. L., Rajendiran, S., Regar, K. L., & Lata, M. (2016). Use of sugarcane industrial by-products for improving sugarcane productivity and soil health. *International Journal of Recycling of Organic Waste in Agriculture*, *5*, 185–194.

Ferreiro, O. B., Kronemberger, F. A., & Borges, C. P. (2021). Sugarcane stillage treatment using direct contact membrane distillation. *Waste and Biomass Valorization*, *12*, 3987–3999.

Food and Agriculture Organization of the United Nations. (2020). FAOSTAT. https://www.fao.org/faostat/en/#data/QC

Fregolente, L. G., dos Santos, J. V., Mazzati, F. S., Miguel, T. B. A. R., de C. Miguel, E., Moreira, A. B., Ferreira, O. P., & Bisinoti, M. C. (2021). Hydrochar from sugarcane industry by-products: Assessment of its potential use as a soil conditioner by germination and growth of maize. *Chemical and Biological Technologies in Agriculture*, *8*, 16.

Frías, M., Villar, E., & Savastano, H. (2011). Brazilian sugar cane bagasse ashes from the cogeneration industry as active pozzolans for cement manufacture. *Cement and Concrete Composites*, *33*, 490–496. https://doi.org/10.1016/j.cemconcomp.2011.02.003.

Gentile, A., Dias, L. I., Mattos, R. S., Ferreira, T. H., & Menossi, M. (2015). MicroRNAs and drought responses in sugarcane. *Frontiers in Plant Science*, *6*, 58.

Gholve, S. G., Kumbhar, S. G., & Bhoite, D. S. (2001). Recycling of various conventional and non-conventional organic sources in adsali sugarcane (*Saccharum officinarum* L.) planted with different patterns. *Indian Sugar*, *1*(1), 23–27.

Ghulam, S., Khan, M. J., Usman, K., et al. (2012). Effect of different rates of press mud on plant growth and yield of lentil in calcareous soil. *Sarhad Journal of Agriculture*, *28*(2), 249–252.

Gilbert, R. A., Glynn, N. C., Comstock, J. C., & Davis, M. J. (2009). Agronomic performance and genetic characterization of sugarcane transformed for resistance to sugarcane yellow leaf virus. *Field Crops Research*, *111*(1-2), 39–46. http://dx.doi.org/10.1016/j.fcr.2008.10.009.

Gujja, B., Loganandhan, N., Goud, V. V., et al. (2009). Sustainable Sugarcane Initiative-improving sugarcane cultivation in India. Training manual by ICRISAT-WWF Project.

Gupta, N., Tripathi, S., & Balomajumder, C. (2011). Characterization of press mud: A sugar industry waste. *Fuel*, *90*(1), 389–394.

Harender, Nimbrayan, P. K., Sunil, et al. (2021). Sustainability of sugarcane cultivation and production in India – A review. *Agricultural Mechanization in Asia, Africa & Latin America*, *51*(1), 1051–1071.

Hoareau, W., Oliveira, F. B., Grelier, S., & Siegmun, B. (2006). Fibreboards based on sugarcane bagasse lignin and fibres. *Macromolecular Materials and Engineering*, *291*(7), 829–839.

Hofsetz, K., & Silva, M. A. (2012). Brazilian sugarcane bagasse: Energy and non-energy consumption. *Biomass and Bioenergy*, *46*(6), 564–573.

Huang, H. R., Gao, Y. J., Malviya, M. K., Verma, K. K., Solanki, M. K., Huang, Y. X., & Xu, L. (2021). Genetic diversity analysis of sugarcane (Saccharum spp. Hybrids) among high-sucrose clones of GT series and commonly used parents by using microsatellite markers in Guangxi, China. *Sugar Tech.*, *24,* 397–407. https://doi.org/10.1007/s12355-021-01033-y.

ICRISAT. (2009). Training manual on sustainable sugarcane initiative: Improving sugarcane cultivation in India, an initiative of ICRISAT-WWF Project, ICRISAT.

Indian Sugar Mills Association. (2019). Presentation to Task Force on Sugarcane and Sugar Industry. Retrieved from: www.indiansugar.com/uploads/Niti_Aayog.pdf (accessed on 27 May 2024).

Inman-Bamber, N. G., & McGlinchey, M. G. (2003). Crop coefficients and water-use estimates for sugarcane based on long-term Bowen ratio energy balance measurements. *Field Crops Research*, *83*, 125–138.

Izaguirre, Y. P., Regal, M. R., Lema, E. R., et al. (2018). Diseases and pests affecting sugarcane: Methods of control. In P. Singh & A. K. Tiwari (Eds.), *Sustainable Sugarcane Production* (pp. 217–240). Apple Academic Press Inc.

Jackson, P. A. (2005). Breeding for improved sugar content in sugarcane. *Field Crop Research*, *92*(2-3), 277–290. https://dx.doi.org/10.1016/j.fcr.2005.01.024

Kamboj, B. R., Malik, R. K., Garg, R., & Yadav, A. (2008). Bed planting – A novel technique to encourage multiple land use. *Technical Bulletin*, *29*, 24.

Kawade, U. R., Rathi, V. R., & Vaishali, D. G. (2013). Effect of use of bagasse ash on strength of concrete. *International Journal of Innovative Research in Science, Engineering and Technology*, *2*, 2997–3000.

Khandagave, R. B. (2003). Influence of organic and inorganic manure on sugarcane and sugar yield. *Indian Sugar*, *52*, 981–989.

Kumar, R., Mohd Yaseen, A. Y. B., Shafiq, N., & Jalal, A. (2017). Influence of metakaolin, fly ash and nano silica on mechanical and durability properties of concrete. *Key Engineering Materials*, *744*, 8–14. https://doi.org/10.4028/www.scientific.net/kem.744.8.

Lakshmanan, P., & Robinson, N. (2013). Stress physiology: Abiotic stresses. *In Sugarcane: Physiology, Biochemistry, and Functional Biology* (pp. 411–434).

Li, Y. R., & Yang, L. T. (2015a). Sugarcane agriculture and sugar industry in China. *Sugar Tech*, *14*, 1–8.

Loganandhan, N., Gujja, B., Goud, V. V., & Natarajan, U. S. (2012). Sustainable Sugarcane Initiative (SSI): A Methodology of 'More with Less.' *Sugar Tech.*, *15,* 98–102. https://doi.org/10.1007/s12355-012-0180-y.

Macedo, I. C., Seabra, J. E. A., & Silva, J. E. A. R. (2008). Greenhouse gases emissions in the production and use of ethanol from sugarcane in Brazil: The 2005/2006 averages and a prediction for 2020. *Biomass and Bioenergy*, *32*, 582–595. https://doi.org/10.1016/j.biombioe.2007.12.006.

Magrini, F. E., de Almeida, G. M., da Maia, S. D., Fuentes, L., Ecthebehere, C., Beal, L. L., da Silveira, M. M., & Paesi, S. (2021). Effect of different heat treatments of inoculum on the production of hydrogen and volatile fatty acids by dark fermentation of sugarcane vinasse. *Biomass Conversion and Biorefinery*, *11*, 2443–2456.

Malyadri, T., & Supriya, J. (2015). Experimental study on bagasse ash in concrete by partial replacement with cement. International Journal of Computational Engineering *Research and Trends*, *2*, 995–1001.

Montiel-Rosales, A., Montalvo-Romero, N., García-Santamaría, L. E., Sandoval-Herazo, L. C., Bautista-Santos, H., & Fernández-Lambert, G. (2022). Post-Industrial use of sugarcane ethanol vinasse: A systematic review. *Sustainability*, *14*, 11635. https://doi.org/10.3390/su141811635.

Narayanamoorthy, A. (2005). Economics of drip irrigation in sugarcane cultivation: Case study of a farmer from Tamil Nadu. *Indian Journal of Agricultural Economics*, *60*(2), 235–248.

OECD. (2016). Safety assessment of transgenic organisms in the environment, vol 6: OECD Consensus Documents, Harmonization of Regulatory Oversight in Biotechnology. OECD Publishing. http://dx.doi.org/10.1787/9789264253421-en.

Parajuli, T., Kurre, D. K., & Jangde, S. K. (2019). Sustainable sugarcane initiative (SSI): An approach to enhance sugarcane cultivation and input use efficiency and sustainable yield of sugarcane in India. *International Journal of Agricultural Sciences*, *15*(1), 222–226. https://doi.org/10.15740/HAS/IJAS/15.1/222-226.

Patil, N. N., Jadhav, S., Ghorpade, S. S., & Sharma, A. B. (2013). Isolation and enrichment of sugar press mud (SPM) adapted microorganism for production of biofertilizer by using sugar press mud. *International Journal of Advanced Biotechnology Research*, *14*(1), 96–104.

PAU. (2018-2019). Package of practices for crops of Punjab. Rabi 2018-19, released by Punjab Agricultural University, Ludhiana.

Prasanna, S. M., & Pattar, P. S. (2014). Successful ratoon management in sugarcane. *Research and Reviews: Journal of Agriculture and Allied Sciences*, *3*(4), 39–47.

Qureshi, M. E., Mallawaarachchi, T., Wegener, M. K., Bristow, K. L., Charlesworth, P. B., & Lisson, S. (2001). Economic evaluation of alternative irrigation practices for sugarcane production in the Burdekin Delta (No. 412-2016-25824).

Ramesh, P. (2000). Effect of different levels of drought during the formative phase on growth parameters and its relationship with dry matter accumulation in sugarcane. *Journal of Agronomy and Crop Science*, *185*(2), 83–89.

Ray, B., & Dasgupta, M. (2006). Sugarcane crop loss due to wilt caused by parasitic angiosperm *Aeginetia pedunculata* (Roxb.) Wall. (*Orobanchaceae*). *Journal of Mycology and Plant Pathology, 36*(1), 31–34.

Raza, Q-U-A., Bashir, M. A., Rehim, A., & Raza, H. M. A. (2021). Role of sugarcane industrial byproducts on soil physicochemical properties and metal accumulation in rice. *Environmental Science and Pollution Research*, *29*, 24726–24736.

Razzaq, A. (2001). Assessing sugarcane filtercake as crop nutrients and soil health ameliorant. *Pakistan Sugar Journal*, *21*(3), 15–18.

Robertson, M. J., Inman-Bamber, N. G., Muchow, R. C., & Wood, A. W. (1999). Physiology and productivity of sugarcane with early and mid-season water deficit. *Field Crops Research*, *64*(3), 211–227.

Rodella, A. A., Silva, L. C. F. D. A., & Filho, J. O. (1990). Effect of filter cake application on sugarcane yields. *Turrialba*, *40*, 323–326.

Rosegrant, M. W. (1997). Water Resources in the Twenty-First Century: Challenges and Implications for Action, Food and Agriculture, and the Environment. Discussion Paper 20, International Food Policy Research Institute, Washington, D.C., U.S.A., March.

Rosegrant, M. W., & Meinzen-Dick, R. S. (1996). Water resources in the Asia-Pacific region: Managing scarcity. *Asian-Pacific Economic Literature*, *10*(2), 32–53.

Rulli, M. M., Villegas, L. B., Barcia, C. S., & Colin, V. L. (2021). Bioconversion of sugarcane vinasse into fungal biomass protein and its potential use in fish farming. *Journal of Environmental Chemical Engineering*, *9*, 106136.

Sanghera, G. S., Singh, R. P., Singh, O., & Tyagi, V. (2018). Initiatives for sustainable sugarcane farming in Punjab. *Journal of Plant Sciences Research*, *34*(2), 137–152.

Santos, N. L., de Oliveira, R. G., Cerri, B. C., Soares, M. R., Kieckbusch, T. G., & da Silva, M. A. (2020). Physicochemical properties of konjac glucomannan/alginate films enriched with sugarcane vinasse intended for mulching applications. *International Journal of Biological Macromolecules*, *165*, 1717–1726.

Sarwar, G., Schmeisky, H., Hussain, N., Muhammad, S., Ibrahim, M., & Safdar, M. E. (2008). Improvement of soil physical and chemical properties with compost application in rice-wheat cropping system. *Pakistan Journal of Botany*, *40*, 275–282.

Sawhney, V., & Singh, D. P. (2002). Effect of chemical desiccation at the post-anthesis stage on some physiological and biochemical changes in the flag leaf of contrasting wheat genotypes. *Field Crops Research*, *77*(1), 1–6.

Shafiq, N., Hussein, A. A. E., Nuruddin, M. F., et al. (2018). Effects of sugarcane bagasse ash on the properties of concrete. *Proceedings of the Institution of Civil Engineers – Engineering Sustainability*, *171*, 123–132.

Shankaraiah, C., & Nagaraju, M. S. (2002). Economical and ecofriendly practices for sustainable sugarcane. *Cooperative Sugar*, *33*(8), 647–653.

Shanthy, T. R., & Ramanjaneyulu, S. (2014). Socio-economic performance analysis of sugarcane cultivation under sustainable sugarcane initiative method. *Indian Research Journal of Extension Education*, *14*(3), 93–98.

Shrivastava, A. K., Srivastava, A. K., & Solomon, S. (2011). Sustaining sugarcane productivity under depleting water resources. *Current Science*, *101*(6), 748–754.

Shukla, M., Patel, R. H., Verma, R., Deewan, P., & Dotaniya M. L. (2013). Effect of bio-organics and chemical fertilizers on growth and yield of chickpea (Cicer arietinum L.) under middle Gujarat conditions. *Vegetos*, *26*(1), 183–187.

Shukla, S. K., Sharma, L., Awasthi, S. K., et al. (2017). Sugarcane in India: Package of practices for different agro-climatic zones, pp. 1–64.

Shukla, S. K., Yadav, R. L., Awasthi, S. K., & Gaur, A. (2016). Soil microbial biomass nitrogen, in situ respiration and crop yield influenced by deep tillage, moisture regimes and N nutrition in sugarcane based system in subtropical India. *Sugar Tech*, *19*, 125–135. https://doi.org/10.1007/s12355-016-0442-1.

Silva, L. C. M., Moreira, R. A., Pinto, T. J. S., Vanderlei, M. R., Athayde, D. B., Lopes, L. F. P., Ogura, A. P., Yoshii, M. P. C., Freitas, J. S., Montagner, C. C., Goulart, B. V., Schiesari, L., Daam, M. A., & Espíndola, E. L. G. (2021). Lethal and sublethal toxicity of pesticides and vinasse used in sugarcane cultivation to *Ceriodaphnia silvestrii* (Crustacea: Cladocera). *Aquatic Toxicology*, *241*, 106017.

Singh, A. (1993). Economics of intercropping in autumn planted sugarcane. *Journal of Sugarcane Technology*, *8*(1), 71–72.

Singh, A. K., Lal, M., & Singh, E. (2018). Headways in agro-techniques for heightened yield of sugarcane: Indian perspective. In: Singh P, Tiwari A K (eds.) Sustainable Sugarcane Production, Apple Academic Press Inc. pp. 17–75.

Singh, I., Verma, R. R., & Srivastava, T. K. (2018). Growth, yield, irrigation water use efficiency, juice quality and economics of sugarcane in Pusa Hydrogel application under different irrigation scheduling. *Sugar Tech*. *20*(1), 29–35.

Singh, V. P., Pareek, N., Singh, S. P., Raverkar, K. P., Satyawali, K., Bisht, N., Joshi, N., Kumar, A., & Kaushik, S. (2017). Halosulfuron+metribuzin effect on weed control in sugarcane and their carry over effect on succeeding lentil. *Indian Journal of Weed Science*, *49*(4), 364–369.

Sivanappan, R. K. (1994). Prospects of micro-irrigation in India. Irrigation and Drainage System, 8, 49–58.

Song, X. P., Mo, F., Yan, M., Zhang, X., Zhang, B., Huang, X., Huang, D., Pan, Y., Verma, K. K., & Li, Y. R. (2022). Effect of smut infection on the photosynthetic physiological characteristics and related defense enzymes of sugarcane. *Life*, *12*, 1201. https://doi.org/10.3390/life12081201.

Song, X. P., Verma, K. K., Tian, D. D., Zhang, X. Q., Liang, Y. J., Huang, X., Li, C. N., & Li, Y. R. (2021). Exploration of silicon functions to integrate with biotic stress tolerance and crop improvement. *Biological Research*, *54*, 19. https://doi.org/10.1186/s40659-021-00344-4.

Sousa, S. P., Lovato, G., Albanez, R., Ratusznei, S. M., & Rodrigues, J. A. D. (2019). Improvement of sugarcane stillage (vinasse) anaerobic digestion with cheese whey as its co-substrate: Achieving high methane productivity and yield. *Applied Biochemistry and Biotechnology*, *189*, 987–1006.

Suganthi, M., & Muthukrishnan, P. (2017). Effect of chemical weed management on quality parameters, yield parameters and yield in sugarcane in western zones of Tamil Nadu. *Indian Journal of Weed Science*, *32*, 46–49.

Suguna, M., & Jayanthi, M. (2021). Sugarcane production in India – an overview. *Journal of Emerging Technology and Innovative Research*, *8*(10), 702–705.

Sundar, A. R., Ashwin, N. M. R., Barnabas, E. L., Malathi, P., & Viswanathan, R. (2015). Disease resistance in sugarcane – an overview. *Scientia Agraria Paranaensis*, *14*(4), 200–212.

Sundara, B. (2011). Agrotechnologies to enhance sugarcane productivity in India. *Sugar Tech. 13*(4), 281–298.

Tamashiro, J. R., Kinoshita, A., Pereira Silva, L. H., de Paiva, F. F. G., Antunes, P. A., & Simões, R. D. (2022). Compressive resistance of concrete produced with recycled concrete aggregate and sugarcane vinasse waste-water. *Clean Engineering and Technology*, *6*, 100362.

United States Department of Agriculture (USDA). (2020). Foreign Agricultural Service. Sugar: World Markets and Trade. https://apps.fas.usda.gov/psdonline/circulars/Sugar.pdf.

Ventorino, V., Nicolaus, B., di Donato, P., Pagliano, G., Poli, A., Robertiello, A., Lavarone, V., & Pepe, O. P. (2019). Bioprospecting of exopolysaccharide-producing bacteria from different natural ecosystems for biopolymer synthesis from vinasse. *Chemical Biology and Technology of Agriculture*, *6*, 18.

Verma, H. (2019). Organic cultivation in sugarcane. In: Rawat A. K., & Tripathi U. K. (eds.) Advances in Agronomy, Akinik Publication, pp. 31–43.

Verma, K. K., Song, X. P., Budeguer, F., Nikpay, A., Enrique, R., Singh, M., Zhang, B., Wu, J. M., & Li, Y. R. (2022b). Genetic engineering: an efficient approach to mitigating biotic and abiotic stresses in sugarcane cultivation. *Plant Signaling & Behavior*, *17*(1), e2108253.

Verma, K. K., Song, X. P., Li, D. M., Singh, M., Wu, J. M., Singh, R. K., Sharma, A., Zhang, B. Q., & Li, Y. R. (2022c). Silicon and soil microorganisms improve rhizospheric soil health with bacterial community, plant growth, performance, and yield. *Plant Signaling & Behavior*, *17*(1), e2104004.

Verma, K. K., Song, X. P., Lin, B., Guo, D. J., Singh, M., Rajput, V. D., Singh, R. K., Singh, P., Sharma, A., Malviya, M. K., Chen, G. L., & Li, Y. R. (2021d). Silicon-induced drought tolerance in crop plants: physiological adaptation strategies. *Silicon*, *14*, 2473–2487. DOI: 10.1007/s12633-021-01071-x.

Verma, K. K., Song, X. P., Rajput, V. D., Boldyreva, V., Zhang, B. Q., Minkina, T., & Li, Y. R. (2022d). Morpho-physiological, biochemical, and ultrastructural modifications on sugarcane to prolonged water deficit. In Verma, K. K., et al. (eds.), Agro-industrial perspectives on sugarcane production under environmental stress. https://doi.org/10.1007/978-981-19-3955-6_11

Verma, K. K., Song, X. P., Tian, D. D., Guo, D. J., Chen, Z. L., Zhong, C. S., Nikpay, A., Singh, M., Rajput, V. D., Singh, R. K., Minkina, T., & Li, Y. R. (2021a). Influence of silicon on biocontrol strategies to manage biotic stress for crop protection, performance, and improvement. *Plants*, *10*, 2163. https://doi.org/10.3390/plants10102163.

Verma, K. K., Song, X. P., Tian, D. D., Singh, M., Verma, C. L., Rajput, V. D., Singh, R. K., Sharma, A., Singh, P., Malviya, M. K., & Li, Y. R. (2021b). Investigation of the defensive role of silicon during drought stress induced by irrigation capacity in sugarcane: physiological and biochemical characteristics. *ACS Omega*, *6*, 19811–19821.

Verma, K. K., Song, X. P., Verma, C. L., Chen, Z. L., Rajput, V. D., Wu, K. C., Liao, F., Chen, G. L., & Li, Y. R. (2021c). Functional relationship between photosynthetic leaf gas exchange in response to silicon application and water stress mitigation in sugarcane. *Biology Research*, *54*, 15. https://doi.org/10.1186/s40659-021-00338-2.

Verma, K. K., Song, X. P., Verma, C. L., Malviya, M. K., Guo, D. J., Rajput, V. D., Sharma, A., Wei, K. J., Chen, G. L., Solomon, S., & Li, Y. R. (2021f). Prediction of photosynthetic leaf gas exchange of sugarcane (Saccharum spp.) leaves in response to leaf positions to foliar spray of potassium salt of active phosphorus under limited water irrigation. *ACS Omega*, *6*, 2396–2409.

Verma, K. K., Song, X. P., Yadav, G., Degu, H. D., Parvaiz, A., Singh, M., Huang, H. R., Mustafa, G., Xu, L., & Li, Y. R. (2022a). Impact of agroclimatic variables on proteogenomics in sugarcane (Saccharum spp.) plant productivity. *ACS Omega*, *7*, 22997–23008.

Verma, K. K., Song, X. P., Zeng, Y., Guo, D. J., Sing, M., Rajput, V. D., Malviya, M. K., Wei, K. J., Sharma, A., Li, D. P., Chen, G. L., & Li, Y. R. (2021e). Foliar application of silicon boosts growth, photosynthetic leaf gas exchange, antioxidative response, and resistance to limited water irrigation in sugarcane (*Saccharum officinarum* L.). *Plant Physiology and Biochemistry*, *166*, 582–592.

Vuyyuru, M., Sandhu, H. S., McCray, M. J., Raid, R. N., & Erickson, J. E. (2019). Effects of nitrogen fertilization and seed piece applied fungicides on establishment, tiller dynamics, and sucrose yields in successively planted sugarcane. *Agronomy*, *9*(387), 1–22. https://doi.org/10.3390/agronomy9070387

Xiao, Z., Liao, X., & Guo, S. (2017). Analysis of sugarcane juice quality indexes. *Journal of Food Quality*, *3*, 1–6. https://doi.org/10.1155/2017/1746982

Xu, L., Deng, Z. N., Wu, K. C., Malviya, M. K., Solanki, M. K., Verma, K. K., Pang, T., Li, Y. J., Liu, X. Y., Kashyap, B. K., Dessoky, E. S., Wang, W. Z., & Huang, H. R. (2022). Transcriptome analysis reveals a gene expression pattern that contributes to sugarcane bud propagation induced by Indole-3-Butyric Acid. *Frontiers in Plant Science*, *13*, 852886. doi: 10.3389/fpls.2022.852886.

CHAPTER 2

Influence of Environment on Propagation and Germination of Sugarcane

MOHSEEN RIAZ UD DEAN,[1] MARISSA CORINNA ASEN,[2] and RAHUL RAVNEEL PRASAD[3]

[1]*Anthropology Program, Faculty of Arts and Social Sciences, University of Waikato, Hamilton, New Zealand*

[2]*School of Law and Social Sciences, University of the South Pacific, Suva, Fiji*

[3]*Pacific Center for Environment and Sustainable Development, University of the South Pacific, Suva, Fiji*

ABSTRACT

Sugarcane is an environmentally sensitive crop. Its spatial distribution and growth are affected by various environmental parameters. Different sugarcane crop cultivars have been found to react and respond differently to varying ecological conditions. Environmental factors that affect plant growth include light, temperature, water, humidity, and nutrition. For instance, environmental factors such as scorching sunlight or heavy downpours can affect germinating crop buds. In some cases, poor environmental conditions, e.g., drought, can impair a germinating crop. Likewise, environmental stressors can weaken a plant and make it more susceptible to disease or insect attack. Understanding how these environmental factors affect sugarcane plant growth and development is integral. With a thorough understanding of these factors, one may be able to advance or manipulate cropping techniques, including the conditions under which crops are sown, as well as devising

Sugarcane Cultivation and Management: Challenges and Opportunities. Krishan K. Verma, Xiu-Peng Song, Munna Singh, Renato de Mello Prado, Jian-Ming Wu, and Yang-Rui Li (Eds.)

crop management plans for increasing overall crop yields. In this chapter, we focus on different environmental parameters and how they affect the overall growth and development of sugarcane crops.

2.1 INTRODUCTION

The propagation and germination of sugarcane are highly influenced by its external environment. Environmental factors that influence the production and yield of sugarcane crops such as temperature, light, water, humidity, nutrition, pests, and diseases are all affected by the impacts of climate change. With external stressors becoming more complex due to these evolving factors, it is important for sugarcane farmers to be aware of the most suitable cropping techniques for their changing environmental conditions, as well as be aware of the latest research in relation to increasing production and yield of sugarcane. Although sugarcane crops are susceptible to external stressors, there are a variety of measures that can be taken to increase production, and farmer awareness and access to resources will be essential in ensuring global sugarcane production can withstand climate change impacts on propagation and germination (Verma et al., 2020a, b, c, 2022a, b, 2023).

2.2 CLIMATE

Sugarcane crop cultivars usually take up to 12 months to grow, with some taking up to 24 months (Riajaya et al., 2022). The germinating crop undergoes varying weather seasons during this period, including different micro- and macro-climatic conditions. Climate plays a significant role in the development of sugarcane in all stages of the crop's life, with rainfall, humidity, sunshine, temperature, pests, and diseases all having various effects on sugarcane production, germination, propagation, sugar yield, and sucrose quality (Zhao & Li, 2015; Verma et al., 2021a; Song et al., 2022). With the impacts of climate change projected to continue to affect agriculture and coupled with various stressors altering the germination and propagation process in a fast-changing environment with extreme climatic occurrences, farmers must adopt all necessary measures to ensure their sugarcane crops are adaptive to the changing environmental factors.

Sugarcane is already susceptible to many environmental factors, and it will be important for farmers to find the ideal combination of various crop management techniques to optimize production and yield (Meena et al., 2020; Verma, 2021b, c). It is integral to note that what may help increase germination, propagation, yield, and production in one sugarcane-producing country may not work for another, as each country has its combination of different factors. Challenges and interventions will also be micro-location specific, which is why building the capacity of farmers is crucial so that they can identify issues and challenges specific to their situations and determine the most appropriate interventions. Knowledge sharing in this space will increasingly become a key factor in building awareness and best practices in the sugar industry (Verma et al., 2021d, 2023).

2.3 TEMPERATURE

While sugarcane crops are asexually or vegetatively propagated for commercial use, the seeds are important for hybridization and genetically modifying crops for more productive and resilient sugarcane (Budeguer et al., 2021). Genetic improvement has become a major priority in the sugarcane industry to produce crops that are more resilient to changing external factors, as well as to maintain and increase overall crop yield. Hybridization allows the combination of important and specific traits of various sugarcane species to produce superior progeny with high production and resilience to abiotic and biotic stressors (Ganapati et al., 2016).

The propagation and germination of sugarcane are susceptible to abiotic factors, with temperature being a major determinant of crop production (Galon et al., 2013). Given the rapidly fluctuating temperatures and global warming caused by climate change, there have been serious concerns globally about the sugarcane crop's ability to withstand this catastrophe of changing temperatures (Hussain et al., 2019). While sugarcane seeds can germinate at temperatures ranging from 11°C to 42°C, the optimal temperature for sugarcane germination is still being debated, with many studies indicating that it falls between approximately 27°C and 36°C, depending on the sugarcane variety (Pierre et al., 2014).

While some varieties of sugarcane have been found to be more resilient as well as vulnerable to varying temperatures, many of the commercial sugarcane varieties grown today are a cross between two or more cultivar species (Cheavegatti-Gianotto et al., 2011). Cultivars high in sucrose but vulnerable to diseases are often crossed with species with increased resilience to varying temperatures, moisture, stress, pests, and diseases but with lower sucrose levels (Goodall, 1998). Therefore, although more resilient to environmental stressors such as extreme temperature, some hybrid sugarcane varieties would have lower sucrose levels and flavor. Temperatures below the ideal range have been found to negatively affect sugarcane crop growth and decelerate cultivation due to limitations on photosynthesis. In contrast, temperatures above the ideal range for sugarcane growth hinder sprouting, leading to fewer plant shoots (Srivastava, 2012). Higher temperatures have also resulted in shorter internodes, higher moisture content, stalk fiber, and lower sucrose levels (Bonnett et al., 2006).

Higher temperatures at night can lead to increased flowering, leading to a cease in internode growth, resulting in decreased sucrose and cane yield (Mhlanga-Ndlovu & Nhamo, 2017). The sugarcane flowering process influences the crop ripening process, and temperature is an external factor affecting sugarcane flowering. The reduction of sucrose and increased stalk fiber are the major results of the flowering process, which is why it is considered undesirable in sugarcane production. Rising temperatures also affect the natural ripening process and decrease the sugarcane quality (Cardozo & Sentelhas, 2013). An increase in the irrigation cycle is often required to compensate for the increased temperature and evaporation brought about by climate change, which poses the risk of over-irrigation, waterlogging, salinity challenges, and reduction in sugarcane yield (Hussain et al., 2019). High temperatures affect all stages of plant growth. Still, germination and propagation are particularly vulnerable stages and can also lead to cellular damage, cell death, and the downfall of cellular organization (Hu et al., 2020). Some common features of sugarcane plants adapting to increasing temperatures include the closure of stomata, enhanced stomatal densities, and changes in membrane lipid compositions.

On the other hand, low temperatures affect sugarcane crop planting season and geographical distribution (Hussain et al., 2019). Slow plant growth and development are common in sugarcane crops dealing with low-temperature stress due to oxidative and osmotic stress caused by the decreased metabolic process. Apart from the direct effects of temperature on sugarcane germination and propagation, it also has indirect impacts that further hinder sugarcane growth by supporting the germination of some weeds, pests, and diseases (ibid). For example, smut disease—a fungal disease known to cause major economic losses in the sugarcane industry—has been shown to increase, most likely parallel to temperatures (Liu et al., 2017). Additionally, summer weeds that often become dormant during winter months may reappear due to climate change-induced increased temperatures (Ramesh et al., 2017).

Temperature changes also enable a more favorable environment for introducing new pests (Skendzic et al., 2021). The efficacy of herbicides is also affected by temperature since stomata conductance is stimulated at higher temperatures and increases permeability, inhibiting herbicide uptake (Matzenbacher et al., 2014). Abiotic ailments such as sunburn and cold/banded chlorosis resulting from extreme hot or cold temperatures can also affect sugarcane physiology and biochemistry and lead to decreased quality and yield (Sanghera, 2020). Increasing temperatures are more likely to negatively affect sugarcane grown in tropical regions where cooler winter months are required to slow plant growth to increase sucrose storage. Given the extreme effects of climate change and poor commitments to reduce or stop global warming, it is extremely important to develop innovative sugarcane cultivars adaptable to the changing climate and extreme temperature fluctuations. This can be done through enhanced research and discovery of suitable genes for agronomic trait development and utilizing physiological screening, basic breeding, as well as new molecular biology technologies for sugarcane germination and propagation as well as increasing resilience against climate change, thus leading to improved overall productivity and sustainability (Zhao & Li, 2015).

2.4 WATER

The amount of water germinating crop buds receive is integral to their health and overall sugarcane crop productivity (Singh et al., 2019). Because of drought, water deficit is one of the most common abiotic stressors of sugarcane cropping and severely affects yield, productivity, and growth. Water deficit can reduce crop productivity by more than half (Hussain et al., 2019). Sugarcane is a high-water-consuming plant, so drought is the major stressor for this crop. Sugarcane cropping is usually focused in areas with sufficient rainfall for optimal growth and productivity. Where this is not the case, sugarcane requires supplemental water supply through irrigation. Approximately 8–12 cane crops are estimated to be produced per milliliter of irrigation water (Ferreira et al., 2017). While sugarcane plants adopt various coping mechanisms to environmental stressors such as drought and water deficiency, the response of sugarcane to these abiotic stressors depends on the length of exposure, degree, area of the plant affected, and variety of plants (Verma et al., 2020d, e, f, 2022b). Stomatal closure, cell elongation, leaf rolling, decreased leaf surface area, and reduced stalk and leaf growth are all common indicators of sugarcane plants affected by water stress or water deficits (ibid).

Rates of photosynthesis are also affected by water deficiency, which severely inhibits sugarcane plant germination and growth. Like abiotic temperature stress, water deficit also negatively affects the sugarcane crop's ability to absorb herbicides effectively and therefore makes the sugarcane crop more susceptible to invasive weeds (Alizade et al., 2020; Verma et al., 2022c). As previously mentioned, climate change-induced global warming severely impacts sugarcane productivity as crops are affected by increasing temperatures. These rising temperatures also lead to changes in frequency and micro- and macro-precipitation levels. Sucrose levels are directly affected by the changing levels of rainfall and precipitation. Similarly, an increasing number of droughts have been found to have negative effects on sensitive sugarcane-germinating crops.

To combat water deficits, farmers naturally turn to increased irrigation. However, this can often lead to other issues, such as salinity problems (Shrivastava & Kumar, 2015; Verma et al., 2023). Waterlogging also results from increased irrigation or extreme rainfall changes, often leading to flooding. Flooding and waterlogging can prevent sugarcane plants' root system's access to oxygen, as well as limit their nutrient uptake (Azhar et al., 2022). How water deficit or drought stress affects sugarcane also depends on the germination stage that the plant is in when exposed to the abiotic stressor. For example, drought may reduce the sucrose levels in the crop at the early to mid-stages of germination.

On the other hand, it has been found that moderate water stress during the mature stage of sugarcane growth can increase sucrose yield, as CO_2 will be channeled to sucrose production and accumulation in the cane stalk since photosynthesis is less susceptible to water deficit than stem growth (Ferreira et al., 2017). Therefore, maintaining sufficient water levels during germination must be effectively managed through efficient water use. High irrigation efficiency and the use of sugarcane plant varieties that are more resistant or tolerant to drought and drier conditions will be crucial for mitigation and adaptation as response measures to increased occurrence of droughts, especially as this also leads to increasing competition between various agricultural sectors for water resources (Sanghera, 2020).

Waterlogging—the flooding or over-irrigation of crops that brings underground water levels to the surface—is also an abiotic stressor to sugarcane and has been found to affect productivity, up to approximately 45% cane yield reduction (Gomathi et al., 2014). Sugarcane tolerance to waterlogging depends on genotypes, plant development stage, length of exposure to waterlogging stress, water type (stagnant or moving), soil type, and drainage. Waterlogging hinders sugarcane leaf and stem growth, and tiller production, and affects the direction of shoot extension (ibid). Sugarcane is more vulnerable to waterlogging in the early stages of plant growth, during the initial 3–4 months. During the latter 5–9 months, exposure to water stress at later stages tends to accelerate maturity (Salter et al., 2014). When exposed to water stress at earlier stages, crop stalk weight and numbers are notably affected, resulting in reduced yield. Stalk mortality, decreased crop growth, decreased nutrient and water uptake, and cane breakage are some effects of waterlogging, reducing sugarcane yield.

Waterlogging can cause hypoxia in the soil environment and affect soil as it turns from an aerobic environment—soil containing oxygen—to anaerobic—soil without oxygen—due to a lack of aeration, which severely affects the crops' root growth and health (Gomathi et al., 2014). Without a regular supply of oxygen due to waterlogged soil, the root hairs die and rot, leading to complete underground root system impairment. This further causes nutrient deficiency as the root system functions are inhibited and cannot uptake the necessary nutrients from the soil. As a coping mechanism for flooding, some sugarcane crops have been found to develop aerial roots (Misra et al., 2020). The aerial roots assist in preserving root activity during waterlogging by absorbing essential oxygen, as well as the nutrients derived from the dry matter accumulations. During flooding, the sensitivity of adventitious root-forming tissues—roots produced from non-root tissues such as stems or leaves—is enhanced due to a higher ethylene concentration.

While long-term flooding can severely hinder the production and yield of sugarcane, shorter-term flooding is more common, and many varieties can withstand these shorter lengths of exposure, particularly in the later stages of plant growth. Water use efficiency is important for both the productivity and profitability of sugarcane cropping. Many sugarcane crop improvement programs and research are heavily focused on and invested in water-use efficiency and water-stress-tolerant varieties/cultivars and crop production systems to find the most effective combination of solutions to combat the increasing challenges associated with water.

2.5 HUMIDITY

Humidity levels can affect the germination and propagation of sugarcane (Abou-Salama, 1990). Sugarcane is a long-duration crop, and several factors, such as atmospheric humidity, day and night temperatures, soil moisture content, and crop age, determine the maturity or ripening of the crop. Unfavorable humidity levels can delay the maturity of sugarcane crops. Hence, to increase sugarcane cropping production, ripening must be induced either by chemical ripening or withholding irrigation. Ideal relative humidity levels for sugarcane cropping are approximately 70–85% (Ashagre & Ashagre, 2020).

Humidity levels can also influence the incidence of disease in sugarcane crops (ibid). As a major environmental stressor and concern for farmers, sugarcane crop diseases threaten crop production and yield. They must be constantly monitored, alongside utilizing mitigation practices to reduce the occurrence of diseases. One way this can be controlled is through precision agriculture, an innovative engineering practice that uses information technology image processing techniques to enhance sugarcane production and yield (Kumar et al., 2017). Through precision agriculture, external factors such as humidity, temperature, and soil health can be analyzed to determine vital data to help inform the necessary steps for the prevention and treatment of diseases and consequently maximize crop yield. Sensor-based systems can help farmers make more accurately informed decisions about preventive or corrective measures to ensure the crop is grown in ideal environments by better understanding the quality and condition of crop fields regarding humidity, water supply, temperature, and soil. Humidity, temperature, and moisture sensors can all play an integral role in more precise agricultural practices to allow for enhanced germination and propagation of sugarcane, as well as more efficient techniques that will enable farmers to reduce costs and increase production, yield, and profits (Khan et al., 2021).

2.6 LIGHT

Light exposure is also important for the development of germinating seeds (Pierre et al., 2014). It is important to provide ample lighting to maximize seed germination in a controlled setting. In the field, where exposure to sunlight cannot be controlled, the effect on crop germination and production is significant. Sugarcane thrives in full sun and will become dormant in the colder winter months. The duration and intensity of lighting are also integral for sugarcane growth and for the crops to establish high photosynthesis capacities (Glover, 1974). Generally, higher radiation favors adequate cane production and sugar yield. On the other hand, extreme sunlight can hinder the germination and propagation of sugarcane (Karuppaiyan et al., 2014). Sunburn injury or sun scald on leaves of sugarcane crops is another abiotic stressor caused by extreme weather and hot temperatures, particularly hot/heat waves during the summer months. While leaves will appear burned, sugarcane growth and juice quality generally do not significantly decline. Therefore, extreme sunlight exposure is not as drastic as other abiotic stressors that currently affect sugarcane germination, propagation, and production.

2.7 NUTRITION

Sugarcane is a crop that requires large amounts of nutrients, which it collects and stores in the plant because of its large mass germination attributes (Oliveira et al., 2018; Verma et al., 2020a, 2021c). To maintain high crop yields and production, farmers must ensure they supply sufficient nutrients to the germinating sugarcane crop. Sugarcane varieties have different levels of efficiency in their nutrient absorption and use. While some varieties show good production levels even with low levels of available nutrients, other varieties that may result in better production yields have been found to have much more demanding nutrient uptake (ibid). When comparing different varieties of sugarcane, the most efficient one would be the variety that can absorb more nutrients in the same climatic condition and soil and produce higher sucrose or biomass related to nutrients absorbed (Verma et al., 2022a; Song et al., 2022). However, it may not be easy to avail varieties that are efficient in both these processes (ibid). Harvesting of sugarcane results in high levels of nutrient removal from soil. Therefore, it is important to regularly conduct chemical soil analysis to understand the nutrient supply capacity of soil—this will better inform levels of chemical and organic fertilizer that should be added or identify the toxic levels of elements which can then be reduced with the addition of lime and gypsum.

Farmers who know their farm soil history, for example, the types of fertilizers utilized previously in the farm, will also have better experiential wisdom of soil health and crop needs. Studies have also shown that continuous sugarcane planting reduces crop production due to a negative impact on soil microbial communities and soil fertility (Pang et al., 2021). There is a substantial negative relationship between the number of years of planting and soil nutrients. Soil acidification because of continued mineral fertilizer, which leads to decreased soil organic matter over time, as well as the buildup of sugarcane residue and allelochemicals, contributes to declining sugarcane production in fields with 10 to 30 years of monoculture history. While corrective soil preparation, such as the introduction of lime, gypsum, and organic fertilizers, helps to prepare the soil for effective germination and propagation, the traditional practice of disking operations and soil plowing encourages microbial communities to develop in the soil environment, which in turn break down organic matter in the soil (Wagner de Oliveira et al., 2020).

These organisms eliminate carbon dioxide into the atmosphere by breaking down the organic matter. Therefore, when traditional farming practices such as disk operation and soil plowing are applied on land with native vegetation to prepare it for use as agricultural plots for sugarcane, it results in the loss of soil carbon and soil organic matter and further contributes to greenhouse gas emissions.

Farmers are encouraged to practice more sustainable agricultural practices, such as adding organic matter back into the soil. This allows for healthy and fertile soil for effective sugarcane germination and propagation while simultaneously practicing sustainable agriculture to minimize contributions toward climate change impacts. Crop rotation is another sustainable methodology to ensure soils remain healthy and fertile and provide a suitable environment for the germination and propagation of sugarcane crops. Farmers will also be able to increase their income through the diversification of crops while simultaneously reducing the chances of soil erosion due to ample coverage during the rainy season.

Reduction of weeds and pests such as the notorious cane borer and lesser cornstalk borer can also be effectively managed through practicing crop rotation (Dhaliwal et al., 2021). Germination and propagation of sugarcane can also be boosted by using green fertilizer. Fertilization of soil helps to increase yields and replace nutrients extracted by crops, to manage soil nutrients over a long period. Sugarcane yields gradually decrease with each new harvest, with a significant difference in yield levels from the first ratoon compared to the later fifth or sixth harvest, even under good climatic conditions. The degree to which yield declines depends on the sugarcane variety, soil nutrition, and climatic and atmospheric environments. With proper fertilization, crop yields can continue to be maximized and ensure decline in subsequent harvests is kept to a minimum, thus increasing the overall sustainability of sugarcane farms.

Wastes generated because of sugarcane crop production are of high value and play an important role in maintaining nutrient levels in soil (Raza et al., 2021). For example, bagasse, vinasses, and filter cake can be reused in sugarcane production to recycle soil nutrients and are major contributors to keeping sugarcane cropping sustainable. Introducing vinasses and filter cake back into the soil aids in recycling organic matter and soil nutrients. The filter cake, a byproduct of syrup clarification resulting from sugar production in modern distilleries, is utilized in sugarcane farms to add calcium, phosphorus, organic matter, and supplementary nutrients, which help to boost soil fertility and increase crop yields. The most efficient use of filter cake is in plantation furrows, where phosphorus released helps stimulate root growth (de Aquino Vidal Lacerda Soares et al., 2021). Additionally, the water in the filter cake supports cane sprouting, boosting sugarcane germination and propagation. Technological interventions, advances, and research improving soil fertility, recovery, and overall sustainability of sugarcane cropping through enhanced germination and propagation continue. Challenges associated with crop nutrition are thus overcome.

2.8 PEST AND DISEASES

Insects and pests can cause devastating losses in sugarcane farming, with sugarcane crops being highly susceptible to some of these (Bhatt et al., 2022). Changes in seasons, weather patterns, and climatic conditions may provide favorable conditions, such as ideal rainfall and temperatures for the increased growth of pests. With more than 200 species of pests known to affect sugarcane crops, a few are of major concern and more common across the sugarcane farming industry. Termites, white grub, and black bugs are some of the many common insects affecting sugarcane worldwide, each appearing at different times of the year and requiring varying control measures (Srikanth et al., 2012). For example, termites are underground-dwelling insects that mainly target the planting stalk. These are particularly threatening during the germination phases of sugarcane crops because termites find their way from the cut ends of the seed, ruining soft tissue. They are active all season but more so during the drought months.

Farmers use pesticides to control termite infestation and ensure they have sprayed over the stalks in the furrow. Adding chemicals in the furrow when planting, placing decaying farm manure, and removing previous crops' stubbles and debris are ways to manage termite infestation (Rana et al., 2021). Borers are also common insects found to wreak havoc in sugarcane crops. As the name suggests, they bore through young shoots, cane stalks, and roots. Borer larvae tunnel their way through the cane stalk, significantly affecting the weight of the stalk and levels of sucrose yield. Another issue with this insect is that the tunnels they create in the cane stalk become points of entry for other invaders.

Weakened stalks are susceptible to breakages and lodging, and severe invasion by borer insects can ruin the sugarcane plant (Reagan & Mulcahy, 2019). There is also a specific type of borer – the shoot borers, which are a major threat to the germination of sugarcane as they often attack the crops during the beginning stages of sugarcane plant growth (Kumar et al., 2019). Early stages will see shoot borers damaging the entire stem, usually tunneling in the mother stem. Some measures that can be taken to combat borer infestations are through chemical treatment of crops or practicing earlier planting of cane before shoot borers have multiplied to peak numbers. Many different types of bacterial, fungal, viral, and phytoplasma pathogen sugarcane diseases have been discovered, with some more common than others. Various diseases affect sugarcane differently and can affect crops at all stages of the crop cycle – from germination to mature stages. The susceptibility of sugarcane to diseases largely depends on a combination of factors such as crop variety, climate, soil environment, and atmospheric environment (Zhao & Li, 2015; Song et al., 2019, 2022).

Common diseases caused by bacteria include red stripe and leaf scald (Christopher & Edgerton, 1930). Some fungal diseases have eye spots, yellow spots, brown spots, pineapple disease, red rot, smut, wilt, and banded sclerotial. Mosaic, leaf yellowing, and grassy shoots are common viral diseases affecting sugarcane crops (Nithya et al., 2020). It is important for farmers to be able to identify early signs of disease in their crops before they cause significant production losses. For example, smut is a common fungal disease affecting sugarcane (Liu et al., 2017; Song et al., 2019, 2022).

Some symptoms of crops affected by smut include the appearance of a grayish-black whip and pencil-thick growth, which develops from the top of the sugarcane plant. It usually targets the terminal bud or lateral shoots and can grow up to about a meter in length. It is composed of both fungal tissue and host plant tissue. Some preventive measures that can be taken to reduce sugarcane affected by smut are treating seed canes with hot water, removing diseased plants, using sugarcane varieties that are resistant or tolerant to the disease, and applying fungicides (Zhao & Li, 2015). Adopting control measures and early detection is crucial in managing insect pests and diseases to prevent significant production and economic losses in sugarcane.

2.9 SUGARCANE PLANT HORMONES

Sugarcane plant hormones also influence sugarcane germination (Croft, 2000). Plant hormones' fluctuations can prevent sugarcane buds' germination on stalks during normal growth. If normal development is disturbed, for example, if the growing point is damaged, then generally, buds will germinate from the top of the stalk due to a hormonal imbalance. After the top buds have germinated, plant hormones will be produced in the growing points of side shoots to prevent further growth of the lower buds. This trend has also been discovered in planted setts where one bud on a set germinates quickly and inhibits the growth of other buds on the setts. Some farmers use ethephon, the chemical that regulates plant growth, to boost the ripening of sugarcane crops (ibid). The ethephon chemical can also cause side effects, such as disturbing plant hormonal balance, which triggers the side-shooting of stalks. When applied to setts, it is noted to improve germination. Plant growth is controlled by five hormones naturally occurring in sugarcane—auxins, gibberellins, cytokinins, ethylene, and inhibitors (Goodall, 1998).

Apart from playing a role in germination, plant hormones also affect cell elongation, cell division, and maturation. By understanding the functions of each plant hormone, one can determine the necessary plant growth regulators required for enhancing sugarcane germination and production. The plant growth hormone auxin boosts shoot cell elongation, cell division, and root formation and can be found in stem apical tissue and roots of sugarcane plants (Dilworth et al., 2017). Gibberellins also enhance cell division and elongation and have been discovered in the stem apical tissue, roots, and lateral buds of cane crops (Castro-Camba et al., 2022). Commercial use of gibberellic acid is prevalent in sugarcane cropping. External application of this hormone boosts the elongation of sugarcane internodes and can maximize the sugar levels in the stalk (Chen et al., 2020). It can also be used as a treatment to help decapitated shoots to support lateral bud growth. Cytokinin is another hormone that promotes cell division and is linked to quick-growing tissue. Naturally occurring cytokinins such as zeatin and kinetin are produced in sugarcane roots, transported to shoots, and help to free lateral buds from apical dominance (Le Bris, 2017).

Equally important in germination and plant growth is the plant hormone ethylene, which is associated with releasing buds from dormancy. Ethylene destroys auxin by boosting peroxidase activity and consequently inhibits main shoot growth, allowing for the increase in the development of side shoots, strut roots, and tillers (Goodall, 1998). Another class of plant hormones is inhibitors, which slow down cell division and elongation in shoot tissues, helping to control plant height (Cunha et al., 2017). Plant hormonal balance is a determining factor in the development of lateral bud germination (Baracat Neto et al., 2017). While auxins inhibit germination and plant growth, ethylene, gibberellins, and cytokinin boost germination and plant growth. Modern-day technological advances have allowed innovative solutions to reduce sugarcane susceptibility to environmental stressors, improving germination, propagation, and production (Meena et al., 2020).

Plant growth regulators include natural plant hormones extracted from plant tissue and synthetic compounds that imitate naturally occurring plant hormones (Dias & da Silva, 2020). Through these modern agricultural practices, farmers can use plant growth regulators to manage the production of sugarcane germination better and influence fast early-stage growth, as well as reduce vulnerability to insect pests and diseases and increase overall sugarcane yield.

2.10 GROWING MEDIUM

The type and quality of material or growing medium play very important roles in the development of sugarcane crops, particularly for germination and growth (Singh et al., 2019). For transplants to grow into healthy crops, it's integral to have a good balance of various elements such as water holding capacity, drainage, and air-filled porosity for the crops to germinate. Typically, the ideal air-filled porosity level for the crops would be around 15–25%. Striking a perfect composition of all these factors is extremely important. For instance, while a coarse growing medium has a robust air-filled porosity, it also has low water-holding capacity (Goodall, 1998). This can be reversed if the ever-increasing medium is comprised of finer particles. This would mean crops are grown in lower air-filled porosity and higher water-holding capacity conditions. If air-filled porosity is below ideal levels, the growing medium is waterlogged and will be subject to a lack of oxygen and algal growth. These factors are important in sugarcane production because a lack of oxygen means root and nutrient uptake is inhibited, which leads to a decline in root weight, affecting germination and growth.

Improving the productivity of sugarcane crops requires healthy seeds. To ensure the growth of disease-free sugarcane seeds, farmers must create the ideal conditions and environments for successful germination by providing a suitable growing medium (Salokhe, 2021). Other factors determine the ideal medium. Apart from water-holding capacity, drainage, and air-filled porosity, the growing medium must also have the right balance of temperature, nutrients, and plant hormones or plant growth regulators. Germinating setts in a controlled environment with an ideal growing medium have shown positive results in sugarcane production. Additionally, several advantages are associated with planting pre-germinated plants, such as less planting material being needed and more effective control over the planting rate due to the controlled environment (Angelini et al., 2021).

2.11 CULTIVATION

Another factor that affects the germination and propagation of sugarcane is cultivation (Zulu et al., 2019). As with many other factors that affect sugarcane propagation, cultivation works conjointly with these factors in determining sugarcane production. For example, farms located on dry land and requiring major spacing of the rows will benefit immensely from single bud chips transplanting – a determining factor in sugarcane growth (Sulistiono et al., 2017). Therefore, the planting method would work together with effective irrigation practices as groundwater and soil fertility are limited in dry conditions, as noted during droughts which affect crop production, growth, and yield.

Wider spacing of sugarcane has been shown to allow for more light to be absorbed and increase photosynthesis activity in plants, resulting in increased growth of sugarcane crops (Ullah et al., 2016). The effects of using the spacing technique when growing sugarcane can be assessed by measuring the leaf area, leaf area index, and crop growth rate. These are the main growth parameters to be evaluated. The tillering and grand growth phases are both critical stages in sugarcane cropping; hence, cultivation methods must be carefully selected, with special consideration for the environment and fast-changing extreme climatic conditions. Therefore, effective crop management practices become essential for improving sugarcane germination, production, and yield. Apart from the cultivated land size, human labor, physical assets such as fertilizers, weedicides, and pesticides are all important variables in sugarcane production (Zulu et al., 2019). Some common cultivation practices, such as crop rotation and tillage, are already widely practiced for maximizing sugarcane production. Additionally, while cultivation practices are important factors in improving sugarcane production, they should not be treated as a substitute for weed control using chemicals, as the roots of the sugarcane plant are highly vulnerable to tillage and compaction (Smith et al., 2008).

2.12 ENHANCING GERMINATION POTENTIAL

The germination and propagation of sugarcane are affected by various factors. Climate change, temperature, water, humidity, light, nutrition, insect pests, plant hormones, growing medium, and cultivation are some major influencing factors that can affect sugarcane crop production. Cane and sugar yields can be maximized through the effective germination of cane setts and for the quick early growth of the crop. Seed cane quality and treatments of setts have great potential to enhance the germination of sugarcane plants (Singh et al., 2019). Seed cane must be strong and healthy to ensure sustainable and maximized production from the onset. The type or variety of seed cane will be determined by the area in which it is planted, with different varieties taking varying amounts of time to germinate and their ability to adjust to abiotic and biotic stressors. To ensure the best quality seed cane, some studies suggest that best practice would involve using the top section of matured cane as this produces the highest and quickest germination compared with rates of mid and lower stalk buds (Goodall, 1998).

Treatment of setts is another way to improve the germination and propagation of sugarcane. Some common treatments include long hot water treatments, short heat treatments, chemical treatments, and fungicides (Wondu, 2015). Long hot water treatment is used to treat ratoon stunting disease or eliminate carbon. Shorter heat treatments are easier on buds and less lethal, encouraging rapid germination of roots and shoots, increased excretion of sugar, and decreased apical dominance. As already mentioned, chemical treatments, including plant growth regulators, boost germination and increase resilience to environmental stressors. Fungicides are integral for facilitating the germination of sugarcane as they help to prevent and fight off any disease that could harm roots, setts, or plant growth. Intercropping may also help to enhance the germination of sugarcane crops (Vuyyuru et al., 2018).

Intercropping is highly encouraged with sugarcane farming (growing another crop between sugarcane rows simultaneously). One reason is that since sugarcane is a long-duration crop, farmers must wait almost a year or longer before reaping the benefits of their sugarcane harvest. The intercropping of shorter-duration crops allows farmers additional means of income and food security while they wait for their main sugarcane harvest (Shukla et al., 2022). In addition, intercropping has been found to assist with sugarcane growth as the plants intercropped act as a form of weed control and reduction (Kaur et al., 2016). In some parts of the world, intercropping is often applied alongside mulching, which uses sugarcane residue or waste to suppress weed growth. Additionally, intercropping helps to improve soil fertility by introducing nutrients from the decomposition of crop residues (Singh et al., 2019). Intercropping also boosts sugarcane growth by increasing soil enzyme activity and a higher bacterial community compared to monoculture (Malviya et al., 2021). Since intercropping reduces production costs and increases income, farmers also have increased means to invest in other agricultural practices that promote the germination and propagation of sugarcane.

2.13 CONCLUSION AND FUTURE OUTLOOKS

As environmental conditions continue to vary with climate change, sugarcane crop vulnerability to these environmental stressors will continue to be a major concern. Research on how best to support the sugarcane industry through these climatic issues is ongoing. Given the importance of sugarcane production globally, it will continue to be a high-demand focus area. The use of modern technology and advances in the agricultural process will become more crucial over the years to support the germination and propagation of sugarcane despite unfavorable environmental conditions. Research and testing of new varieties of sugarcane cultivars are crucial to finding the most adaptive varieties to the increasing environmental stressors. Germination in controlled settings where growth can be closely monitored before being transported to the field may become a more common practice if natural environments become less viable for the early growth stages of sugarcane. Additionally, since a large variety of factors can affect sugarcane germination and propagation, education, and awareness are of top priority among farmers to ensure they can understand their environmental conditions and the health of their produce. Farmers also need support and capacity building to determine any remedial processes that need to take place to address agricultural challenges and/or maximize sugarcane crop production. As it will be a constantly changing and evolving field due to human-induced climate change, governments of sugarcane-producing countries can provide support by establishing platforms where farmers can refer to for the latest best practices on sugarcane production, share challenges and solutions, and stay up to date on the latest research and technology that support the improvement of production and yield of sugarcane, specific to locations and local circumstances of each country.

KEYWORDS

- **climate variables**
- **germination**
- **humidity**
- **sugarcane cropping technique**
- **sugarcane germination**
- **sugarcane propagation**
- **sugarcane yield**

REFERENCES

Abou-Salama, A. M. (1990). Sugarcane pollen viability and seed setting as affected by daylength decline rates and relative humidity. *LSU Historical Dissertations and Theses, 5029*, 1–116, https://repository.lsu.edu/gradschool_disstheses/5029.

Alizade, S., Keshtkar, E., Mokhtasi-Bidgoli, A., Sasanfar, H., & Streibig, J. C. (2020). Effect of water deficit stress on benzoylprop-ethyl performance and physiological traits of winter wild oat (*Avena sterilis* subsp. ludoviciana). *Crop Protection*, *137,* 105292. https://doi.org/10.1016/j.cropro.2020.105292.

Angelini, L. G., Clemente, C., & Tavarini, S. (2021). Pre-germination treatments, temperature, and light conditions improved seed germination of Passiflora incarnata L. *Agriculture*, *11*(10), 937. https://doi.org/10.3390/agriculture11100937.

Ashagre, S. Y., & Ashagre, Y. (2020). Determining optimum period of withholding irrigation for inducing maturity of sugarcane (Saccharum spp. hybrid) in Southern Ethiopia. *Journal of Applied Biotechnology and Bioengineering*, *7*, 16–25. https://doi.org/10.15406/jabb.2020.07.00211.

Azhar, A., Asano, K., Sugiura, D., Kano-Nakata, M., & Ehara, H. (2022). Waterlogged conditions influence the nitrogen, phosphorus, potassium, and sugar distribution in sago palm (*Metroxylon sagu* Rottb.) at seedling stages. *Plants*, *11*(5), 710. https://doi.org/10.3390/plants11050710.

Baracat, N. J., Scarpare, F. V., Araújo, R. B., & Scarpare-Filho, J. A. (2017). Initial development and yield in sugarcane from different propagules. *Pesquisa Agropecuária Tropical*, *47*(3), 273–278. https://doi.org/10.1590/1983-40632016v4744472.

Bhatt, R., Kumar, R., Kashyap, L., Alataway, A., Dewidar, A. Z., & Mattar, M. A. (2022). Growth, yield, quality and insect-pests in sugarcane (Saccharum officinarum) as affected by differential regimes of irrigation and potash under stressed conditions. *Agronomy*, *12*(8), 1982. https://doi.org/10.3390/agronomy12081942.

Bonnett, G. D., Hewitt, M. L., & Glassop, D. (2006). Effects of high temperature on the growth and composition of sugarcane internodes. *Australian Journal of Agricultural Research*, *57*(10), 1087–1095. https://doi.org/10.1071/ar06042.

Budeguer, F., Enrique, R., Perera, M. F., Racedo, J., Castagnaro, A. P., Noguera, A. S., & Welin, B. (2021). Genetic transformation of sugarcane, current status and future prospects. *Frontiers in Plant Science*, *12*, 768609. https://doi.org/10.3389/fpls.2021.768609.

Cardozo, N. P., & Sentelhas, P. C. (2013). Climatic effects on sugarcane ripening under the influence of cultivars and crop age. *Scientia Agricola*, *70*(6), 449–456.

Castro-Camba, R., Sánchez, C., Vidal, N., & Vielba, J. M. (2022). Plant development and crop yield: The role of gibberellins. *Plants*, *11*(19), 2650. https://doi.org/10.3390/plants11192650

Cheavegatti-Gianotto, A., de Abreu, H. M., Arruda, P., Bespalhok Filho, J. C., Burnquist, W. L., Creste, S., & Ulian, E. C. (2011). Sugarcane (*Saccharum X officinarum*): A reference study for the regulation of genetically modified cultivars in Brazil. *Tropical Plant Biology, 4*(1), 62–89. https://doi.org/10.1007/s12042-011-9068-3.

Chen, R., Fan, Y., Yan, H., Zhou, W., Zhou, Z., Weng, M., & Wu, J. (2020). Enhanced activity of genes associated with photosynthesis, phytohormone metabolism and cell wall synthesis is involved in gibberellin-mediated sugarcane internode growth. *Frontiers in Genetics, 11*, 570094. https://doi.org/10.3389/fgene.2020.570094.

Christopher, W. N., & Edgerton, C. W. (1930). Bacterial stripe diseases of sugarcane in Louisiana. *Journal of Agricultural Research, 41*(3), 259–267.

Croft, B. J. (2000). Review of methods of improving the germination of sugarcane. Report. Queensland, Australia: Bureau of Sugar Experiment Stations.

Cunha, C. P., Roberto, G. C., Vicentini, R., Lembke, C. G., Souza, G. M., Ribeiro, R. V., & Menossi, M. (2017). Ethylene-induced transcriptional and hormonal responses at the onset of sugarcane ripening. *Scientific Reports, 7*(1), 43364. https://doi.org/10.1038/srep43364.

de Aquino Vidal Lacerda Soares, A., de Mello Prado, R., Caione, G., Rodrigues, M., Pavinato, P. S., & Naudi Silva Campos, C. (2021). Phosphorus dynamics in sugarcane fertilized with filter cake and mineral phosphate sources. *Frontiers in Soil Science, 1*, 719651. https://doi.org/10.3389/fsoil.2021.719651.

Dhaliwal, S. S., Sharma, V., Mandal, A., Naresh, R. K., & Verma, G. (2021). Improving soil micronutrient availability under organic farming. In *Advances in Organic Farming.* 93–114. https://doi.org/10.1016/b978-0-12-822358-1.00002-x.

Dias, J. P. T., & da Silva, M. A. (2020). Effect in greenhouse of commercial biostimulant on sugarcane seedlings growth. *Biotecnología Vegetal, 20*(1), 23–31.

Dilworth, L. L., Riley, C., & Stennett, D. K. (2017). Plant constituents: carbohydrates, oils, resins, balsams, and plant hormones. *Pharmacognosy*, pp. 61–80. https://doi.org/10.1016/B978-0-12–802104-0.00005-6.

Ferreira, T. H., Tsunada, M. S., Bassi, D., Araújo, P., Mattiello, L., Guidelli, G. V., & Menossi, M. (2017). Sugarcane water stress tolerance mechanisms and its implications on developing biotechnology solutions. *Frontiers in Plant Science, 8*, 01077.https://doi.org/10.3389/fpls.2017.01077.

Galon, L., Conceno, G., Evander, A., Aspiazu, I., da Silva, A. F., Clevison, L., & Andres, A. (2013). Influence of biotic and abiotic stress factors on physiological traits of sugarcane varieties. *Photosynthesis* [Preprint]. InTech. (https://doi.org/10.5772/55255.

Ganapati, R. K., Alam, R., & Rahman, M. M. (2016). Abiotic factor in germination of sugarcane seeds (*Saccharum spp*). *International Journal of Plant Biology and Research, 1*, 1053.

Glover, J. (1974). The rate of apparent photosynthesis of whole sugarcane plants. *Annals of Botany, 38*(4), 909–920. https://doi.org/10.1093/oxfordjournals.aob.a084885.

Gomathi, R., Rao, P. N. G., Chandran, K., & Selvi, A. (2014). Adaptive responses of sugarcane to waterlogging stress: An overview. *Sugar Tech, 17*, 325–338. https://doi.org/10.1007/s12355-014-0319-0.

Goodall, J. L. (1998). Factors affecting germination and growth of sugarcane transplants. Unpublished master's thesis, University of Natal.

Hu, S., Ding, Y., & Zhu, C. (2020). Sensitivity and responses of chloroplasts to heat stress in plants. *Frontiers in Plant Science, 11*, 375. https://doi.org/10.3389/fpls.2020.00375.

Hussain, S., Khaliq, A., Mehmood, U., Qadir, T., Saqib, M., Amjed Iqbal, M., & Hussain, S. (2019). Sugarcane production under changing climate: Effects of environmental vulnerabilities on sugarcane diseases, insects and weeds. *Climate Change and Agriculture* [Preprint]. https://doi.org/10.5772/intechopen.81131.

Karuppaiyan, R., Ram, B., Pooja, Kumar, S., Ali, M., Meena, M. R., & Kumar, R. (2014). Sugarcane sunburn: An emerging abiotic disorder in subtropical India. *Sugar Tech, 17*(3), 283–290. https://doi.org/10.1007/s12355-014-0314-5.

Kaur, N., Bhullar, M., & Gill, G. (2016). Weed management in sugarcane-canola intercropping systems in northern India. *Field Crops Research, 188, 1–9.* https://doi.org/10.1016/j.fcr.2016.01.009.

Khan, Z., Zahid Khan, M., Ali, S., Abbasi, I., Ur Rahman, H., Zeb, U., Khattak, H., & Huang, J. (2021). Internet of Things-based smart farming monitoring system for bolting reduction in onion farms. *Scientific Programming, 2021*. 1–15. https://doi.org/10.1155/2021/7101983.

Kumar, K., Jaglan, M. S., & Verma, T. (2019). Biology of early shoot borer, *Chilo Infuscatellus* Snellen on sugarcane, *Saccharum Officinarum* L. *Journal of Entomology and Zoology Studies, 7*(5), 262–268.

Kumar, S., Mishra, S., Khanna, P., & Pragya. (2017). Precision sugarcane monitoring using SVM classifier. *Procedia Computer Science, 122, 881–887*. https://doi.org/10.1016/j.procs.2017.11.450.

Le Bris, M. (2017). Hormones in growth and development. *Reference Module in Life Sciences* [Preprint]. https://doi.org/10.1016/b978-0-12-809633-8.05058-5.

Liu, S., Lin, N., Chen, Y., Liang, Z., Liao, L., Lv, M., & Zhang, L. (2017). Biocontrol of sugarcane smut disease by interference of fungal sexual mating and hyphal growth using a bacterial isolate. *Frontiers in Microbiology, 8*, 778. https://doi.org/10.3389/fmicb.2017.00778.

Malviya, M. K., Solanki, M. K., Li, C., Wang, Z., Zeng, Y., Verma, K., & Li, Y. (2021). Sugarcane-legume intercropping can enrich the soil microbiome and plant growth. *Frontiers in Sustainable Food Systems, 5*, 606595. https://doi.org/10.3389/fsufs.2021.606595.

Matzenbacher, F. O., Vidal, R. A., Merotto Jr., A., & Trezzi, M. M. (2014). Environmental and physiological factors that affect the efficacy of herbicides that inhibit the enzyme protoporphyrinogen oxidase: A literature review. *Planta Daninha, 32*(2), 457–463. https://doi.org/10.1590/s0100-83582014000200024.

Meena, M. R., Kumar, R., Chinnaswamy, A., Karuppaiyan, R., Kulshreshtha, N., & Ram, B. (2020). Current breeding and genomic approaches to enhance the cane and sugar productivity under abiotic stress conditions. *Biotechnology, 10*(10), 440. https://doi.org/10.1007/s13205-020-02416-w.

Mhlanga-Ndlovu, B. F., & Nhamo, G. (2017). An assessment of Swaziland sugarcane farmer associations' vulnerability to climate change. *Journal of Integrative Environmental Sciences, 14*(1), 39–57. https://doi.org/10.1080/1943815x.2017.1335329.

Misra, V., Solomon, S., Mall, M. K., Prajapati, C. P., Hashem, A., Abd_Allah, E., & Ansari, M. (2020). Morphological assessment of water stressed sugarcane: A comparison of waterlogged and drought affected crop. *Saudi Journal of Biological Sciences, 27*(5), 1228–1236. https://doi.org/10.1016/j.sjbs.2020.02.007.

Nithya, K., Parameswari, B., & Viswanathan, R. (2020). Mixed infection of sugarcane yellow leaf virus and grassy shoot phytoplasma in yellow leaf affected Indian sugarcane cultivars. *The Plant Pathology Journal, 36*(4), 364–377. https://doi.org/10.5423/ppj.oa.06.2020.0092

Oliveira, M. W., Macêdo, G., Martins, J., Silva, V., & Oliveira, A. (2018). Mineral nutrition and fertilization of sugarcane. *Sugarcane – Technology and Research* [Preprint]. https://doi.org/10.5772/intechopen.72300.

Pang, Z., Tayyab, M., Kong, C., Liu, Q., Liu, Y., Hu, C., & Yuan, Z. (2021). Continuous sugarcane planting negatively impacts soil microbial community structure, soil fertility, and sugarcane agronomic parameters. *Microorganisms, 9*(10), 2008. https://doi.org/10.3390/microorganisms9102008.

Pierre, J. S., Rae, A. L., & Bonnett, G. D. (2014). Abiotic limits for germination of sugarcane seed in relation to environmental spread. *Tropical Plant Biology, 7*(3–4), 100–110. https://doi.org/10.1007/s12042-014-9141-9.

Ramesh, K., Matloob, A., Aslam, F., Florentine, S., & Chauhan, B. S. (2017). Weeds in a changing climate: Vulnerabilities, consequences, and implications for future weed management. *Frontiers in Plant Science, 8*, 95. https://doi.org/10.3389/fpls.2017.00095.

Rana, A., Chandel, R. S., Verma, K. S., & Joshi, M. J. (2021). Termites in important crops and their management. *Indian Journal of Entomology, 83*(3), 486–504. https://doi.org/10.5958/0974-8172.2021.00001.8.

Raza, Q. U. A., Bashir, M. A., Rehim, A., Sial, M. U., Ali Raza, H. M., Atif, H. M., & Geng, Y. (2021). Sugarcane industrial byproducts as challenges to environmental safety and their remedies: A review. *Water, 13*(24), 3495. https://doi.org/10.3390/w13243495.

Reagan, T. E., & Mulcahy, M. M. (2019). Interaction of cultural, biological, and varietal controls for management of stalk borers in Louisiana sugarcane. *Insects, 10*(9), 305. https://doi.org/10.3390/insects10090305.

Riajaya, P. D., Hariyono, B., Cholid, M., Kadarwati, F., Santoso, B., Djumali, & Subiyakto. (2022). Growth and yield potential of new sugarcane varieties during plant and first ratoon crops. *Sustainability, 14*(21), 14396. https://doi.org/10.3390/su142114396.

Salokhe, S. (2021). Development of an efficient protocol for production of healthy sugarcane seed cane through meristem culture. *Journal of Agriculture and Food Research, 4, 100126.* https://doi.org/10.1016/j.jafr.2021.100126.

Salter, B., Park, G., & Kok, E. (2014). A field experiment to evaluate the response of sugarcane varieties to waterlogging. *Report. Proceedings of the Australian Society of Sugar Cane Technology*, *40*, 138–150.

Sanghera, G. S. (2020). Sugarcane disorders associated with temperature extremes and mitigation strategies. *East African Scholars Journal of Agriculture and Life Sciences, 3*(4), 101–114. https://doi.org/10.36349/EASJALS.2020.v03i04.002.

Shrivastava, P., & Kumar, R. (2015). Soil salinity: A serious environmental issue and plant growth promoting bacteria as one of the tools for its alleviation. *Saudi Journal of Biological Sciences, 22*(2), 123–131. https://doi.org/10.1016/j.sjbs.2014.12.001.

Shukla, S. K., Sharma, L., Jaiswal, V. P., Dwivedi, A. P., Yadav, S. K., & Pathak, A. D. (2022). Diversification options in sugarcane-based cropping systems for doubling farmers' income in subtropical India. *Sugar Tech, 24*(4), 1212–1229. https://doi.org/10.1007/s12355-022-01127-1.

Singh, P., Singh, S., Tiwari, A., Pathak, S., Singh, A., Srivastava, S., & Mohan, N. (2019). Integration of sugarcane production technologies for enhanced cane and sugar productivity targeting to increase farmers' income: Strategies and prospects. *Biotechnology, 9*(2), 48. https://doi.org/10.1007/s13205-019-1568-0.

Skendžić, S., Zovko, M., Živković, I. P., Lešić, V., & Lemić, D. (2021). The impact of climate change on agricultural insect pests. *Insects, 12*(5), 440. https://doi.org/10.3390/insects12050440

Smith, D. T., Richard, E. P., & Santo, L. T. (2008). Weed control in sugarcane and the role of triazine herbicides. In *The triazine herbicides: 50 years Revolutionizing Agriculture*; LeBaron, H. M.; McFarland, J. E.; Burnside, O. C., Eds.; Elsevier: San Diego, Amsterdamand Oxford, pp. 185–198. https://doi.org/10.1016/b978–044451167–6.50018–0.

Song, X. P., Mo, F., Yan, M., Zhang, X., Zhang, B., Huang, X., Huang, D., Pan, Y., Verma, K. K., & Li, Y. R. (2022). Effect of smut infection on the photosynthetic physiological characteristics and related defense enzymes of sugarcane. *Life, 12*(8), 1201. https://doi.org/10.3390/life12081201.

Srikanth, J., et al. (2012). *Sugarcane pests and their management*. Coimbatore: Sugarcane Breeding Institute.

Srivastava, A. K. (2012). Sugarcane production: Impact of climate change and its mitigation. *Biodiversitas: Journal of Biological Diversity, 13*(4), 214–227. https://doi.org/10.13057/biodiv/d130408.

Sulistiono, W. (2017). Growth analysis of transplanted sugarcane bud chips seedling in the dry land. *International Journal of Scientific and Technology Research, 6*(1), 87–93.

Ullah, S., Ahmad Anjum, S., Mohsin Raza, M., Riaz, A., Abbas, A., Muhammad Yousif, M., Ma, J., Liu, Y., Zhang, J., Cheng, D., Xu, Y., & Ali, I. (2016). Optimizing row spacing to ameliorate the productivity of spring sugarcane (*Saccharum officinarum* L.). *Agricultural Sciences, 7*(8), 531–538. https://doi.org/10.4236/as.2016.78053.

Verma, K. K., Anas, M., Chen, Z., Rajput, V. D., Malviya, M. K., Verma, C. L., Singh, R. K., Singh, P., Song, X. P., & Li, Y. R. (2020d). Silicon supply improves leaf gas exchange, antioxidant defense system and growth in sugarcane responsive to water limitation. *Plants, 9*(8), 1032. https://doi.org/10.3390/plants9081032.

Verma, K. K., Li, D. M., Singh, M., Rajput, V. D., Malviya, M. K., Minkina, T., Singh, R. K., Singh, P., Song, X. P., & Li, Y. R. (2020b). Interactive role of silicon and plant-rhizobacteria mitigating abiotic stresses: A new approach for sustainable agriculture and climate change. *Plants, 9*(9), 1055. https://doi.org/10.3390/plants9091055.

Verma, K. K., Liu, X. H., Wu, K. C., Singh, R. K., Song, Q. Q., Malviya, M. K., Song, X. P., Singh, P., Verma, C. L., & Li, Y. R. (2020f). The impact of silicon on photosynthetic and biochemical responses of sugarcane under different soil moisture levels. *Silicon, 12*, 1355−1367.

Verma, K. K., Singh, P., Song, X. P., Malviya, M. K., Singh, R. K., Chen, G. L., Solomon, S., & Li, Y. R. (2020e). Mitigating climate change for sugarcane improvement: Role of silicon in alleviating abiotic stresses. *Sugar Tech, 22*(5), 741−749. https://doi.org/10.1007/s12355-020-00831-0.

Verma, K. K., Song, X. P., Budeguer, F., Nikpay, A., Enrique, R., Singh, M., Zhang, B., Wu, J. M., & Li, Y. R. (2022a). Genetic engineering: An efficient approach to mitigating biotic and abiotic stresses in sugarcane cultivation. *Plant Signaling & Behavior, 17*(1), e2108253.

Verma, K. K., Song, X. P., Li, D. M., Singh, M., Wu, J. M., Singh, R. K., Sharma, A., Zhang, B. Q., & Li, Y. R. (2022b). Silicon and soil microorganisms improve rhizospheric soil health with bacterial community, plant growth, performance, and yield. *Plant Signaling & Behavior, 17*(1), e2104004.

Verma, K. K., Song, X. P., Lin, B., Guo, D. J., Singh, M., Rajput, V. D., Singh, R. K., Singh, P., Sharma, A., Malviya, M. K., Chen, G. L., & Li, Y. R. (2021b). Silicon-induced drought tolerance in crop plants: Physiological adaptation strategies. *Silicon, 14*, 2473–2487. https://doi.org/10.1007/s12633-021-01071-x.

Verma, K. K., Song, X. P., Rajput, V. D., Boldyreva, V., Zhang, B. Q., Minkina, T., & Li, Y. R. (2022c). Morpho-physiological, biochemical and ultrastructural modifications on sugarcane to prolonged water deficit. In K. K. Verma et al. (Eds.), *Agro-industrial perspectives on sugarcane production under environmental stress*. 193–213. https://doi.org/10.1007/978-981-19-3955-6_11.

Verma, K. K., Song, X. P., Rajput, V. D., Solomon, S., Li, Y. R., & Rao, G. P. (Eds.). (2023). *Agro-industrial perspectives on sugarcane production under environmental stress*. Springer, Singapore. https://doi.org/10.1007/978-981-19-3955-6_1.

Verma, K. K., Song, X. P., Verma, C. L., Chen, Z. L., Rajput, V. D., Wu, K. C., Liao, F., Chen, G. L., & Li, Y. R. (2021a). Functional relationship between photosynthetic leaf gas exchange in response to silicon application and water stress mitigation in sugarcane. *Biological Research, 54*(1), 15. https://doi.org/10.1186/s40659-021-00338-2.

Verma, K. K., Song, X. P., Verma, C. L., Malviya, M. K., Guo, D. J., Rajput, V. D., Sharma, A., Wei, K. J., Chen, G. L., Solomon, S., & Li, Y. R. (2021d). Predication of photosynthetic leaf gas exchange of sugarcane (*Saccharum spp.*) leaves in response to leaf positions to foliar spray of potassium salt of active phosphorus under limited water irrigation. *ACS Omega, 6*, 2396–2409.

Verma, K. K., Song, X. P., Zeng, Y., Guo, D. J., Singh, M., Rajput, V. D., Malviya, M. K., Wei, K. J., Sharma, A., Li, D. P., Chen, G. L., & Li, Y. R. (2021c). Foliar application of silicon boosts growth, photosynthetic leaf gas exchange, antioxidative response and resistance to limited water irrigation in sugarcane (*Saccharum officinarum* L.). *Plant Physiology and Biochemistry, 166*, 582–592.

Verma, K. K., Song, X. P., Zeng, Y., Li, D. M., Guo, D. J., Rajput, V. D., Chen, G. L., Barakhov, A., Minkina, T. M., & Li, Y. R. (2020c). Characteristics and correlation of leaf stomata and its relationship with photosynthesis on *Saccharum spp.* under different irrigation and silicon application. *ACS Omega, 5*, 24145–24153.

Verma, K. K., Wu, K. C., Singh, P., Malviya, M. K., Singh, R. K., Song, X. P., & Li, Y. R. (2019). The protective role of silicon in sugarcane under water stress: Photosynthesis and antioxidant enzymes. *Biomedical Journal of Scientific & Technical Research,15*, 1–7. https://doi.org/10.26717/BJSTR.2019.15.002685.

Verma, K. K., Wu, K. C., Verma, C. L., Li, D. M., Malviya, M. K., Singh, R. K., Singh, P., Chen, G. L., Song, X. P., & Li, Y. R. (2020a). Developing mathematical model for diurnal dynamics on photosynthetic responses in sugarcane responsive to different irrigation and silicon application. *PeerJ, 8*, e10154. https://doi.org/10.7717/peerj.10154.

Vuyyuru, M., Sandhu, H., McCray, J., & Raid, R. (2018). Effects of soil-applied fungicides on sugarcane root and shoot growth, rhizosphere microbial communities, and nutrient uptake. *Agronomy, 8*(10), 223. https://doi.org/10.3390/agronomy8100223.

Wagner de Oliveira, M., Cláudio Inácio da Silveira, L., Bosco de Oliveira, A., Henrique Pereira Barbosa, M., Gomes Pereira, M., & Bezerra Albino Oliveira, T. (2020). Sugarcane production systems in small rural properties. *Multifunctionality and Impacts of Organic and Conventional Agriculture [Preprint]*. https://doi.org/10.5772/intechopen.84975.

Wondu, E. (2015). Design for fabrication of effective seed cane hot water treatment plant for Ethiopian sugar estates/projects. *International Journal of Science Technology & Society, 3*(5), 225–229. https://doi.org/10.11648/j.ijsts.20150305.11.

Zhao, D., & Li, Y. R. (2015). Climate change and sugarcane production: Potential impact and mitigation strategies. *International Journal of Agronomy, 2015*, 1–10. https://doi.org/10.1155/2015/547386.

Zulu, N. S., Sibanda, M., & Tlali, B. (2019). Factors affecting sugarcane production by small-scale growers in Ndwedwe Local Municipality, South Africa. *Agriculture, 9*(8), 170. https://doi.org/10.3390/agriculture9080170.

CHAPTER 3

Impact of Irrigation Management on Sugarcane During Stress

MUSA OPEYEMI AHMED,[1,2] ADEPOJU MORONKOLA IDRIS,[1,3] FAITH EWELUEGIM ENAHORO-OFAGBE,[1,3] and KINGSLEY O. IWUOZOR[1,4]

[1]*Nigeria Sugar Institute, Ilorin, Kwara State, Nigeria*

[2]*Department of Agricultural and Environmental Engineering, University of Ibadan, Oyo, Nigeria*

[3]*Department of Agronomy, University of Ilorin, P.M.B., Ilorin, Kwara, Nigeria*

[4]*Department of Pure and Industrial Chemistry, Nnamdi Azikiwe University, P.M.B., Akwa, Nigeria*

ABSTRACT

Sugarcane is a C_4 crop and requires the right conditions to thrive. However, climate change has been reported in the literature to affect most agricultural systems through biotic and abiotic stresses. The reoccurrence of abiotic stress is expected to decrease sugarcane yield over the coming decades by 50%. Therefore, to rectify the issues relating to stress conditions, there is a need to understand proper irrigation management. This chapter carefully reviews the literature that has been published to study how sugarcane reacts to various stress conditions. Stresses such as drought, waterlogging, salinity, and heat, which are the major abiotic stresses that contribute adversely to sugarcane growth, development, recovery, and cane yield, are discussed in detail in this chapter. It is observed that some of the stress conditions drastically affected sugarcane at the early and grand growth phases, thereby causing reductions

Sugarcane Cultivation and Management: Challenges and Opportunities. Krishan K. Verma, Xiu-Peng Song, Munna Singh, Renato de Mello Prado, Jian-Ming Wu, and Yang-Rui Li (Eds.)

in sugarcane growth, quality, and yield. The chapter critically discusses the mechanism through which proper irrigation management remediates this problem of unfavorable stress conditions and its impact on sugarcane yield.

3.1 INTRODUCTION

Sugarcane, which is grown primarily for ethanol and sugar production in tropical and subtropical regions of the world, is important to the global economy (Dlamini, 2021). Over the past 10 years, yields from this crop have increased noticeably, accounting for 70% of the sugar produced globally (Azevedo et al., 2011; Verma et al., 2023). As a result, the effect of sugarcane on the economy is undeniable, and its usage in various ways cannot be viewed as prospective but rather as actuality. Sugarcane belongs to the Poaceae family and is mainly grown within latitude 36.7°N to 31°S of the equator (Kumar et al., 2021). It grows erect up to 5 or 6 meters long and produces multiple stems. Sugarcane can store up to 50% of its dry weight in sucrose in the stalks of the plant (Kumar et al., 2021; Verma et al., 2022a, b, c). Apart from being a crucial crop for the manufacturing of sugar and alcohol, its biomass is also used for bio-electricity generation, paper, and fertilizers (Afiomah & Iwuozor, 2020; Rabelo et al., 2020; Dlamini, 2021; Iwuozor et al., 2022).

Sugarcane has a long growth cycle, which presents significant environmental challenges that impact its development and growth (Misra et al., 2020; Verma et al., 2022d; Song et al., 2019, 2022). The decline in sugarcane yield is attributed to the stress that the cane undergoes. Sugarcane is vulnerable to two types of stress: biotic and abiotic. Biotic stress results from harm to plants caused by other living things, such as nematodes, weeds, insect pests, disease-causing organisms, and allelopathic chemicals (Rana et al., 2016). On the other hand, abiotic stress occurs naturally due to changing climatic conditions, thereby causing a reduction in plant yield. Amidst the aforementioned types of stresses, abiotic stress is the most concerning in the light of irrigation (Verma et al., 2021a, b, c; Deng et al., 2022).

Abiotic stress has emerged as a persistent issue as a result of the sudden changes in climatic conditions. Abiotic stressors directly or indirectly influence sugarcane by changing the crop's metabolism, growth, and development (Misra et al., 2020; Verma et al., 2021d, e; Xu et al., 2022). As a result of climate change, abiotic stresses on sugarcane affect the health of the soil, the growth and development of the crop, its synthesis and accumulation of sugar, chemical composition, and the availability of seed cane, as well as aggravating other abiotic and biotic stresses that increase losses (Shrivastava et al., 2017; Song et al., 2019). Abiotic stress includes drought, salinity, waterlogging, and temperature stress (Rana et al., 2016). Various researchers have reported the impact of these abiotic stresses on sugarcane (Misra et al., 2016, 2020; Shrivastava et al., 2017; Meena et al., 2020). For example, drought has been reported by Misra et al. (2020) to have posed a severe threat to the entire Indian nation. Nearly 2.97 lakh acres of land used for sugarcane production are particularly vulnerable to drought, which can hinder the crop at different stages of growth. Waterlogging conditions reported by Gomathi et al. (2015) in sugarcane fields have caused piping in stalks and roots, development of adventitious roots, a high rate of tiller mortality, reduced growth and development, as well as a reduction in stalk weight and a 5–45% loss in cane yield. The problem of salinity has been reported by Meena et al. (2020) as one of the major global problems affecting crop production, with 33% of irrigated land and 20% of cultivated land globally suffering from salt accumulation. Hence, to mitigate the effect of these abiotic stresses on sugarcane, proper irrigation management must be carried out.

The management of irrigation is a critical topic that needs to be extensively discussed. Using various methods and technologies to manage water resources properly is one practical way to address the overwhelming issue of sugarcane under stress. Many studies have examined the potential for effective irrigation management techniques (Cahn & Johnson, 2017; Saccon, 2018; Verma et al., 2019a, b). However, most of these studies focus heavily on drought conditions (Pereira et al., 2002; Singels et al., 2019). Therefore, to the best of the author's knowledge, there has not been any research that has studied the impact of irrigation management on sugarcane under different stress conditions. In light of the aforementioned issues, this chapter focuses on the irrigation management concepts applied to reducing the effects of sugarcane under different stress scenarios. When all of the information provided in the chapter is considered, it will improve knowledge of the irrigation management approaches used to control the different stresses sugarcane experiences in order to increase cane output.

3.2 IRRIGATION MANAGEMENT

A set of technical choices comprising the features of the crop, climate, water, soil, and irrigation system are referred to as irrigation management (de Oliveira et al., 2015). A producer can achieve higher productivity levels, save water and energy, and help preserve the environment by properly managing irrigation and other farming methods. A well-run management strategy includes identifying the ideal time to begin irrigation and calculating the amount of water the crop will need, using information about the timing of the water application or the irrigation equipment's displacement rate.

3.2.1 IMPORTANT PARAMETERS IN THE MANAGEMENT OF IRRIGATION

The soil provides natural water storage for plants. Between the field capacity and the permanent wilting point, the water plants can access the soil is present.

3.2.1.1 SATURATED HUMIDITY

A soil is said to be in "saturated humidity" when all the pores present in the soil are filled with water. Continued water application at this stage will result in a downward movement and air replacement. The growth of the plant is impeded due to a lack of aeration.

3.2.1.2 FIELD CAPACITY

This is the moisture content of the soil after the drainage of gravitational water has ceased (Santra et al., 2018). Hence, the water stored in the soil macropores has been drained out by gravity action and replaced with air. This level of wetness encourages greater nutrient and water uptake by plants. It usually occurs in the field two days after adequate rainfall or irrigation. The retention curve approach is typically used in laboratories to measure field capacity. According to this method, depending on the texture, structure, and amount of organic matter in the soil, the value of the field capacity moisture is represented by the balanced moisture with tension ranging from 6 to 33 kPa.

3.2.1.3 LIMIT OF STRESS

This is the theoretical point where water is less readily available for the plant. At this stage, the plant's roots require more effort to extract water since water from soil micro-pores is not equally easy to access. A balance of moisture with a tension of 2 bar is required at this stage.

3.2.1.4 PERMANENT WILTING POINT

This is the moisture content of the soil that is unavailable for plant use. At this stage, the soil moisture is tightly bound to the soil particle and requires a suction pressure greater than 15 bar (Santra et al., 2018). Ultimately, the plant remains wilted and will die if no water is added. The permanent wilting point is determined in the laboratory using the pressure plate chamber at a suction pressure of 15 bar.

3.2.1.5 SOIL DENSITY

The soil density is the relationship between a dry soil sample's mass and volume (Robinson et al., 2022). An Uhland Sampler, whose cylinder is implanted in the soil, in the medium depth of the soil layer explored by the roots of the plants, can be used to make this assessment. The sample is ready after the cylinder has been removed, and it is then placed in an oven to dry for 24 hours at a temperature of roughly 105°C to calculate its mass. The volume is calculated using the diameter and height measurements of the soil sample. The density is then calculated using the following equation:

$$Ds=\frac{Ms}{Vs} \tag{1}$$

where; 'Ds' is the soil density (g/cm^3); 'ms' is the dry soil mass (g); and 'Vs' soil sample volume (cm^3).

3.2.1.6 SOIL POROSITY

It is the fraction of the bulk volume of soil containing air and water. In other words, it is the percentage of pores (micro and macro) in the soil. Clay is observed to have the highest porosity of 0.60, while sandy soil has the lowest porosity at 0.28.

3.3 IRRIGATION MANAGEMENT STRATEGIES

3.3.1 IRRIGATION SCHEDULING

Irrigation scheduling is the process of coordinating the application of water to crops in order to match their water usage. When done correctly, the right amount of water is applied at the right frequency to produce the best possible yield. Irrigation scheduling involves planning to ensure adequate water is given to plants to help them withstand other external stressors, making it an essential part of management (Singels et al., 1998).

Irrigation scheduling considers factors such as the growth stage of the crops, the weather conditions, and the soil's water storage capacity. By carefully planning and executing irrigation schedules, farmers can ensure that their crops receive the water they need to grow and thrive. This can help maximize yields and minimize the negative effects of water stress on the crops. In irrigation scheduling, the inputs considered include rainfall and irrigation. Outputs to consider include the water used by crops through processes like evapotranspiration and water loss through deep percolation. To manage the impact of water stress on plants, irrigation scheduling will then involve estimating the capacity of soil to hold water, estimating crops' water needs to determine how much water to apply and when, and replacing the soil moisture used with irrigation water.

The water balance method is a common practice used for determining when irrigation is necessary. This method balances the output (water loss) with the input (water supply). Irrigation becomes essential when the soil moisture reaches a predetermined target allowable for depletion, called the management acceptable depletion level. The irrigation applied will be the water needed to bring the soil back to field capacity. Factors that determine irrigation water requirement (IWR) are reference evapotranspiration (ET_o), crop evapotranspiration (ET_c, also known as crop water requirement, CWR), and effective rainfall (P_e).

A tensiometer is also used to schedule irrigation. It measures the soil moisture potential (tension or suction). The tensiometer consists of a hollow tube joined to a porous ceramic tip at the base, a vacuum gauge, and a removable cap at the top (Ors et al., 2021). The mode of operation involves moving water from within the tensiometer through the ceramic tip when the soil is dry. The water loss creates a vacuum in the tensiometer, which is then recorded by the vacuum gauge as a suction reading. For irrigation scheduling, the tensiometer must be calibrated with field capacity (FC) and the permanent wilting point of the soil (PW). Irrigation begins when the tensiometer reads a predetermined level close to the limit of stress (LS) value. After irrigation or rainfall, water moves back through the ceramic tip, and the vacuum is reduced in the tensiometer.

3.3.2 FULL IRRIGATION MANAGEMENT

Full irrigation management is carried out to prevent water stress impacts on a plant throughout its growing season (Irmak & Rudnick, 2014). Scheduling irrigation to avoid water stress necessitates a thorough knowledge of the irrigation system, i.e., system capacity and efficiency as well as regular and precise information on crop status and field characteristics, including current and anticipated crop water demands and plant soil water availability. The volume of irrigation application must be sufficient to meet crop water needs while still being moderate enough to minimize percolation and surface runoff of water and nutrients beneath the crop root zone (Irmak & Rudnick, 2014).

3.3.3 DEFICIT AND LIMITED IRRIGATION MANAGEMENT

As previously indicated, there are instances when applying less water than the crop needs is the only alternative when full irrigation control is not an option. As a result, researchers have developed deficit and/or limited irrigation management systems to improve or optimize yield per unit of applied water. Limited irrigation distributes the whole season's available water in fixed amounts throughout the growing season, regardless of the crop development stage. In contrast, deficit irrigation withholds water at crop growth phases less vulnerable to water stress than others. Due to this, the primary distinction between deficit and limited irrigation approaches is that the former takes crop growth stages into account while the latter does not (Irmak & Rudnick, 2014).

Deficit irrigation is a technique that aims to maximize the efficiency of irrigation in order to produce higher yields with less water. It is usually used when there is a water allocation or a severe water shortage. This is done by carefully regulating the amount of water provided to crops, particularly during critical growth stages, to optimize irrigation resources. Farmers can improve their yields and make the most of their irrigation water by using this method. As for limited irrigation, water must be available at the time of irrigation when using the limited irrigation management method, which involves administering less water than the crop ETa. Limited irrigation management is used for various reasons, including water allocations, delays in irrigation scheduling, and operational savings from pumping water. Insufficient irrigation system capacity to meet agricultural water demands is also among them. Depending on the amount and distribution of precipitation, different levels of water stress might emerge from limited irrigation. Because precipitation timing and volume vary greatly from place to place, it is critical to develop locally specialized deficit or limited irrigation solutions to ensure the effectiveness of these approaches (Irmak & Rudnick, 2014).

3.4 IMPACT OF STRESS ON SUGARCANE

Sugarcane is the most important sugar crop and is highly sensitive to stress. Generally, stress is an unfavorable condition that poses a threat to crop productivity due to its ability to reduce metabolic activities, inhibit physiological processes, and cause declines in growth and development as well as the yield of sugarcane during its active growing phase (Shrivastava et al., 2017; Singh et al., 2019; Misra et al., 2020; Verma et al., 2023). Biotic and abiotic stresses are two major categories of stress that influence sugarcane productivity. Abiotic stresses such as drought, waterlogging, salinity, and heat are major abiotic stresses that contribute adversely to sugarcane growth, development, sucrose accumulation, recovery, and cane yield (Shrivastava et al., 2017). However, water resource management has proven effective in alleviating the detrimental effect of water stress in sugarcane.

3.4.1 IMPACT OF IRRIGATION CONTROL ON SUGARCANE DURING A DROUGHT

Irrigation, which means providing additional water during a drought, has always been a sustainable solution to the problem of drought in sugarcane-producing regions around the world (Santos et al., 2019). Drought is a significant abiotic stress that affects sugarcane growth and development as it affects over 30% of the arable land use in the sugarcane-cultivated region. Drought is mostly established when transpiration exceeds rainfall during sugarcane's active growing period. However, irrigation is employed to supplement the deficit in rainfall to ensure balance (Misra et al., 2020; Verma et al., 2020). According to Inman-Bamber & Smith (2005); and Barbosa et al. (2015), water is essential for proper physiological processes for sugarcane growth, while its deficit will result in low photosynthesis due to impairment in biochemical processes and reduced leaf gas exchange, consequently resulting in a reduction in growth. Inman-Bamber & Smith (2005) stated that irrigation management is crucial for the efficient use of water in a manner that will not be detrimental to the crop, as sugarcane morphology responds differently to irrigation. Hence, a water deficit at the early stages of sugarcane growth does not significantly affect the crop yield and sucrose level. For example, five months of water deficit during the elongation stage of sugarcane when the leaf area index (LAI) is less than two will reduce the leaf area significantly from 1.8 to 0.9 but will not significantly reduce yield and sucrose accumulation due to the ability to recover faster and low evapotranspiration (ET) at the early stage of growth when irrigated (Wiedenfeld, 2000; Inman-Bamber & Smith, 2005).

Furthermore, reduction in total biomass (19%) and sucrose accumulation (27%) have been attributed to water stress, as reported by Inman-Bamber et al. (2008), which implies that water stress can be both beneficial and detrimental depending on the grower's objectives and stage of the crop. Although sugarcane is a high-biomass crop that requires a large amount of water for optimum production, a reduction in cane yield and sugar content has been attributed to the grand growth stage, when the crop canopy is greater than 70% as a result of increased evapotranspiration (ET) and insufficient water. Supplemental water application at this stage will balance the water loss by ET for optimum crop production (Wiedenfeld, 2000; Inman-Bamber & Smith, 2005; Inman-Bamber et al., 2008). Shorter internode (SH), reduction in leaf area index (LAI), reduced stem girth (SG), reduction in photosynthesis, and consequently yield reductions have been attributed to 12–80% moisture depletion during the sugarcane growth stage (Table 3.1). Increases in water stress due to drought will reduce yield by reducing the stalk elongation rate (Thomas et al., 1978; Lowveld, 1996; Wiedenfeld, 2000). Conversely, water stress is primarily related to the level or degree of stress about ET demand rather than the crop growth stages at which the stress occurred; therefore, prior understanding of the ET during crop growth stages will ensure efficient irrigation management (Wiedenfeld, 2000). De Almeida et al. (2011) indicated that sufficient irrigation during crop emergence, tillering, and canopy establishment. However, the consumptive water use by sugarcane crops is more effective at the grand growth stage than the formal. Moreover, effective irrigation management for limited water conditions is beneficial at an early stage of sugarcane growth. Hence, early water application before full canopy cover is necessary to balance water loss by evapotranspiration. Full irrigation management practices can be used in areas susceptible to drought stress. Therefore, sufficient irrigation during crop emergence, tillering, and canopy establishment will help reduce the impact of drought stress on sugarcane, thereby causing stalk elongation, increasing leaf area index, increasing stomatal conductance, and increasing yield.

TABLE 3.1 Drought Level Effect on Physiological Parameters of Sugarcane

Level of Stress	Crop Stage at Which Stress is Applied	Observations	References
60 days of water withholding	Early growth stage (60–120 DAT)	Total biomass, shoot, and root dry mass.	Dinh et al. (2017)
60 days of water withholding	Tilling up to the grand growth phase (60–120 DAP).	Stalk diameter, number of internodes, intermodal length, cane height, leaf length, and width, fresh and dry root weight.	Misra et al. (2020)
Withholding irrigation	Formative growth phase (60–150 DAP)	Internode length, internode number, cane length, and cane diameter.	Hemaprabha et al. (2013)
Withholding water for 20 days	Tilling phase	Numbers of leaves, plant height, and stalk diameter.	Vu et al. (2019)
20% available water content	Tilling phase (88–120 DAP)	Stalk height, stalk diameter, leaf area, and tiller mass.	Endres et al. (2016)
Water withholding	Formative growth phase	Stalk weight, stalk number, stalk diameter, and stalk height.	Silva et al. (2008)

3.4.2 IMPACT OF IRRIGATION CONTROL ON SUGARCANE DURING A WATER LOGGING STRESS

Waterlogging is the major water stress that seriously impacts sugarcane productivity, especially in areas with a high water table. Globally, climate change has caused frequent and severe waterlogging conditions in sugarcane-growing regions of the world (Meena et al., 2014). Waterlogging occurs when the water table in the soil rises to the point where the root zone becomes saturated. This is a condition that results in a metabolic shift in the plant from an aerobic to an anaerobic state due to a significant depletion of oxygen in the root zone, thereby hampering the growth and development of sugarcane (Gilbert et al., 2007, 2008; Sasidharan et al., 2017; Singh et al., 2019). Waterlogging causes an 18–64% reduction in cane yield depending on genotype, stress duration, growth stage, and management practice in terms of drainage (Singh et al., 2019). According to Gomathi et al. (2015); and Singh et al. (2019), sugarcane is mainly susceptible to waterlogged conditions during the early stages of crop development, especially in the first 3–4 months, and considerably tolerant at 5–9 months, depending on the water level, waterlogged duration, or submerged condition, which is also varietal dependent.

Furthermore, the severity of this stress on the morphology and physiological processes of sugarcane is dependent on variety, time, and field waterlogging (Hasan et al., 2003). Waterlogging conditions result in yield reduction, tiller abortion, reduced stalk weight, stalk breakage, decreased photosynthesis, imbalanced nutrition, decreased root development, and reduced leaf area. Furthermore, researchers have attributed poor germination and poor crop establishment to waterlogged conditions that persist up to 3 days after planting due to the reduction of dissolved oxygen around the cane set, significantly affecting bud sprouting. Therefore, the construction of good infield drains is essential to alleviate the effect of waterlogging during this developmental phase. It is critical to consider the height of the water table during irrigation because it is inversely proportional to the irrigation interval, which means that the higher the water table, the less frequent the irrigation, and vice versa. Therefore, to mitigate the impact of waterlogging on sugarcane, deficit irrigation management can be practiced. This is done by carefully regulating the amount of water provided to crops, particularly during critical growth stages, in order to optimize the use of irrigation resources. The water must be applied with a high degree of uniformity to prevent excessive deep percolation in some parts of the field, which could cause waterlogging. Also, the deficit or limited irrigation practice will provide the soil with proper aeration for the sugarcane to thrive. As a result, the impact will increase yield, proper photosynthetic processes, stalk elongation, and tillering ability. Irrigation systems appropriate to this management approach include drip irrigation, linear move sprinkler systems, center-pivot sprinkler systems, and low-energy precise application (LEPA) irrigation machines. Furthermore, the withdrawal of water and the construction of infield drains have been the age-old management approach to waterlogging (Gomathi et al., 2015; Jain et al., 2017; Singh et al., 2019).

3.5 IMPACT OF IRRIGATION CONTROL ON SUGARCANE DURING A HEAT STRESS

Heat stress is another form that impacts sugarcane productivity in sugarcane-producing regions worldwide. Heat stress is simply a rise in temperature beyond the favorable threshold level for plant growth, which may result in a 2.5–10% yield decline, inhibit plant physiological processes, and therefore result in poor morphological performance (Hasanuzzaman et al., 2013; Zhao & Li, 2015). According to Hasanuzzaman et al. (2013), sugarcane's formative and grand growth stages were found to be more sensitive to high temperatures. Heat stress processes have little or no effect on non-induced stress genotypes in terms of chlorophyll fluorescence FV/Fm, leaf RWC (relative water content), Nitrogen reductase NR activities, proline accumulation, and other metabolic activities. In this report, heat stress increases sugarcane's total soluble solids (TSS) during these two growth phases. Sugarcane yield and its components were reported to have been affected by heat stress, with a significant decrease in tillering, elongation, and stalk characteristics at temperatures greater than 35°C, which consequently resulted in the reduction of yield decrease across the sugarcane-producing region (Mall et al., 2016). At different sugarcane developmental phases, temperatures above or below the optimum temperature threshold of 20–40°C may slow the growth and development of sugarcane and drastically reduce production.

Similarly, increased photorespiration due to high temperatures may reduce sugar accumulation (Gawander, 2007; Teixeira et al., 2013; Rao et al., 2015; Jaiphong et al., 2016; Mall et al., 2016). However, effective irrigation or rainfall before or during heat stress could help reduce the effect on sugarcane productivity as the soil moisture content reduces under heat stress due to increased evaporation (Hunsigi, 2012; Santos et al., 2015). Furthermore, full irrigation management is usually advised to be practiced to mitigate the effect of heat stress on the sugarcane. This practice will help to reduce the rising temperatures of both soil and plants due to increasing heat stress. The relative water content of the leaf will rise, as a result, enhancing morphological processes and yield.

3.6 IMPACT OF IRRIGATION CONTROL ON SUGARCANE DURING A SALINITY STRESS

Abiotic stress potentially results in soil salinity due to concentrations of soluble salts (e.g., sodium) on the soil surface, which leads to an increase in the electrical conductivity (EC) of the soil (Sharma & Uppal, 1999; Wiedenfeld, 2008). Salinity has been reported to have caused a 25–50% reduction in sugarcane germination at 45 DAP, a situation caused by inadequate water around the buds or the direct effect of sodium salt on germination. The number of leaves, leaf area index (LAI), and tillering are reduced by 25–50%, 35–75%, and 20–60%, respectively. Hence, this significant effect on the sugarcane growth component will impact the photosynthetic activities of the plant and ultimately result in yield loss (Sharma & Uppal, 1999; Vasantha et al., 2010). Low transpiration rate, stomata closure, and reduced sucrose accumulation have reportedly been caused by salinity; moreover, research has revealed that declines in sucrose percentage, yield, and its components (stalk length, stalk height, girth, and internode length) are the resultant effects of salinity during the actively growing stages of sugarcane. Also, cane quality, such as sucrose, purity, pol, and CCS (commercial cane sugar), has been significantly affected by salinity (Sharma & Uppal, 1999; Wiedenfeld, 2008; Vasantha et al., 2010). To prevent the adverse effects of salinity, a proper irrigation management plan must be carried out. Irrigation management recommendations include the use of subsurface drainage to avoid the accumulation of salt, and periodic rainfall, the introduction of leaching factors to wash away soil from the root zone, and the use of the electrostatic precipitation method when irrigation water quality analysis indicates electrical conductivity and TDS exceed 3.0 dS/m and 2,000 mg/l, respectively (Wiedenfeld, 2000, 2008; Enciso & Wiedenfeld, 2005; de Oliveira et al., 2015; Culas & Baig, 2020).

3.7 CONCLUSION

According to the above analysis, abiotic stressors are the main obstacles to increasing sugarcane production globally. Hence, to address the problem of the nation's sugar security, reducing the negative effects of these pressures on sugarcane crops' productivity and production is vital. This chapter presented the impact of different abiotic stresses on sugarcane, such as drought, waterlogging, heat, and salinity. The adverse effects of the stress include a reduction of the leaf area index, decreased photosynthesis rate, reduced cane elongation, poor germination, and decreased yield and quality. Sugarcane is a particularly vulnerable plant to water stress. Under these circumstances, the amount of irrigation water, when to irrigate, and how to irrigate are questions that must be considered carefully. Proper irrigation management is vital to maximize sugarcane quality, yield, and water use efficiency while minimizing environmental impacts. It was concluded that the use of irrigation management practices such as irrigation scheduling and deficit or limited irrigation would help reduce the adverse effects of these stresses and increase sugarcane quality and yield.

ACKNOWLEDGMENTS

The author would like to thank the entire management of the Nigeria Sugar Institute for their support during the writing of this manuscript.

KEYWORDS

- **drought**
- **high-temperature**
- **irrigation management**
- **limited water supply**
- **salinity**
- **soil-flooding**
- **sugarcane**
- **sugarcane quality**

REFERENCES

Afiomah, C. S., & Iwuozor, K. O. (2020). Nutritional and phytochemical properties of Beta vulgaris Linnaeus (Chenopodiaceae) – A review. *Nigerian Journal of Pharmaceutical and Applied Science Research*, *9*(4), 38–44.

Azevedo, R. A., Carvalho, R. F., Cia, M. C., & Gratão, P. L. (2011). Sugarcane under pressure: An overview of biochemical and physiological studies of abiotic stress. *Tropical Plant Biology*, *4*(1), 42–51.

Barbosa, A., Guidorizi, K., Catuchi, T., Marques, T., Ribeiro, R., & Souza, G. (2015). Biomass and bioenergy partitioning of sugarcane plants under water deficit. *Acta Physiologiae Plantarum, 37*(8), 1–8.

Cahn, M. D., & Johnson, L. F. (2017). New approaches to irrigation scheduling of vegetables. *Horticulturae*, *3*(2), 28.

Culas, R. J., & Baig, I. A. (2020). Impacts of irrigation water user allocations on water quality and crop productivity: The LCC irrigation system in Pakistan. *Irrigation and Drainage*, *69*(1), 38–51.

de Almeida Silva, M., Jifon, J. L., Sharma, V., da Silva, J. A., Caputo, M. M., Damaj, M. B., Guimarães, E. R., & Ferro, M. I. (2011). Use of physiological parameters in screening drought tolerance in sugarcane genotypes. *Sugar Tech*, *13*(3), 191–197.

de Oliveira, R. A., Ramos, M. M., & de Aquino, L. A. (2015). Irrigation management. *In Sugarcane* (pp. 161–183). Elsevier.

Deng, Z. N., Huang, H. R., Huang, C. M., Verma, K. K., Luo, H. B., Jiang, S. L., Li, Y. J., Cao, H. Q., Wu, K. C., & Xu, L. (2022). Short-term limited water irrigation influences the photosynthetic pigments, enzymatic and non-enzymatic activities in Saccharum

spontaneum L. at vegetative stage. *Sugar Tech. 25*, 110–118. https://doi.org/10.1007/s12355-022-01175-7.
Dinh, T. H., Watanabe, K., Takaragawa, H., Nakabaru, M., & Kawamitsu, Y. (2017). Photosynthetic response and nitrogen use efficiency of sugarcane under drought stress conditions with different nitrogen application levels. *Plant Production Science*, *20*(4), 412–422.
Dlamini, P. J. (2021). Drought stress tolerance mechanisms and breeding effort in sugarcane: A review of progress and constraints in South Africa. *Plant Stress*, *2*, 100027.
Enciso, J., & Wiedenfeld, B. (2005). Irrigation guidelines based on historical weather data in the Lower Rio Grande Valley of Texas. *Agricultural Water Management*, *76*(1), 1–7.
Endres, L., Oliveira, N. G., Ferreira, V. M., Silva, J. V., Barbosa, G. V., & Maia Junior, S. O. (2016). Morphological and physiological response of sugarcane under abiotic stress to neonicotinoid insecticides. *Theoretical and Experimental Plant Physiology*, *28*(4), 347–355.
Gawander, J. (2007). Impact of climate change on sugar-cane production in Fiji. *World Meteorological Organization Bulletin*, *56*(1), 34–39.
Gilbert, R. A., Rainbolt, C. R., Morris, D. R., & Bennett, A. C. (2007). Morphological responses of sugarcane to long-term flooding. *Agronomy Journal*, *99*(6), 1622–1628.
Gilbert, R., Rainbolt, C., Morris, D., & McCray, J. (2008). Sugarcane growth and yield responses to a 3-month summer flood. *Agricultural Water Management*, *95*(3), 283–291.
Gomathi, R., Gururaja Rao, P., Chandran, K., & Selvi, A. (2015). Adaptive responses of sugarcane to waterlogging stress: An overview. *Sugar Tech*, *17*(4), 325–338.
Hasan, M. F., Alam, M., Jabber, M., Begum, M., & Miah, M. (2003). Effects of water-logging on juice quality and yield of sugarcane. *Pakistan Journal of Biological Sciences*, *6*(13), 1151–1155.
Hasanuzzaman, M., Nahar, K., Alam, M. M., Roychowdhury, R., & Fujita, M. (2013). Physiological, biochemical, and molecular mechanisms of heat stress tolerance in plants. *International Journal of Molecular Sciences*, *14*(5), 9643–9684.
Hemaprabha, G., Swapna, S., Lavanya, D. L., Sajitha, B., & Venkataramana, S. (2013). Evaluation of drought tolerance potential of elite genotypes and progenies of sugarcane (Saccharum sp. hybrids). *Sugar Tech*, *15*(1), 9–16.
Hunsigi, G. (2012). Production of sugarcane: Theory and practice. Springer Science & Business Media.
Inman-Bamber, N., & Smith, D. M. (2005). Water relations in sugarcane and response to water deficits. *Field Crops Research*, *92*(2–3), 185–202.
Inman-Bamber, N., Bonnett, G., Spillman, M., Hewitt, M., & Jackson, J. (2008). Increasing sucrose accumulation in sugarcane by manipulating leaf extension and photosynthesis with irrigation. *Australian Journal of Agricultural Research*, *59*(1), 13–26.
Irmak, S., & Rudnick, D. R. (2014). Corn irrigation management under water-limiting conditions. University of Nebraska-Lincoln, Extension.
Iwuozor, K. O., Anyanwu, V. U., Olaniyi, B. O., Mbamalu, P. S., & Adeniyi, A. G. (2022). Adulteration of sugar: A growing global menace. *Sugar Tech*, *24*, 914–919.
Jain, R., Singh, A., Singh, S., Singh, S. P., Srivastava, V. K., Chandra, A., Pathak, A. D., & Solomon, S. (2017). Physio-biochemical characterization of sugarcane genotypes for waterlogging tolerance. *World Journal of Agricultural Sciences*, *13*, 90–97.
Jaiphong, T., Tominaga, J., Watanabe, K., Nakabaru, M., Takaragawa, H., Suwa, R., Ueno, M., & Kawamitsu, Y. (2016). Effects of duration and combination of drought and flood

conditions on leaf photosynthesis, growth and sugar content in sugarcane. *Plant Production Science, 19*(3), 427–437.

Kumar, D., Malik, N., & Sengar, R. S. (2021). Physio-biochemical insights into sugarcane genotypes under water stress. *Biological Rhythm Research, 52*(1), 92–115.

Lowveld, E. (1996). Effect of irrigation regime on cane and sugar yields of variety Nco376 in the south-east Lowveld of Zimbabwe. *Proceedings of the South African Sugar Technologists' Association, 70*, 59–62.

Mall, R., Sonkar, G., Bhatt, D., Sharma, N., Baxla, A., & Singh, K. (2016). Managing impact of extreme weather events in sugarcane in different agro-climatic zones of Uttar Pradesh. *Mausam, 67*(1), 233–250.

Meena, M. R., Kumar, R., Chinnaswamy, A., Karuppaiyan, R., Kulshreshtha, N., & Ram, B. (2020). Current breeding and genomic approaches to enhance the cane and sugar productivity under abiotic stress conditions. *3 Biotech, 10*(10), 1–18.

Meena, V., Dotaniya, M., Coumar, V., Rajendiran, S., Kundu, S., & Subba Rao, A. (2014). A case for silicon fertilization to improve crop yields in tropical soils. *Proceedings of the National Academy of Sciences, India Section B: Biological Sciences, 84*(3), 505–518.

Misra, V., Solomon, S., & Ansari, M. I. (2016). Impact of drought on post-harvest quality of sugarcane crop. *Advances in Life Sciences, 20*(5), 9496–9505.

Misra, V., Solomon, S., Mall, A., Prajapati, C., Hashem, A., Abd_Allah, E. F., & Ansari, M. I. (2020). Morphological assessment of water stressed sugarcane: A comparison of waterlogged and drought affected crop. *Saudi Journal of Biological Sciences, 27*(5), 1228–1236.

Ors, S., Sahin, U., Ekinci, M., Turan, M., & Yildirim, E. (2021). Principles of irrigation management for vegetables. In Vegetable Crops-Health Benefits and Cultivation. IntechOpen.

Pereira, L. S., Oweis, T., & Zairi, A. (2002). Irrigation management under water scarcity. *Agricultural Water Management, 57*(3), 175–206.

Rabelo, S. C., de Paiva, L. B. B., Pin, T. C., Pinto, L. F. R., Tovar, L. P., & Nakasu, P. Y. S. (2020). Chemical and energy potential of sugarcane. *In Sugarcane Biorefinery: Technology and Perspectives* (pp. 141–163). Elsevier.

Rana, D., Dass, A., Rajanna, G., & Kaur, R. (2016). Biotic and abiotic stress management in pulses. *Indian Journal of Agronomy, 61*, S238–S248.

Rao, B. B., Chowdary, P. S., Sandeep, V., Pramod, V., & Rao, V. (2015). Spatial analysis of the sensitivity of wheat yields to temperature in India. *Agricultural and Forest Meteorology, 200*, 192–202.

Robinson, D., Thomas, A., Reinsch, S., Lebron, I., Feeney, C. J., Maskell, L. C., Wood, C. M., Seaton, F. M., Emmett, B. A., & Cosby, B. J. (2022). Analytical modelling of soil porosity and bulk density across the soil organic matter and land-use continuum. *Scientific Reports, 12*(1), 1–13.

Saccon, P. (2018). Water for agriculture, irrigation management. *Applied Soil Ecology, 123*, 793–796.

Santos, F., Borém, A., & Caldas, C. (2015). Sugarcane: Agricultural production, bioenergy and ethanol. Academic Press.

Santos, L. C., Coelho, R. D., Barbosa, F. S., Leal, D. P. V., Júnior, E. F. F., Barros, T. H. S., Lizcano, J. V., & Ribeiro, N. L. (2019). Influence of deficit irrigation on accumulation and partitioning of sugarcane biomass under drip irrigation in commercial varieties. *Agricultural Water Management, 221*, 322–333.

Santra, P., Kumar, M., Kumawat, R., Painuli, D., Hati, K., Heuvelink, G., & Batjes, N. (2018). Pedotransfer functions to estimate soil water content at field capacity and permanent wilting point in hot arid Western India. *Journal of Earth System Science*, *127*(3), 1–16.

Sasidharan, R., Bailey-Serres, J., Ashikari, M., Atwell, B. J., Colmer, T. D., Fagerstedt, K., Fukao, T., Geigenberger, P., Hebelstrup, K. H., Hill, R. D., Holdsworth, M. J., Ismail, A. M., Licausi, F., Mustroph, A., Nakazono, M., Pedersen, O., Perata, P., Sauter, M., Shih, M., Sorrell, B. K., Striker, G. G., van Dongen, J. T., Whelan, J., Xiao, S., Visser, E. J. W., Voesenek, L. A. C. J. & Hill, R. D. (2017). Community recommendations on terminology and procedures used in flooding and low oxygen stress research. *New Phytologist*, *214*(4), 1403–1407.

Sharma, S., & Uppal, K. S. K. (1999). Influence of salt stress on growth and quality on sugarcane. *Indian Sugar*, *22*, 1–4.

Shrivastava, A., Pathak, A., Misra, V., Srivastava, S., Swapna, M., & Shukla, S. (2017). Sugarcane crop: Its tolerance towards abiotic stresses. *In Abiotic Stress Management for Resilient Agriculture* (pp. 375–397). Springer.

Silva, M. d. A., Silva, J. A. G. d., Enciso, J., Sharma, V., & Jifon, J. (2008). Yield components as indicators of drought tolerance of sugarcane. *Scientia Agricola*, *65*, 620–627.

Singels, A., Kennedy, A., & Bezuidenhout, C. (1998). Irricane: A simple computerised irrigation scheduling method for sugarcane. Proceedings of the South African Sugar Technologists' Association.

Singels, A., Paraskevopoulos, A., & Mashabela, M. (2019). Farm level decision support for sugarcane irrigation management during drought. *Agricultural Water Management*, *222*, 274–285.

Singh, S., Singh, S., Pathak, A., & Pandey, N. (2019). Assessment of waterlogging induced physio-biochemical changes in sugarcane varieties and its association with waterlogging tolerance. *Journal of Environmental Biology*, *40*(3), 384–392.

Song, X. P., Mo, F., Verma, K. K., Wei, J., Zhang, X., Yang, L., & Li, Y. R. (2019). Effect of sugarcane smut (Ustilago scitaminea Syd.) on ultrastructure and biochemical indices of sugarcane. *Biomedical Journal of Science & Technical Research*, *17*(1), 12546–12551. Advance online publication. DOI: 10.26717/BJSTR.2019.17.002950.

Song, X. P., Mo, F., Yan, M., Zhang, X., Zhang, B., Huang, X., Huang, D., Pan, Y., Verma, K. K., & Li, Y. R. (2022). Effect of smut infection on the photosynthetic physiological characteristics and related defense enzymes of sugarcane. *Life*, *12*, 1201. https://doi.org/10.3390/life12081201.

Teixeira, E. I., Fischer, G., Van Velthuizen, H., Walter, C., & Ewert, F. (2013). Global hot-spots of heat stress on agricultural crops due to climate change. *Agricultural and Forest Meteorology*, *170*, 206–215.

Thomas, J., Salinas, F., & Namken, L. (1978). Growth and yield of sugarcane as affected by row spacing and irrigation regime. Proceedings-New Series. American Society of Sugar Cane Technologists (USA).

Vasantha, S., Venkataramana, S., Gururaja Rao, P., & Gomathi, R. (2010). Long term salinity effect on growth, photosynthesis and osmotic characteristics in sugarcane. *Sugar Tech*, *12*(1), 5–8.

Verma, K. K., Singh, P., Song, X. P., Malviya, M. K., Singh, R. K., Chen, G. L., Solomon, S., & Li, Y. (2020). Mitigating climate change for sugarcane improvement: Role of silicon in alleviating abiotic stresses. *Sugar Tech*, *22*(5), 741–749.

Verma, K. K., Singh, R. K., Song, Q. Q., Singh, P., Zhang, B. Q., Song, X. P., Chen, G. L., & Li, Y. R. (2019a). Silicon alleviates drought stress of sugarcane plants by improving antioxidant responses. *Biomedical Journal of Science & Technical Research. 17*(1), 12580–12586. Advance online publication. DOI: 10.26717/BJSTR.2019.17.002957.

Verma, K. K., Song, X. P., Budeguer, F., Nikpay, A., Enrique, R., Singh, M., Zhang, B. Q., Wu, J. M., & Li, Y. R. (2022b). Genetic engineering: An efficient approach to mitigating biotic and abiotic stresses in sugarcane cultivation. *Plant Signaling & Behavior*, *17*(1), e2108253.

Verma, K. K., Song, X. P., Li, D. M., Singh, M., Wu, J. M., Singh, R. K., Sharma, A., Zhang, B. Q., & Li, Y. R. (2022c). Silicon and soil microorganisms improve rhizospheric soil health with bacterial community, plant growth, performance and yield. *Plant Signaling & Behavior*, *17*(1), e2104004.

Verma, K. K., Song, X. P., Lin, B., Guo, D. J., Singh, M., Rajput, V. D., Singh, R. K., Singh, P., Sharma, A., Malviya, M. K., Chen, G. L., & Li, Y. R. (2021d). Silicon induced drought tolerance in crop plants: physiological adaptation strategies. *Silicon*, *14*(6), 1–15. DOI: 10.1007/s12633-021-01071-x.

Verma, K. K., Song, X. P., Rajput, V. D., Boldyreva, V., Zhang, B. Q., Minkina, T., & Li, Y. R. (2022d). Morpho-physiological, biochemical and ultrastructural modifications on sugarcane to prolonged water deficit. In K. K. Verma et al. (Eds.), Agro-industrial perspectives on sugarcane production under environmental stress. https://doi.org/10.1007/978-981-19-3955-6_11.

Verma, K. K., Song, X. P., Rajput, V. D., Solomon, S., Li, Y. R., & Rao, G. P. (Eds.). (2023). Agro-industrial Perspectives on Sugarcane Production under Environmental Stress. Springer, Singapore. https://doi.org/10.1007/978-981-19-3955-6_1.

Verma, K. K., Song, X. P., Tian, D. D., Guo, D. J., Chen, Z. L., Zhong, C. S., Nikpay, A., Singh, M., Rajput, V. D., Singh, R. K., Minkina, T., & Li, Y. R. (2021a). Influence of silicon on biocontrol strategies to manage biotic stress for crop protection, performance and improvement. *Plants*, *10*, 2163. https://doi.org/10.3390/plants10102163.

Verma, K. K., Song, X. P., Tian, D. D., Singh, M., Verma, C. L., Rajput, V. D., Singh, R. K., Sharma, A., Singh, P., Malviya, M. K., & Li, Y. R. (2021b). Investigation of defensive role of silicon during drought stress induced by irrigation capacity in sugarcane: Physiological and biochemical characteristics. *ACS Omega*, *6*, 19811–19821.

Verma, K. K., Song, X. P., Verma, C. L., Chen, Z. L., Rajput, V. D., Wu, K. C., Liao, F., Chen, G. L., & Li, Y. R. (2021c). Functional relationship between photosynthetic leaf gas exchange in response to silicon application and water stress mitigation in sugarcane. *Biological Research*, *54*, 15. https://doi.org/10.1186/s40659-021-00338-2.

Verma, K. K., Song, X. P., Yadav, G., Degu, H. D., Parvaiz, A., Singh, M., Huang, H. R., Mustafa, G., Xu, L., & Li, Y. R. (2022a). Impact of agroclimatic variables on proteogenomics in sugarcane (Saccharum spp.) plant productivity. *ACS Omega*, *7*, 22997–23008.

Verma, K. K., Song, X. P., Zeng, Y., Guo, D. J., Sing, M., Rajput, V. D., Malviya, M. K., Wei, K. J., Sharma, A., Li, D. P., Chen, G. L., & Li, Y. R. (2021e). Foliar application of silicon boosts growth, photosynthetic leaf gas exchange, antioxidative response and resistance to limited water irrigation in sugarcane (*Saccharum officinarum* L.). *Plant Physiology and Biochemistry*, *166*, 582–592.

Verma, K. K., Wu, K. C., Singh, P., Malviya, M. K., Singh, R. K., Song, X. P., & Li, Y. R. (2019b). The protective role of silicon in sugarcane under water stress: photosynthesis and antioxidant enzymes. *Biomedical Journal of Science & Technical Research. 15*(2), 11301–11307. Advance online publication. DOI:0.26717/ BJSTR.2019.15.002685.

Vu, N. T., Vu, D. C., & Pham, T. T. (2019). Growth and physiological responses of sugarcane to drought stress at an early growth stage. *Tap chi Khoa hoc Nong nghiep Viet Nam/Vietnam Journal of Agricultural Sciences*, *2*(4), 451–460.

Wiedenfeld, B. (2008). Effects of irrigation water salinity and electrostatic water treatment for sugarcane production. *Agricultural Water Management*, *95*(1), 85–88.

Wiedenfeld, R. P. (2000). Water stress during different sugarcane growth periods on yield and response to N fertilization. *Agricultural Water Management*, *43*(2), 173–182.

Xu, L., Deng, Z. N., Wu, K. C., Malviya, M. K., Solanki, M. K., Verma, K. K., Pang, T., Li, Y. J., Liu, X. Y., Kashyap, B. K., Dessoky, E. S., Wang, W. Z., & Huang, H. R. (2022). Transcriptome analysis reveals a gene expression pattern that contributes to sugarcane bud propagation induced by Indole-3-Butyric Acid. *Frontiers in Plant Science*, *13*, 852886. doi: 10.3389/fpls.2022.852886.

Zhao, D., & Li, Y. R. (2015). Climate change and sugarcane production: Potential impact and mitigation strategies. *International Journal of Agronomy*, *2015*, 1–10. https://link.gale.com/apps/doc/A462272803/AONE?u=anon~4518575c&sid=sitemap&xid=46aff863 (accessed on 07 June 2024).

CHAPTER 4

Sugarcane: An Overview of Physiological and Biochemical Studies Under Drought and Salt Stress

POOJA DHANSU,[1] SAPNA YADAV,[1] KUMAR PREET,[1,2] SUNITA DEVI,[3] RAVINDER KUMAR,[1] ASHWANI KUMAR,[3] ADHINI S. PAZHANI,[4] and S. K. PANDEY[1]

[1]*ICAR–Sugarcane Breeding Institute, Regional Center, Karnal, Haryana, India*

[2]*Maharishi Markandeshwar (Deemed to be) University, Mullana, Haryana, India*

[3]*ICAR–Central Soil Salinity Research Institute, Karnal, Haryana, India*

[4]*ICAR–Sugarcane Breeding Institute, Coimbatore, Tamil Nadu, India*

ABSTRACT

Sugarcane is a major crop for biofuel and sugar production, adding to the gross domestic product (GDP) of India. Climatic sensitivities are key limits to sugarcane growth, development, and output. In this chapter, we aimed to concentrate on two main abiotic stresses, drought, and salinity, to which sugarcane is most likely to be exposed. Sugarcane plants develop a variety of adaptation and escape strategies to combat different environmental stresses. Plant responses to drought and salinity are often additive, intermingled, and very complex. Changes in plant morphology, growth, and physio-chemical processes are generally triggered by difficulties in water absorption and toxicity of certain ions, which reduce plant growth, development, quality, and yield or cause a total loss of output. Given the anticipated expansion of sugarcane crop acreage in the coming years, the selection of more suited

Sugarcane Cultivation and Management: Challenges and Opportunities. Krishan K. Verma, Xiu-Peng Song, Munna Singh, Renato de Mello Prado, Jian-Ming Wu, and Yang-Rui Li (Eds.)

plant types to this climate is a major necessity for the long-term sustainability of this sector. Surprisingly, the literature on the study of abiotic stresses on sugarcane is quite sparse, emphasizing that much more research is needed and investigations into this domain should be encouraged. Thus, this is the need of the hour to evaluate the physiological and biochemical responses of sugarcane varieties to drought and salt stress.

4.1 INTRODUCTION

Sugarcane (*Saccharum* spp.) is a member of the Poaceae family and is one of the world's 10 most cultivated crops, with a wide distribution in tropical and subtropical climates (Sánchez-Ken & Clark, 2010; Kumar et al., 2022; Verma et al., 2022a, b, 2023), also including regions where water availability is limited or highly inconsistent (Azevedo et al., 2011; Pooja et al., 2017). From a global perspective, sugarcane (*Saccharum* spp. hybrid complex) is an important cash crop cultivated to produce sugar and ethanol. It was adopted in Brazil during the colonial period and is now one of the primary economic crops (Huang et al., 2021; Song et al., 2021). India is the world's second-largest producer and consumer of sugar. Sugarcane is grown on 5.0 million hectares in India, with an average production of 68 tonnes ha^{-1}. It is anticipated that India will demand 33.0 million tonnes of sugar by 2030, with an average sugar recovery of 10.75%. This will entail productivity of 110 tons ha^{-1} as the area under sugarcane may stabilize at 5.0 mHa. Sugarcane is a glycophyte with high sensitivity to salt stress at varied crop growth stages. Like the majority of other crops, it cannot withstand higher salinity at the vegetative phase, influencing its physiology and metabolic balance at the cell level via osmotic adjustments (Zhu et al., 2011; Verma et al., 2021a, b; Anas et al., 2021a, b; Song et al., 2022).

The salts interfere with sugarcane yield in a twofold manner, first by affecting the growth rate and yield of the cane and secondly by affecting the sucrose accumulation pattern of the stalk. Sometimes it may show stunted growth or no development, with yields falling to 50% or less than its real potential limitations under salinity stress, even though sugarcane has developed complex mechanisms to adapt to the ionic and osmotic stresses due to salinity. However the sensitivity of the crop towards high salt concentration at various crop stages causes a subsequent reduction in crop growth, height, number of internode formations, and number of millable cane (NMC), and finally, crop yield and productivity. Physiologically, it reduces the photosynthetic activity of plants and the sucrose accumulation pattern in the stalks.

4.2 DROUGHT STRESS

Sugarcane, a C_4 plant, needs a lot of water to flourish properly. Water is reportedly a rare entity among other abiotic pressures in many places of the world. Future conditions will become worse due to decreased precipitation frequency and an increase in global temperature. Water stress is a major impediment to sugarcane production (Misra et al., 2022; Dhansu et al., 2021; Verma et al., 2021c). Several cane-growing countries are dealing with this issue, causing cane yield losses of up to 60%. It is a common abiotic disaster affecting the sugarcane-producing states of Tamil Nadu, Maharashtra, Andhra Pradesh, Bihar, Haryana, Rajasthan, and Uttar Pradesh (Mall et al., 2022). Drought conditions have recently resulted in a 40% loss of cane production in the Indian states of Maharashtra and Karnataka. Plants evolve mechanisms to detect external stress signals and adapt by altering their physiological and biochemical responses (Hussain et al., 2018; Verma et al., 2020a, b, c; Dhansu et al., 2022). Drought escape, drought avoidance, and drought tolerance are the three main drought-resistance strategies adopted by plants, including sugarcane (Figure 4.1). The escape strategy is predicated on appropriate reproduction via a reduced life cycle far in advance of the incidence of stress, as well as accelerated growth rates and effective storage (Campos et al., 2004) (Figure 4.1).

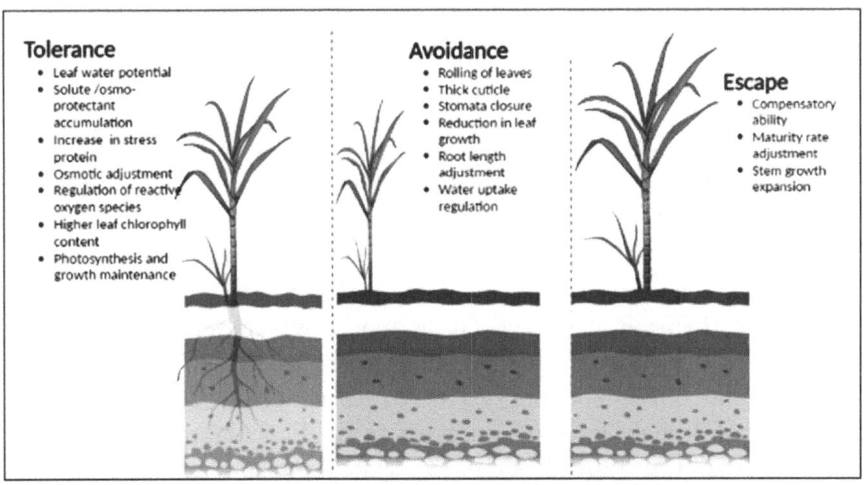

FIGURE 4.1 The three primary responses of plants, such as sugarcane, to drought, are characterized by interacting mechanisms (Mall et al., 2022).

Source: Adapted from Mall et al., 2022

The avoidance strategy entails preserving moisture content without considerable loss in the plant under stressful events that are helpful to the plant's growth and development. This approach implies that the plant loses either minimum water through senescence of older leaves, stomatal closure, or reduction in leaf area by leaf rolling, or maximal water loss through expanding the length of plant roots to a depth where water is accessible (Araus et al., 2008; Blum, 2009). The tolerance strategy permits a plant to sustain metabolism despite reducing tissue water potential when subjected to water deficiency stress (Mitra, 2001). This approach involves numerous plant physiological, biochemical, and regulatory processes (Khan et al., 2011). During water stress, an intricate, complex mechanism disrupts many physio-biochemical pathways. The plant has a broad repertoire of morpho-physiological and biochemical alterations, adaptations, and reactions for dealing with water deficit (Seki et al., 2003; Yamaguchi-Shinozaki, 2006; Pooja et al., 2019, 2020).

4.2.1 EFFECTS OF DROUGHT STRESS ON SUGARCANE

4.2.1.1 GROWTH ANALYSIS

Sugarcane responds to water stress mainly by rolling its leaves, lowering leaf surface area for incident light and evaporation (Inman & Smith, 2005). It is accompanied by cell wall shrinkage and leaf aging (Waraich et al., 2011; Verma et al., 2022a, b). Sugarcane roots spread laterally in moist soils but expand deeply and unevenly in soils with low moisture content (Waraich et al., 2011). Since lateral roots' main purpose is to absorb nutrients, a lack of nutrients alters the structure of the roots, lowering yield (Yamaguchi & Blumwald, 2005; Hoang et al., 2019). The development of the plant's shoots is also impacted by the relocation of nutrients from other plant parts (like the stem and leaves to the roots (Inman & Smith, 2005; Khonghintaisong et al., 2018; Hoang et al., 2019). Furthermore, genotypes of plants that can re-root *in vitro* have been categorized as drought-tolerant (Snyman et al., 2016; Dlamini, 2020). The root systems of susceptible genotypes are shorter than those of drought-tolerant genotypes in field conditions (Kumar et al., 2017). Bidoia et al. (2006) examined sugarcane plants in various growth stages subjected to drought. They found that dry mass accumulation was significantly reduced during the vegetative stage, with growth reduced by 61% and 46% under moderate and severe water stress. Moreover, Ramesh

(2000) discovered that five sugarcane genotypes responded differently to dry conditions versus irrigated conditions during the vegetative growth phase in terms of leaf number and dry mass distribution. The cultivars' ability to adapt to their surroundings regarding climate and type of soil can also be greatly influenced by the distribution of their roots (Vasconcelos et al., 2003). Therefore, plants that survive drought tend to produce more absorbing hairs (thinner roots) to maximize water absorption (Aiken & Smucker, 1996; Faroni & Trivelin, 2006). Similarly, in another study, severe water stress (drought) causes morphological changes in plants due to cellular component damage or simply as a defensive response (Murphy, 2011; Dinh et al., 2017; Ogbaga et al., 2020) (Table 4.1).

TABLE 4.1 Different Sugarcane Growth Stages and the Impact of Drought Stress

Stages	Impacts		Crop Water Requirement
Germination/ ratooning	Poor germination or ratooning, root dieback.	Less sensitive to drought.	Restore soil to field capacity after harvesting, low water requirement at 8–16 weeks.
Tillering	Decreased tillering, peak population, and root growth.	–	–
Stalk elongation	Reduced peak population, and leaf area expansion resulting in downregulated photosynthetic efficiency and stalk growth.	Most sensitive to drought.	High water requirement
Maturation	If drought occurred before the maturation stage, then reduced water content, increased DMC and sucrose content, sucrose capacity, and storage also decreased.		
Flowering	Decreased green leaf area, sucrose produces storage and enhanced pithiness in the stalk.	Less sensitive to drought.	Low water requirement, drying off period, and schedule depend on total available moisture and seasons.

Source: Adapted from: Philani et al. (2021).

4.2.2 PHYSIOLOGICAL RESPONSES

4.2.2.1 PLANT-WATER RELATIONS

Under stress, these relationships help plants control or maintain the ideal level of hydration in their cells, which has critical repercussions for physiological and metabolic functions. It regulates almost all cellular physiological activities that depend on a sufficient amount of water being available. A measure of the water balance in plants is relative water content (Gonzalez et al., 2001). It provides information about the level of both cell and tissue hydration, which is crucial for physio-biochemical systems (Silva et al., 2007). During water deficiency conditions, plant RWC decreased, resulting in stomatal closure and decreased CO_2 uptake (Buckley, 2005). In general, the tolerant genotype of sugarcane has a higher RWC than the susceptible clones, and Silva et al. (2007) report that the tolerant clones have a higher RWC (87%) than the susceptible genotypes (80%). Almeida et al. (2013) also noted that after 30 days of being subjected to water deficit, the drought-tolerant clone RB 72910 experienced no significant effect on RWC, while the sensitive genotype RB 943365 experienced a 50% decrease. Medeiros et al. (2013) noticed a substantial drop in RWC owing to the decline in soil moisture, with average values of 90.7% and 88.7% in the RB 962962 and RB 867515 varieties, respectively. Water potential (ψ) and Osmotic potential (Ψs) are the physiological parameters employed to measure the level of plant stress (Wahid et al., 2005; Verma et al., 2023).

Reduced leaf water potential brought on by a water shortage made it more difficult for plants to draw moisture from the soil and keep their turgor. Stomatal conductance, which controls transpiration, photosynthesis, and root water uptake, is regulated by water potential based on the variation in potential between leaf and soil water. Due to the drought stress, the leaf water potential (ψ) dropped by 34.50%. Though the decrease in water potential was greater in sensitive genotypes CoVc 93,136 and Co 99,014 (greater than 50%), it was less in resistant genotypes Co 99,004 and Co 99,012 (26.96 and 30.15%) (Gomathi et al., 2011). The RB 72910 variety maintained higher ψ values than the RB 943365 variety under water deficit conditions. During the stress treatment, the water potential (ψ) of the RB 867515 and RB 962962 varieties decreased by an average of 1.19 and 0.78 MPa, respectively, compared to values found in the control treatment, which were approximately 11-fold larger. The rewatering treatment restored the plant's water balance, as expected (Medeiros et al., 2013). Another physiological measure used to assess plant stress is osmotic potential (Ψs). Under drought stress, the osmotic potential of sugarcane cultivar leaves has decreased (Inman et al., 1986). By subjecting drought-sensitive (BL-4) and drought-tolerant (CP 43/33) sugarcane cultivars to a 200 mol m^3 mannitol solution, Basra et al. (1999) examined the water relations of the plants. They found that over time, the turgor of sugarcane BL-4 decreased more quickly than that of CP 43/33. Compared to CP 43/33, BL-4 had higher bulk leaf osmotic pressures and cell wall solutes. In sugarcane, Biancos et al. (2000) found that the maximum osmotic adjustment (i.e., the gap in osmotic potential between well-watered and stressed plants) was roughly 0.5 and 0.6 MPa for mature leaves and shoot tips, respectively. Four sugarcane cultivars were found to have significantly lower leaf RWC, osmotic potential, and water potential when subjected to drought stress (Pooja et al., 2019). In a different study, Pooja et al. (2021) discovered that the sugarcane clones Co 05011 (78.77%) and Co 0238 (76.88%) displayed greater RWC under stress conditions, with an average decrease of 16.70% compared to Co-canes RWC.

4.2.2.2 OSMOTIC ADJUSTMENTS

A crucial component of osmotic adjustment is the accumulation of osmoprotectors in plant vacuoles or the cytosol as a good way to maintain water stability and preserve the cellular integrity of proteins, enzymes, and cellular membranes (Jaleel et al., 2007). The accumulation of these substances in plant tissues is a method for establishing osmoprotectants in cultivars resistant to drought. The disaccharide trehalose and the amino acid proline are the most notable osmoprotectors (Bartels & Sunkar, 2005; Paul et al., 2008). Carlin & Santos (2009) evaluated the sugarcane variety IAC91-5155 and found that trehalose concentrations had increased to 2.54 mol g^{-1} of fresh weight, a 25.9% accumulation (an increase of 0.54 mol g^{-1} fresh mass weight). Trehalose is unique among disaccharides in that it can replace the water lost under dehydrating conditions, allowing hydrophilic structures to remain hydrated (Crowe, 2007). Sucrose would inevitably be more effective than trehalose in protecting these structures in the majority of species, as trehalose accumulation in superior angiosperms under unfavorable environments is uncommon. Sucrose and other disaccharides, like maltose, induce cellular stability to a lesser extent when compared to trehalose.

Moreover, this carbohydrate has a three times greater impact than sucrose in terms of stabilizing membranes in water shortage situations (Patist & Zoerb, 2005). There is insufficient evidence to indicate trehalose's actual role in plant development because its metabolism in superior plants was only recently discovered (El-Bashiti et al., 2005). Free proline preserves cellular turgescence and provides an immediate energy supply for tissue growth and proliferation (Kishor et al., 2005). For this to happen, proline would need to be translocated from the phloem to the roots of the plant. In water-stressed plants, proline levels were higher in the phloem and elevated in the root apical meristem than in other plant parts (Raymond & Smirnoff, 2002). Considering this perspective, it would be crucial to ascertain how other energy production and carbon metabolism pathways are affected by the buildup of free proline throughout drought conditions (Kishor et al., 2005). Consequently, this osmotic adjustment may also help to maintain leaf cell expansion and growth (Ashraf & Foolad, 2007).

4.2.2.3 GAS EXCHANGE TRAITS

Photosynthesis is a fundamental physiological process for plant carbon uptake and crop production (Chaves et al., 2009; Verma et al., 2019). The stomatal constraint is generally accepted as the primary factor that reduces photosynthetic efficiency under environmental stress, directly impacting sub-stomatal CO_2 concentration (Cornic et al., 2000; Song et al., 2016). It appears that the photosynthetic apparatus of plants is more susceptible to water stress (Petsas et al., 2009; Zivcak et al., 2014). Improving photosynthetic traits is the foundation for increasing harvestable plant yield. Drought slows photosynthesis and lowers photosynthetic pigment levels by closing stomata, changing the electron transport rate (ETR), and exacerbating the effects of excessive light intensity-induced photoinhibition. These changes could contribute to a further decrease in crop yield (Chen et al., 2017; Li et al., 2019; Gazanchian et al., 2006; Verma et al., 2019). According to Casson & Hentherington (2010), the key elements involved in plant environmental adaptation were two indications: adjusting stomatal motion and optimizing stomatal density and appearance. Stomata morphology or size directly influenced plant leaves' photosynthetic CO_2 assimilation and transpiration rate. The closure of stomata in drought-stressed environments led to a reduction in transpiration rate, which decreased photosynthetic rates and CO_2 fixation (Taiz & Zeiger, 2002). However, additional factors influenced the photosynthetic rate. Some studies found no significant difference in photosynthetic rates for plant growth under drought stress conditions and revealed anatomical traits of decreasing leaf thickness, increasing stomatal density, or decreasing stomatal size (Bosabalidis & Kofidis, 2002; Nawazish et al., 2006).

Fluorescence of leaves has been used as a proxy for leaf photochemical capacity. When absorbed by photosystem II (PSII), light energy can be converted to photochemical energy (in the form of ATP and NADPH) or dissipated as heat or fluorescence. Thus, all photochemical reactions influence the emission of fluorescence. For instance, decreased electron transport increases fluorescence emission (Guo & Tan, 2015). Several studies (Silva et al., 2007, 2013, 2014; Da Graca et al., 2010) have documented a decrease in Fv/Fm (Figure 4.3) in sugarcane plants under water stress. Therefore, the Fv/Fm ratio indicates the conversion of light energy to chemical energy or photochemical quenching (Silva et al., 2014). Figure 4.3 depicts the reduction of photosystem II quantum yield (PSII yield or Fv/Fm) as a water stress response in sugarcane (Cha-Um & Kirdmanee, 2008; Ribeiro et al., 2013). Higher F'v/F'm may indicate a genotype that performs better under moderate water deficit (Silva et al., 2007; Verma et al., 2021c).

When leaves are dark-adapted, QA (plastoquinone) is maximally oxidized, and the PSII is referred to as "open." Exposure of the leaf to a weak measuring light results in a minimal level of measured fluorescence (F0). A saturation pulse is emitted, and Fm, or maximum fluorescence, is measured. The distinction between Fm and F0 is referred to as Fv or variable fluorescence. Fv/Fm is the maximum quantum yield of QA reduction or PSII photochemistry. The minimum fluorescence level rises above the original background (F') when a leaf is light-adapted. In this case, less QA is oxidized, and when a light pulse is emitted, the maximum fluorescence for light-adapted leaves (F'm) is measured, and its level is lower than Fm. This is because when plants are subjected to stress, the photochemical quenching is diminished due to the photo-inactivation of PSII, resulting in a higher level of non-photochemical quenching (NPQ) or the dissipation of energy as heat. Fv is derived from F'm-F.' Maximum PSII efficiency is denoted as Fv'/Fm. This parameter calculates the NPQ's contribution to observing PSII operation changes (Ferreira et al., 2017).

4.2.2.4 ANTIOXIDATIVE DEFENSE MECHANISM OF SUGARCANE UNDER DROUGHT

The photosynthetic process gets diminished under drought because the enzymes involved in it, such as PEPCO (sugarcane mesophyll cell) and RUBISCO, fructose-1,6-bisphosphatase (FBPase), malic enzyme (NADP-ME), and pyruvate orthophosphate dikinase (PPDK) get denatured under severe water stress. Since photosynthesis is essential for plant sucrose production, this reduces carbohydrate accumulation and plant biomass (Sage et al., 2013). Due to the over-assimilation of carbon brought on by a lack of photosynthesis reactants (water) during water deficit and the subsequent inadequate dissipation of electrons generated during the electron transport chain, reactive oxygen species (ROS) build up (Zingaretti et al., 2011). ROS are responsible for the production of superoxide radicals (O_2^-), hydrogen peroxide (H_2O_2), and hydroxyl radicals (OH) (Azevedo et al., 2011; Rai et al., 2011). Drought is expected to result in lessened harvest and harm to biological molecules like chlorophyll, fatty acids, enzymes, and polynucleotides due to the excess production of ROS (Rai et al., 2011; dos Santos & de Almeida Silva, 2015; Ogbaga et al., 2020). Despite the damage it can do, ROS has indeed been explained as a signaling molecule in response to the majority of stressors (Geisler, 2010; Azevedo et al., 2011). Hydroxyl radicals stimulate the Calcium ions and potassium ions (K^+) on the cell membrane of plant roots (Demidchik et al., 2003).

Underwater scarcity, calcium ions control how the stomata widen and shut (Li & Assmann, 2010). H_2O_2 further stimulates the expression of post-translational protein descriptors like the SUMO proteins under heat and drought stress conditions (Kurepa et al., 2003). The negative effects of reactive oxygen species (ROS) are offset by the transcriptional activation of antioxidant compounds (Murphy, 2011). Superoxide dismutase (SOD) has been identified as an antioxidant enzyme with a well-defined function during drought stress in sugarcane (Azevedo et al., 2011). Ascorbate peroxidase (APX), catalase (CAT), glutathione-S-transferase (GST), polyamines, ascorbate (ASC), and phenolic compounds are additional antioxidant enzymes found in plants (Abbas et al., 2014; Tsaniklidis et al., 2021). Plastids, cytosol, apoplasts, and endosomes all contain APX, which uses ascorbate to reduce hydrogen peroxide. Contrarily, CAT, which is only present in peroxisomes, is crucial for ROS detoxification whenever there is stress, and a greater amount of ROS is generated (Scandalios et al., 2000). In addition, peroxisome proliferation is caused by oxidative stress (Lee et al., 1999). Catalase can lower the levels of hydrogen peroxide in peroxisomes, whereas chloroplasts lack this enzyme. Specific ascorbate peroxidase fulfills the function of catalase (Asada, 1994). A marked rise in the activity of antioxidative enzymes such as CAT, POX, APX, and SOD was induced by water stress (PEG treatment). Compared to the control, Co 86032 sugarcane callus had significantly greater SOD activity under drought and salt stress. The CAT activity of stressed and control plants did not differ significantly (Patade et al., 2011). Ngamhui et al. (2012) observed different sugarcane varieties and concluded that higher levels of APX and POX provide tolerance to plants under drought stress. Hydrogen peroxide accumulation negatively affects living cells and is also involved in stress signaling and cellular redox status regulation (Mittler et al., 1993; Neill et al., 2001; Verma et al., 2021b).

Arora et al. (2002) reported that when plants close their stomata in response to a water deficit and reduce their internal CO_2 concentration, the production of reactive oxygen species appears to stimulate mechanisms that reduce oxidative stress; therefore, it may play a significant role in drought tolerance. To protect plant cells from oxidative damage, non-enzymatic antioxidant molecules can collaborate with enzymatic ROS-scavenging mechanisms. Ascorbic acid (AA), reduced glutathione (GSH), tocopherol, carotenoids, phenolics, flavonoids, and proline make up the nonenzymatic system (Gill & Tuteja, 2010; Das & Roychoudhury, 2014). Proline (Pro) is an effective OH and $^1O_2$ scavenger. In addition, proline can serve as a compatible osmolyte, a molecular chaperone, a carbon and nitrogen reserve, and a pH regulator for the cytosol (Das & Roychoudhury, 2014). Pro accumulates in plants during periods of water stress primarily due to increased synthesis and decreased degradation. Pyrroline-5-carboxylate (P5C) synthetase (P5CS) and P5C reductase catalyze the biosynthesis of pro from glutamate (P5CR). Ornithine can also be converted to P5C/GSA by ornithine-aminotransferase (OAT) (Bhaskara et al., 2015). Low ASM levels induce the antioxidative defense system by increasing reactive oxygen species (ROS) and the specific activities of antioxidative enzymes, including peroxidase, catalase, and ascorbate peroxidase (Pooja et al., 2020).

4.3 SALT STRESS

Salinity is among the most serious ecological extremes hindering plant metabolic rate and growth and is considered the most restricting determinant for agriculture. Due to exacerbating agriculture and global climate change, this environmental issue is spreading across the globe. According to the UNEP Report, Nearly 20% of the world's farmland, or even more than 900 mHa, is damaged by salt worldwide (FAO, 2020), making up more than 6% of the planet's total land area. Sodic and saline soil types comprise about 60 and 40% of the nearly 7 mha of SAS (salt-affected soils) in India (Sharma & Singh, 2015). Salt-affected soils are prevalent in arable and moderately arid parts of the world, where irrigation is necessary to boost agricultural production to meet the demand for food (Abrol et al., 1988). The salinity of the soil is a growing issue for crop production, affecting some of the world's most productive crop regions. The main factors preventing the growth of plants in such salty farmland are a lack of water and a high salinity level. Due to the limited agricultural land and lack of salt-resistant crop plant varieties, the agricultural situation is deteriorating, so it is urgent to adapt the crop plants to the new conditions. Regardless of the crop type and the stress duration, these factors limit the crop's yield. The addition and application of new technologies and conventional methods could be employed to address such issues (Figure 4.2).

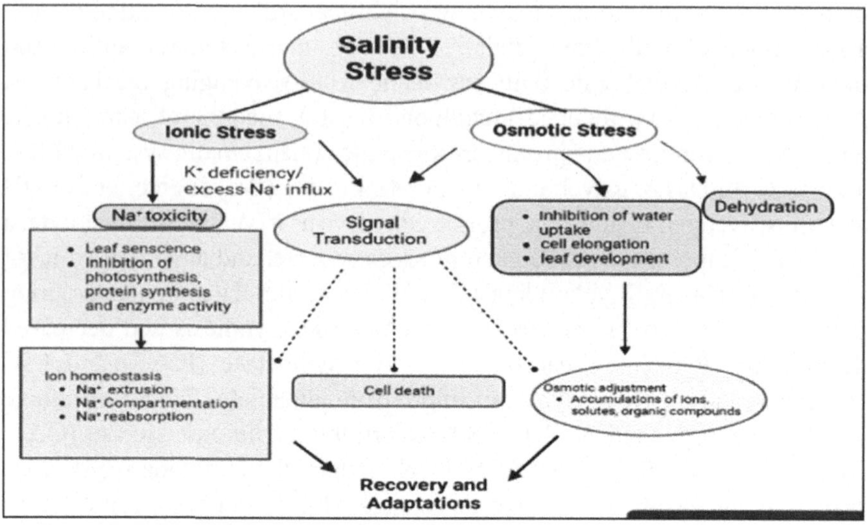

FIGURE 4.2 General effects of salt stress on plants.

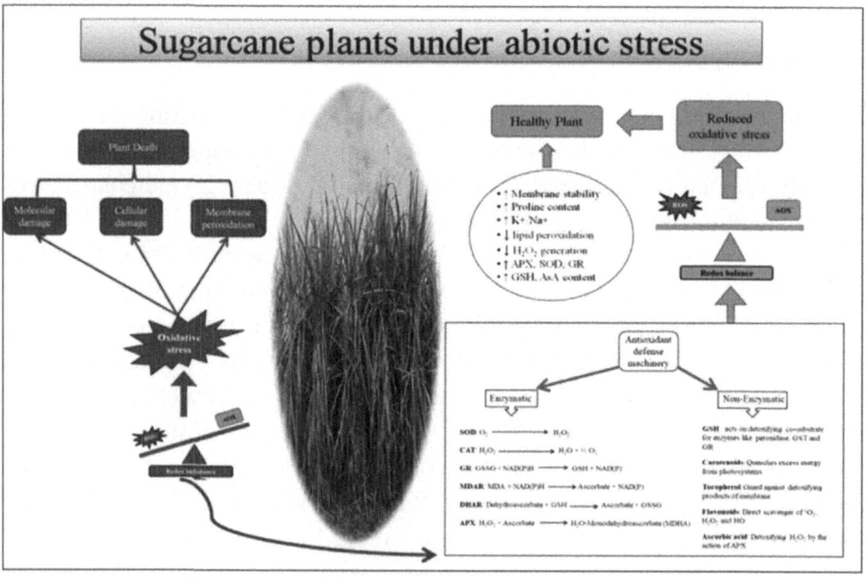

FIGURE 4.3 Physiological and biochemical action of mechanisms of sugarcane in response to drought and saline stress conditions.

The capabilities of plant species to avoid or combat the harmful effects of dissolved salts in their rhizosphere, as well as to sustain overall productivity in salt-affected soils corresponding to their development in non-saline soils, are known as salt tolerance. Hence, it is the relative growth of plants in salinity. Salinity negatively impacts agricultural productivity as it affects plant growth on one side and land use on the other. A particular insufficiency or toxicity may cause salinity stress or be brought on by an accumulation of extra salt in the plant's root system. In India, the sugar industry is the second largest agro-processing industry. About 45 million farmers are directly involved in sugarcane cultivation, and about 7.5% of the rural population directly or indirectly depends on the sugar industry. Recent studies show that about 5% of agricultural land is adversely affected by salinity. In addition, it has also been anticipated that expanding soil salinity in agricultural areas will decrease the arable lands for cultivation by 30% over the next 25 years and approximately 50% by the year 2050 (Patade et al., 2009). Higher salt content in the soil has a negative impact on plant growth and development by interfering with various physiological processes. The harmful consequences of saline soil on the plant growth rate, tillering, reproductive stage, and, finally, limit the crop yield. The world population was mounting startlingly, and agricultural land was lessening due to industrialization and habitual use. Hence, to meet the rising demand for food, it is necessary to use the land that has been affected by salt.

Physiologically, salt tolerance is also complex, as plants show a wide array of adaptations to salt tolerance. To thrive under salinity, plants have evolved and adapted a number of physio-chemical mechanisms. Salinity directly impacts photosynthetic activity by lessening leaf area, total chlorophyll, and stomatal closure, and to a smaller extent, by lowering photosystem II efficiency. Salinity has detrimental effects on plant development but is more severe during the reproductive stage. Salinity restricts microsporogenesis, and stamen filament elongation increases induced apoptosis in different tissues, ovule abortion, and senescence of fertilized embryos, all detrimental to reproductive development (Sengar et al., 2013). One such mechanism to avoid the osmotic effects of salinity is osmotic adjustment (OA; buildup of compatible solutes, which keep water moving into cells from regions of lower potential to higher in plant tissues). For cells to withstand osmotic stresses, compatible organic solutes, such as proline, sugars, polyols, quaternary ammonium compounds like glycine betaine, and other small-molecular-weight metabolites, lower or balance the osmotic potential of intracellular and extracellular ions. Osmotic adjustment is also greatly aided by ionic species, particularly Na, K, Ca, and Cl (Chen & Jiang, 2010). The organic compatible solutes facilitate the stabilization of enzymes and proteins and the preservation of membrane stability (Yancey et al., 1982; Yeo, 1998).

Photosynthesis is suppressed when chloroplasts contain high levels of sodium ions and/or chloride. Salts seem to have little effect on photosynthetic electron transport, but they may impact photophosphorylation or carbohydrate metabolism. Stomata may close due to salinity stress, minimizing the amount of carbon dioxide available in the leaves and preventing photosynthesis (Parida & Das, 2005). It exposes chloroplasts to too many incident photons, which could produce reactive oxygen species (ROS), resulting in oxidative stress (Parvaiz & Satyawati, 2008). Overproduction of ROS can end up causing cell apoptosis or oxidative damage to the cellular components within the cells. In addition to helping to scavenge free radicals and safeguard peptides, some of the osmoprotectants that build up in cells in reaction to stress also have other functions. High soil salinity levels negatively impact ion toxicity and nutrient availability (Serrano et al., 1997). Nutrient imbalance results from high Na+ ion concentrations in the soil, reducing the amount of K^+, Mg^{++}, and Ca^{++} that are readily available (Epstein, 1972). Complex physiological characteristics, metabolic processes, and molecular systems play a role in the adjustment or tolerance of plants to salinity stress. Principal mechanisms include biosynthesis of osmoprotectants and compatible solutes, ion homeostasis and compartmentalization, ion transport and uptake, activation of antioxidative enzymes and synthesis of antioxidant compounds, synthesis of polyamines, etc. (Figure 4.2).

4.4 EFFECTS OF SALT STRESS ON SUGARCANE

4.4.1 GROWTH TRAITS

Salt stress has various complicated implications for plants, which can be seen as alterations in plant morphology, growth, and physiochemical processes. It can be brought on by problems with water uptake and the toxicity of particular ions, which either reduce plant development, quality, and yield or result in the complete loss of the plant's ability to produce. Due to the sensitivity of the growth processes to the impacts of high salinity, growth rate, and photosynthetic capacity are useful metrics to assess these effects and the plant's tolerance to this stress (Dhansu et al., 2022; Kumar et al., 2022). Previous research on sugarcane found that the electrical conductivity of soil greater than 1.7 dS m^{-1} at crucial stages of development remarkably prompts cane length, thickness, NMC, and yield. Each unit change over and above 1.7 dS m^{-1} restricts production to a tune of 5.9% (Rao et al.,

2015). While salinity levels higher than 8 dS m^{-1} substantially decreased growth and physiological traits and yield losses by 50% (Plaut et al., 2000; Santana et al., 2007). Rao et al. (2021) studied the effect of salinity stress on sugarcane genotypes at the elongation stage. They noted that the growth rate was decreased in almost all the sugarcane genotypes under saline conditions, and morphological growth traits like mother shoot height, number of green leaves, and leaf area were lower in the salty environment than normal. Djajadi et al. (2022) noted that the interaction between NaCl concentration and sugarcane varieties significantly impacted plant growth, expressed by the stalk diameter and the number of tillers.

Studies on sugarcane variety to salinity stress depicted that sugarcane growth was inhibited above 50 mM NaCl, including root length, plant height, fresh/dry root, and shoot weight. In contrast, above 100 mM NaCl salt injury symptoms were more pronounced at leaves (Yunita et al., 2020). Saline water irrigation of 10 dS m^{-1} significantly decreased plant height, cane girth, and the number of tillers as well as overall sugarcane growth (Simões et al., 2016; Singh & Sengar, 2020). The buildup of harmful ions may hamper sugarcane development. Sugarcane's sucrose output is reduced by salinity in the root zones due to its effects on biomass and juice quality.

4.4.2 PHYSIOLOGICAL RESPONSES

4.4.2.1 WATER RELATIONS

Water potential (ψw), osmotic potential (ψs), and turgor potential (ψt), which are plant-water relation parameters, are becoming more important to explain water movement in the soil-plant-atmosphere system, the impact of stress on morpho-physiological and biochemical parameters, and the mechanism of stress resistance in plants. Gomathi et al. (2010) found that plants growing in typical soil showed high osmotic and water potential. However, under salt stress conditions, these were drastically decreased. According to Mattioni et al. (1997); and Lutts et al. (1999), genotypes more effective in accumulating compatible solutes linked to greater concentrations of K under stressful situations directly impact the degree and decrease of OP and WP. The tolerant sugarcane genotypes demonstrated greater osmotic adjustment by accumulating more K, free proline, and sugar content, which decreased osmotic potential and water potential compared to salt-sensitive sugarcane lines (Nasir Ahmed et al., 1999). Brindha et al. (2019) observed that relative water content varied from 69 to 80% among the genotypes in control, while under EC_{iw} ~8 dS m^{-1} at the formative stage, it ranged from 48 to 70%, with Co 97010 recording the least turgidity, indicating stress severity on leaf moisture content during formative phase is effective. In plants under salinity, both osmotic and ionic effects occur as the high soil salt concentration decreases water uptake, increases salt uptake, and increases cellular concentrations of Na^+ and Cl^- (Munns & Gilliham, 2015).

4.4.2.2 OSMOTIC ADJUSTMENTS

Some organic molecules or low molecular substances, such as sugars, proline, polyols, and quaternary ammonium compounds like glycine betaine, can lower or balance the osmotic potential of the cell to withstand osmotic stresses (Negro et al., 2017). Osmotic adjustment is greatly aided by inorganic ions, particularly Na^+, K^+, Ca^{2+}, and Cl^- (Chen & Jiang, 2010). However, their synthesis occurs at an energy cost (Raven, 1985), which may be one of the reasons for the reduction in plant growth. Other organic compatible solutes, besides OA, play a role in stabilizing enzymes/proteins and protecting membrane integrity (Yancey et al., 1982; Bohnert & Jensen, 1996; Yeo, 1998). In addition to osmotic regulation, osmolyte accumulation also plays a crucial physiological role in maintaining cellular structure by acting as free radical scavengers (Kumar et al., 2021). Osmolytes also balance the buildup of Na^+ in vacuoles or apoplast (Bohnert & Jensen, 1996). The endogenous level of free proline, free amino acid, and polyols increased to 45, 31, and 15% in sugarcane genotypes, respectively, under salinity conditions, with higher levels noted in tolerant genotypes (Gomathi et al., 2010). Sugarcane genotypes grown under drought and salinity stress accumulated higher total soluble and reducing sugars under drought stress than salt-stressed and control plants (Patade et al., 2011).

In comparison to the control, plants that were stressed by salt or PEG accumulated four times more free proline. Similarly, stressed plants exposed to salt (23%) or PEG (16%) accumulated more glycine betaine than control plants. Compared to varieties cultivated in normal soil conditions, the nitrogenous components, such as soluble proteins, free amino acids, and proline levels, always rose in the saline-grown variety during the crop's formative stage (Saxena et al., 2010). Tolerant variants CoPant 97222, CoS 7250, and Co 98014 were observed to accumulate more proline than genotypes Co 0238, CoSe 8457, and Co 0118 when exposed to salt stress, whereas CoS

8279, Co 5011, and Co 5009 exhibit intermediate behavior (Singh & Sengar, 2020). Thus, to prevent ion toxicity, plants under salt stress may utilize salty ions as an osmoticum, but plants under dehydration stress entirely rely on the synthesis of suitable solutes to maintain cell turgor (Munns & Tester, 2008; Lokhande et al., 2010).

4.4.2.3 PHOTOSYNTHETIC LEAF GAS EXCHANGE

Plant photosynthetic response to salt stress is complicated as it relies on soil salt content as well as stress duration. Therefore, symptoms and effects will vary correspondingly (Chaves et al., 2009). Zhao et al. (2020) observed that sugarcane genotypes grown under 300 mM NaCl showed a reduction of about 35%, 31%, and 29% in photosynthetic rate, stomatal conductance, and transpiration rate, respectively, compared to control plants. Photosynthetic efficiency in terms of gas exchange and chlorophyll fluorescence was significantly reduced under salt treatments. It was found that 2,000 and 3,000 mg NaCl L^{-1} irrigation significantly decreased the photosynthetic efficiency of sugarcane, resulting in considerable reductions of photosynthesis by 7.3 and 9.6 μmol m^{-2} s^{-1} after 1 month and by 8.8 and 14.3 μmol m^{-2} s^{-1} after 2 months of saline irrigation compared to the control, respectively (Watanabe et al., 2020). Under mild salinity conditions of 80 mM NaCl, sugarcane cv. IAC 87-3396 showed higher gas exchange traits than SP 81-3250, while at severe stress of 160 mM NaCl, cv. IAC 87-3396 showed a significant decline compared to cv. SP 81-3250, which maintained equal rates in all treatments (Chiconato et al., 2019). This indicates that the reaction to saline conditions is genetically varied, and the crucial salt concentration may vary based on the varieties used.

Dhansu et al. (2022) observed that continued saline irrigation resulted in a significant diminution in growth, correlated with changes in several gas exchange traits, including chlorophyll content. During the formative phase, the photosynthetic and transpiration rates weren't significantly different among control and salt-treated plants. Still, the tolerant sugarcane clones managed relatively unified photosynthetic rates at grand growth. The sensitive ones decreased dramatically due to different salinity levels (Vasantha et al., 2010). Stomatal closure and feedback inhibition caused by lower sink activity are two causes of reduced photosynthesis. Furthermore, the osmotic effects of salinity may result in a decrease in stomatal conductance. As opposed to this, internal CO_2 concentration is influenced by the activity of the ribulose-1,5-bisphosphate carboxylase oxygenase enzyme, which, in ideal conditions, improves carboxylation and lessens oxygenation of the carbon fixation, resulting in enhanced CO_2 usage (Lima et al., 2021; Verma et al., 2020a, c).

4.4.2.4 ANTIOXIDATIVE DEFENSE MECHANISM OF SUGARCANE UNDER SALT STRESS

Numerous plant reactions to salt stress are triggered, as is the elevated production of reactive oxygen species (ROS) like superoxide and hydrogen peroxide, and hydroxyl radical affects cellular structures and metabolism. The generation of ROS outcomes from metabolic disturbances brought on by the osmotic stress (caused by salt or dehydration stress or ionic toxicity of salt stress) eventually causes membrane leakage, lipid peroxidation, protein degradation, and decreased enzyme activities (Rani et al., 2016; Pooja et al., 2020). Plants have an antioxidant defense mechanism that generates non-enzymatic (phenolics, flavonoids, tocopherols, ASC, and GSH) and enzymatic antioxidants (superoxide dismutase, peroxidase, catalase, glutathione reductase, ascorbate peroxidase, monodehydroascorbate dehydrogenase, and dehydroascorbate reductase) to counteract the oxidative effects of ROS. Willadino et al. (2011) observed an increase in the activity of ascorbate peroxidase (APX) and catalase (CAT) in RB867515 sugarcane genotypes, while in genotype RB863129, there was a reduction in the activity of these two enzymes. The results of the de Morais et al. (2020) study showed that sugarcane variety, as well as stress treatments (water and saline), all had a substantial effect on SOD, CAT, and APX activities. In all sugarcane varieties, a boost in the activity of the oxidative stress defense enzymes SOD, CAT, and APX showed a strongly favorable connection. The coordinated action of SOD with CAT and APX is critical for redox equilibrium maintenance because it ensures the efficient removal of oxidative molecules, avoiding oxidative alterations in various biological components (Shi et al., 2015).

Pote et al. (2019) studied the effect of salinity on the activities of antioxidative enzymes and noticed that APX and SOD activities varied depending on the genetic makeup and stress levels. Karpe et al. (2012) conducted a comparative study to evaluate how different sugarcane varieties responded to salt stress. This analysis revealed that the salt-tolerant genotype had several times more APX activity than the salt-susceptible genotype. Significant genotypic changes in APX activity were seen under salt stress, and sugarcane genotypes C 92038 and Co 85004 exhibited an increasing trend in APX, GR, POX, CAT, and SOD activities (Gomathi & Rakkiyapan, 2011). Salt stress produces an imbalance in the utilization of ROS, which causes a disorderly buildup of these species in plant tissues and can result in severe plant damage. As a result, salt supplementation raises the activity of one or more antioxidative system enzymes such as APX, CAT, and POD. Catalase is a very efficient enzyme that works in cellular detoxification from excessive H_2O_2 during stress conditions (Gill & Tuteja, 2010), whereas the POD enzyme composes antioxidative protection. Another enzyme, APX, actively removes hydrogen peroxide by oxidizing hydrogen peroxide into water (Teixeira et al., 2004). SODs play a crucial role in sequestering oxygen radicals by the dismutation of two molecules of oxygen radicals into oxygen molecules and hydrogen peroxide (Wang et al., 2005). Maintaining redox homeostasis is critical to avoiding oxidative stress caused by a mismatch between the generation and appropriation of ROS. The plant can employ metabolites and enzymes to achieve equilibrium. The degree of stress and the plant's reaction is determined by the duration and intensity of a stressor's effect and the linkage of more than one component.

4.5 CONCLUSIONS AND FUTURE PROSPECTIVE

Environmental vulnerabilities are key limits to sugarcane plant growth, development, and output. Drought and salinity are significant environmental factors that affect sugarcane growth, yield, and metabolism globally. Plant growth involves several physio-biochemical processes that are extremely sensitive to water and salt stress. Plant survival under different conditions is influenced by its ability to generate and transmit signals, as well as biochemical and physiological changes. Since sugarcane is a complex polyploidy crop, no single approach has been proven to be effective for confirming polygenic and phenotypic traits. To better comprehend the use of fundamental omics in sugarcane for agronomic characteristics and industrial quality attributes, as well as responses to varied biotic and abiotic challenges, it is necessary to investigate the physiology, genomic structure, functional integrity, and collinearity of sugarcane with other crops/plants that are more or less comparable. This crop's genetic progress is limited by its complex genome, low fertility ratio, extended production cycle, and vulnerability to a variety of biotic and abiotic stressors. Interventions in biotechnology are anticipated to open up possibilities for overcoming these challenges and enhancing sugarcane production. By using more advanced quantitative and automated selection and screening techniques that concentrate on plant physiology and quality attributes, the genetic diversity of genotypes and phenotypes is connected. These characteristics, when paired with decision-making algorithms, would facilitate the delivery of newly bred varieties to farmers and obviate the need for extensive field surveys and protracted development periods. Furthermore, more focus should be given to CRISPR and epigenetics molecular pathways that are most significant to plant adaptability to changing environments gradually. Finally, the major consideration towards crop improvement is to encourage transparent dialogue between molecular biologists and plant physiologists on the one hand and farmers, breeding companies, and the public on the other hand in order to jointly address the economic, sociological, legal, and ethical hurdles.

KEYWORDS

- **biochemical changes**
- **CRISPR**
- **era of climate change**
- **genetic diversity**
- **genotypes**
- **leaf gas exchange**
- **physiology**
- **sugarcane**

REFERENCES

Abbas, S., Ahmad, S., Sabir, S., & Shah, A. (2014). Detection of drought-tolerant sugarcane genotypes (Saccharum officinarum) using lipid peroxidation, antioxidant activity, glycine-betaine and proline contents. *Journal of Soil Science and Plant Nutrition*, *14*, 233–243.

Abrol, I. P., Yadav, J. S. P., & Massoud, F. I. (1988). Salt Affected Soils and Their Management. Soil Resources Management and Conservation Series. FAO Land and Water Development Division Bulletin, pp. 39.

Aiken, R. M., & Smucker, A. J. M. (1996). Root system regulation of whole plant growth. *Annual Review of Phytopathology*, *34*, 325–346.

Almeida, C. M. A., Silva, T. D., Malafaia, C. B., Amaral, D. O. J., Arruda, I., & Brito, G. G. (2013). Proteomic and physiological analysis of response to water deficit in sugarcane. *Wudpecker Journal of Agricultural Research*, *2*(1), 1–7.

Anas, M., Verma, K. K., Haq, I. U., Naeem, M., Li, Q., Liao, F., Xiang, Y. X., & Li, Y. R. (2021). Characterization of exotic, native and wild-type genotypes of sugarcane (Saccharum spp. Hybrids) for internal nitrogen use efficiency under different nitrogen levels. *Sugar Tech. 23*, 1258–1267. https://doi.org/10.1007/s12355-021-01014-1.

Anas, M., Verma, K. K., Riaz, M., Qiang, L., Liao, F., Liu, Y., & Li, Y. R. (2021). Physio-morphological and biochemical mechanism of nitrogen use efficiency in sugarcane (Saccharum spp.) genotypes under different growth stages and nitrogen levels. *Journal of Plant Interactions*, *16*, 332–343.

Araus, J. L., Slafer, G. A., Royo, C., & Serret, M. D. (2008). Breeding for yield potential and stress adaptation in cereals. *Critical Reviews in Plant Sciences*, *27*, 1–36.

Arora, A., Sairam, R. K., & Srivastava, G. C. (2002). Oxidative stress and antioxidative systems in plants. *Current Science*, *82*, 1227–1238.

Ashraf, M., & Foolad, M. R. (2007). Roles of glycine betaine and proline in improving plant abiotic stress resistance. *Environmental and Experimental Botany*, *59*, 206–216.

Azevedo, R. A., Carvalho, R. F., Cia, M. C., & Gratao, P. L. (2011). Sugarcane under pressure: An overview of biochemical and physiological studies of abiotic stress. *Tropical Plant Biology*, *4*, 42–51.

Bartels, D., & Sunkar, R. (2005). Drought and salt tolerance in plants. *Critical Reviews in Plant Sciences*, *24*, 23–58.

Basra, S. A. M., Ahmed, N., & Khalid, A. (1999). Water relations studies in water stressed sugarcane (Saccaharum officinarum L.). *International Journal of Agricultural Biology*, *1*, 1–4.

Bhaskara, G. B., Yang, T. H., & Verslues, P. E. (2015). Dynamic proline metabolism: Importance and regulation in water limited environments. *Frontiers in Plant Science*, *6*, 484. https://doi.org/10.3389/fpls.2015.00484.

Biancos, R. L., Rieger, M., & Sung, S. S. (2000). Effect of drought on sorbitol and sucrose metabolism in sinks and sources of peach. *Physiologia Plantarum*, *108*, 71–78.

Bidoia, M. A. P., Santos, D. M. M., Marin, A., Landell, M. G. A., Banzatto, D. A., & Cazetta, J. O. (2006). Effect of water deficiency on free proline accumulation in sugar cane at different developmental periods. *Stab – Açúcar, Álcool e Subprodutos*, *24*, 6–9.

Blum, A. (2009). Effective use of water (EUW) and not water-use efficiency (WUE) is the target of crop yield improvement under drought stress. *Field Crops Research*, *112*, 119–123.

Boaretto, L. F., Carvalho, G., Borgo, L., Creste, S., Landell, M. G. A., & Mazzafera, P. (2014). Water stress reveals differential antioxidante responses of tolerant and non-tolerant sugarcane genotypes. *Plant Physiology and Biochemistry*, *7*, 165–175.

Bohnert, H. J., & Jensen, R. G. (1996). Strategies for engineering water stress tolerance in plants. *Trends in Biotechnology*, *14*, 89–97.

Bosabalidis, A. M., & Kofidis, A. (2002). Comparative effects of drought stress on leaf anatomy of two olive cultivars. *Plant Science*, *163*, 375–379.

Campos, H. A., Cooper, J. E., Habben, G. O., Edmeades, J. R., & Schussler, S. (2004). Improving drought tolerance in maize: a view from industry. *Field Crops Research*, *90*, 19–34.

Carlin, S. D., & Santos, D. M. M. (2009). Physiological indicators of the interaction between water deficit and soil acidity in sugarcane. *Pesquisa Agropecuária Brasileira, 44*, 1106–1113. (In Portuguese with abstract in English).

Casson, S. A., & Hetherington, A. M. (2010). Environmental regulation of stomatal development. *Current Opinion in Plant Biology, 13*, 90–95.

Cha-Um, S., & Kirdmanee, C. (2008). Effect of osmotic stress on proline accumulation, photosynthetic abilities and growth of sugarcane plantlets (*Saccharum officinarum* L.). *Pakistan Journal of Botany, 40*, 2541–2552.

Chaves, M. M., Flexas, J., & Pinheiro, C. (2009). Photosynthesis under drought and salt stress: regulation mechanisms from whole plant to cell. *Annals of Botany, 103*, 551–560.

Chaves, M. M., Maroco, J. P., & Pereira, J. S. (2003). Understanding plant responses to drought-from genes to the whole plant. *Functional Plant Biology, 30*, 239–264.

Chen, H., & Jiang, J. G. (2010). Osmotic adjustment and plant adaptation to environmental changes related to drought and salinity. *Environmental Reviews, 18*, 309–319.

Chen, Z. K., Niu, Y. P., Ma, H., Hafeez, A., Luo, H. H., & Zhang, W. F. (2017). Photosynthesis and biomass allocation of cotton as affected by deep-layer water and fertilizer application depth. *Photosynthetica, 55*, 638–647.

Chiconato, D. A., da Silveira Souza Junior, G., dos Santos, D. M. M., & Munns, R. (2019). Adaptation of sugarcane plants to saline soil. *Environmental and Experimental Botany. 162*, 201–211. https://doi.org/10.1016/j.envexpbot.2019.02.021.

Cornic, G. (2000). Drought stress inhibits photosynthesis by decreasing stomatal aperture—not by affecting ATP synthesis. *Trends in Plant Science, 5*, 187–188.

Crowe, J. H. (2007). Trehalose as a "chemical chaperone": fact and fantasy. *Advances in Experimental Medicine and Biology, 594*, 143–158.

Da Graça, J., Rodrigues, F., Farias, J., Oliveira, M. d., Hoffmann-Campo, C., & Zingaretti, S. (2010). Physiological parameters in sugarcane cultivars submitted to water deficit. *Brazilian Journal of Plant Physiology, 22*, 189–197.

Das, K., & Roychoudhury, A. (2014). Reactive oxygen species (ROS) and response of antioxidants as ROS-scavengers during environmental stress in plants. *Frontiers in Environmental Science, 2*, 53. https://doi.org/10.3389/fenvs.2014.00053.

de Morais, M. B., Azevedo, R. A., Camara, T. R., Ulisses, C., & de Albuquerque, C. C. (2020). Antioxidative metabolism in sugarcane (Poaceae) varieties subjected to water and saline stress. *Revista Brasileira de Engenharia Agrícola e Ambiental, 24*(11), 776–782.

Demidchik, V., Shabala, S. N., Coutts, K. B., Tester, M. A., & Davies, J. M. (2003). Free oxygen radicals regulate plasma membrane Ca^{2+} and K^+ permeable channels in plant root cells. *Journal of Cell Science, 16*, 81–88.

Dhansu, P., Kulshrestha, N., Kumar, R., Raja, A. K., Pandey, S. K., & Goel, V. (2021). Identification of drought-tolerant co-canes based on physiological traits, yield attributes and drought tolerance indices. *Sugar Tech, 23*, 747–761.

Dhansu, P., Kumar, R., Kumar, A., Vengavasi, K., Raja, A. K., Vasantha, S., Meena, M. R., Kulshreshtha, N., & Pandey, S. K. (2022). Differential Physiological Traits, Ion Homeostasis and Cane Yield of Sub-Tropical Sugarcane Varieties in Response to Long-Term Salinity Stress. *Sustainability, 14*, 132–146.

Dhansu, P., Nandwal, A. S., Kumar, S., Chand, M., Rani, B., & Kulshreshtha, N. (2022). Comparative evaluation of growth, yield and yield attributing traits in sugarcane (Saccharum officinarum) under different soil moisture regimes. *Indian Journal of Agricultural Sciences, 92*(8), 942–946.

Dinh, T. H., Watanabe, K., Takaragawa, H., Nakabaru, M., & Kawamitsu, Y. (2017). Photosynthetic response and nitrogen use efficiency of sugarcane under drought stress conditions with different nitrogen application levels. *Plant Production Science*, *20*, 412–422.

Djajadi, D., Syaputra, R., & Hidayati, S. N. (2022). Effect of salt stress on nutrients content in soil and leaves of three varieties sugarcane. *In: 2nd International Conference on Sustainable Plantation, IOP Conference Series: Earth and Environmental Science*, 974, 012048.

Dlamini, P. J. (2020). In vitro Mutagenesis and Preliminary Characterization of Potential Drought-Tolerant Sugarcane Lines. Department of Agriculture, University of Zululand, Empangeni. https://hdl.handle.net/10530/2239.

dos Santos, C. M., & de Almeida Silva, M. (2015). Physiological and biochemical responses of sugarcane to oxidative stress induced by water deficit and paraquat. *Acta Physiologiae Plantarum*, *37*, 1–14.

El-Bashiti, T., Hamamci, H., Oktem, H. A., & Yucel, M. (2005). Biochemical analysis of trehalose and its metabolizing enzymes in wheat under abiotic stress conditions. *Plant Science*, *169*, 47–54.

Epstein, E. (1972). Mineral nutrition of plants: Principles and perspectives. John Wiley.

Faroni, C. E., & Trivelin, P. C. O. (2006). Quantification of sugar cane active metabolism roots. *Pesquisa Agropecuária Brasileira*, *41*, 1007–1013. (In Portuguese, with abstract in English).

Ferreira, T. H. S., Tsunada, M. S., Bassi, D., Araújo, P., & Mattiello, L. (2017). Sugarcane Water Stress Tolerance Mechanisms and Its Implications on Developing Biotechnology Solutions. *Frontiers in Plant Science*, *8*, 1–18.

Gazanchian, A., Hajheidari, M., Sima, N. K., & Salekdeh, G. H. (2006). Proteome response of Elymus elongatum to severe water stress and recovery. *Journal of Experimental Botany*, *58*, 291–300.

Geisler, M. (2010). Transcription and signaling factors in the drought response regulatory network. In M. A. Jenks & A. J. Wood (Eds.), *Genes for Plant Abiotic Stress* (pp. 55–111). Wiley-Blackwell.

Gill, S. S., & Tuteja, N. (2010). Reactive oxygen species and antioxidant machinery in abiotic stress tolerance in crop plants. *Plant Physiology and Biochemistry*, *48*, 909–930.

Gomathi, R., & Rakkiyapan, P. (2011). Comparative lipid peroxidation, leaf membrane thermostability, and antioxidant system in four sugarcane genotypes differing in salt tolerance. *International Plant Physiology and Biochemistry*, *34*, 67–74.

Gomathi, R., Vasantha, S., & Thandapani, V. (2010). Mechanism of Osmo Regulation in Response to Salinity Stress in Sugarcane. *Sugar Tech*, *12*(34), 305–311.

Gonzalez, L., & Gonzalez-Vilar, M. (2001). Determination of relative water content. In M. J. Reigosa (Ed.), *Handbook of Plant Ecophysiology Techniques* (pp. 207–212). Kluwer Academic Publishers.

Guo, Y., & Tan, J. (2015). Recent advances in the application of Chlorophyll a fluorescence from Photosystem II. *Photochemistry and Photobiology*, *91*, 1–14.

Hoang, D. T., Hiroo, T., & Yoshinobu, K. (2019). Nitrogen use efficiency and drought tolerant ability of various sugarcane varieties under drought stress at the early growth stage. *Plant Production Science*, *22*, 250–261.

Huang, H. R., Gao, Y. J., Malviya, M. K., Verma, K. K., Solanki, M. K., Huang, Y. X., Li, X., Deng, Y., Yan, J., Tang, S. Y., Wang, L. W., & Xu, L. (2021). Genetic diversity analysis of sugarcane (Saccharum spp. Hybrids) among high-sucrose clones of GT series and commonly used parents by using microsatellite markers in Guangxi, China. *Sugar Tech. 24*, 397–407. https://doi.org/10.1007/s12355-021-01033-y.

Hussain, S., Khaliq, A., Mehmood, U., Qadir, T., Saqib, M., Amjed Iqbal, M., & Hussain, S. (2019). Sugarcane production under changing climate: Effects of environmental vulnerabilities on sugarcane diseases, insects and weeds. In: *Climate Change and Agriculture,* ed. Saddam Hussain. IntechOpen, *2019*, 1–16. http://dx.doi.org/10.5772/intechopen.81131.

Inman, N. G., & Smith, D. M. (2005). Water relations in sugarcane and response to water deficits. *Field Crops Research*, *92*, 185–202.

Inman-Bamber, N. G., & de Jager, I. M. (1986). Effect of water stress on growth, leaf resistance and canopy temperature in field-grown sugar cane. *Proceedings of the South African Sugar Technologists' Association*, *60*, 156–161.

Jaleel, C. A., Manivannan, P., Kishorekumar, A., Sankar, B., Gopi, R., Somasundaram, R., & Paneerselvam, R. (2007). Alterations in osmoregulations, antioxidant enzymes and indole alkaloid levels in Catharanthus roseus exposed to water deficit. *Colloids and Surfaces B: Biointerfaces*, *59*, 150–157.

Karpe, A., Nikam, A. A., Chimote, K. P., Kalwade, S. B., Kawar, P. G., Babu, H., & Suprasanna, P. (2012). Differential responses to salinity stress of two varieties CoC 671 and Co 86032 of sugarcane Saccharum officinarum L. *African Journal of Biotechnology*, *1137*, 9028–9035.

Khan, N., Mishra, A., Chauhan, P. S., & Nautiyal, C. S. (2011). Induction of Paenibacillus lentimorbus biofilm by sodium alginate and CaCl2 alleviates drought stress in chickpea. *Annals of Applied Biology*, *159*, 372–386.

Khonghintaisong, J., Songsri, P., Toomsan, B., & Jongrungklang, N. (2018). Rooting and physiological trait responses to early drought stress of sugarcane cultivars. *Sugar Tech*, *20*, 396–406.

Kishor, P. B. K., Sangam, S., Amrutha, R. N., Sri Laxmi, P., Naidu, K. R., Rao, S. S., Rao, S., Reddy, K. J., Theriappan, P., & Sreenivasulu, N. (2005). Regulation of proline biosynthesis, degradation, uptake and transport in higher plants: Its implications in plant growth and abiotic stress tolerance. *Current Science*, *88*, 424–438.

Kumar, P. A., Reddy, N., & Lakshmi, N. J. (2017). PEG induced screening for drought tolerance in tomato genotypes. *International Journal of Current Microbiology and Applied Sciences*, *6*, 168–181.

Kumar, R., Dhansu, P., Kulshreshtha, N., Meena, M. R., Kumaraswamy, M. H., Appunu, C., Chhabra, M. L., & Pandey, S. K. (2023). Identification of Salinity Tolerant Stable Sugarcane Cultivars Using AMMI, GGE and Some Other Stability Parameters Under Multi Environments of Salinity Stress. *Sustainability*, *15*, 1119. https://doi.org/10.3390/su15021119.

Kurepa, J., Walker, J. M., Smalle, J., Gosink, M. M., Davis, S. J., Durham, T. L., Sung, D. Y., & Vierstra, R. D. (2003). The small ubiquitin-like modifier SUMO protein modification system in Arabidopsis: accumulation of SUMO 1 and -2 conjugates is increased by stress. *Journal of Biological Chemistry*, *278*, 6862–6872.

Lee, H., Xiong, L., Ishitani, M., Stevenson, B., & Zhu, J. K. (1999). Cold-regulated gene expression and freezing tolerance in an Arabidopsis thaliana mutant. *Plant Journal*, *17*, 301–308.

Li, J. H., Wang, Y. Y., Li, N. N., Zhao, R. H., Khan, A., Wang, J., & Luo, H. H. (2019). Cotton leaf photosynthetic characteristics, biomass production, and their correlation analysis under different irrigation and phosphorus application. *Photosynthetica*, *57*, 1066–1075.

Li, S., & Assmann, S. M. (2010). Genetic determinants of stomatal function. In M. A. Jenks & A. J. Wood (Eds.), *Genes for Plant Abiotic Stress* (pp. 5–33). Wiley-Blackwell.
Lima, A. F. D. S., Santos, M. F. D., Oliveira, M. L., Sousa, G. G. D., Mendes, P. F., & Luz, L. N. D. (2021). Physiological responses of inoculated and uninoculated peanuts under saline stress. *Revista AmbienteÁgua*, *16*, 1–10.
Lokhande, V. H., Nikam, T. D., & Penna, S. (2010). Differential osmotic adjustment to iso-osmotic NaCl and PEG stress in the in vitro cultures of *Sesuvium portulacastrum* L. *Journal of Crop Science and Biotechnology*, *134*, 251–256.
Lutts, S., Majerus, V., & Kinet, J. M. (1999). NaCl effects on proline metabolism in rice Oryza sativa seedlings. *Physiology of Plants*, *105*, 450–458.
Mall, A. K., Misra, V., Pathak, A. D., & Srivastava, S. (2022). Breeding for Drought Tolerance in Sugarcane: Indian Perspective. *Sugar Tech*, *24*(6), 1625–1635.
Mattioni, C., Lacernza, N. G., Troccoli, A., De Leonardis, A. M., & Fonzo, D. (1997). Water and salt stress-induced alterations in proline metabolism of *Triticum durum* seedlings. *Physiology of Plants*, *101*, 787–792.
Medeiros, D. B., da Silva, E. C., Nogueira, R. J. M. C., Teixeira, M. M., & Buckeridge, M. S. (2013). Physiological limitations in two sugarcane varieties under water suppression and after recovering. *Theoretical and Experimental Plant Physiology*, *25*(3), 213–222.
Meena, M. R., Tabish, A., Appunu, C., Raja, A. K., Vasantha, S., Dhanshu, P., Kumar, R., & Pandey, S. K. (2022). Impact of Salinity Stress on Sugarcane Yield and Quality: Management Approaches for Higher Cane Sugar Productivity. In Springer.
Misra, V., Solomon, A. K., Mall, S., Abid, M., & Mohammad, I. A. (2022). Drought Stress and Sustainable Sugarcane Production. *Microbe Biotechnology for Sustainable Agriculture*, *1*, 353–368.
Mitra, J. (2001). Genetics and genetic improvement of drought resistance in crop plants. *Current Science*, *80*, 758–763.
Mittler, R., & Zilinskas, B. (1993). Detection of ascorbate peroxidase activity in native gels by inhibition of the ascorbate dependent reduction of nitroblue tetrazolium. *Analytical Biochemistry*, *212*, 540–546.
Munns, R., & Gilliham, M. (2015). Salinity tolerance of crops – what is the cost? *New Phytologist*, *208*, 668–673.
Munns, R., & Tester, M. (2008). Mechanisms of salinity tolerance. *Annual Review of Plant Biology*, *59*, 651–681.
Murphy, D. (2011). Crop improvement in the 20th century IV: plants, society and the future. *In Plants, Biotechnology and Agriculture* (pp. 53–71). CABI, University of Glamorgan, UK.
Nasir, A. S., Ravichandran, V. K., Selvan, N. T., Sankaran, V. M., Sathyavelu, A., & Giridharan, S. (1999). Performance of sugarcane varieties in problem soils. *Sugar Tech*, *1*(1), 47–49.
Nawazish, S., Hameed, M., & Naurin, S. (2006). Leaf anatomical adaptations of Cenchrus ciliaris L. from the Salt Range, Pakistan against drought stress. *Pakistan Journal of Botany*, *38*, 1723–1730.
Neill, S. J., Desikan, R., Clarke, A., Hurst, R., & Hancock, J. T. (2001). Hydrogen peroxide and nitric oxide as signalling molecules in plants. *Journal of Experimental Botany*, *53*, 1237–1247.
Ogbaga, C. C., Athar, H. R., Amir, M., Bano, H., Chater, C. C., & Jellason, N. P. (2020). Clarity on frequently asked questions about drought measurements in plant physiology. *South African Journal of Botany*, *8*, 1–11.

Oliveira, M. A., Verma, K. K., Bhatt, R., & Oliveira, T. B. A. (2022). Impact of green and organic fertilizers on soil fertility and sugarcane productivity. In K. K. Verma et al. (Eds.), Agro-industrial perspectives on sugarcane production under environmental stress.

Parida, A. K., & Das, A. B. (2005). Salt tolerance and salinity effects on plants. *Ecotoxicology and Environmental Safety*, *60*, 324–349.

Parvaiz, A., & Satyawati, S. (2008). Salt stress and phytobiochemical responses of plants. *Plant, Soil and Environment*, *54*, 89–99.

Patade, V. Y., Bhargava, S., & Suprasanna, P. (2011). Salt and drought tolerance of sugarcane under iso-osmotic salt and water stress: growth, osmolytes accumulation, and antioxidant defense. *Journal of Plant Interactions*, *6*, 275–282.

Patist, A., & Zoerb, H. (2005). Preservation mechanisms of trehalose in food and biosystems. *Colloids and Surfaces B: Biointerfaces*, *40*, 107–113.

Paul, M. J., Primavesi, L. F., Jhurreea, D., & Zhang, Y. (2008). Trehalose metabolism and signaling. *Annual Review of Plant Biology*, *59*, 417–441.

Petsas, A., & Grammatikopoulos, G. (2009). Drought resistance and recovery of photosystem II activity in a Mediterranean semideciduous shrub at the seedling stage. *Photosynthetica, 47*, 284–292.

Philani, J., & Dlamini, J. (2021). Drought stress tolerance mechanisms and breeding effort in sugarcane: A review of progress and constraints in South Africa. *Journal of Plant Stress Physiology, 100027*(2), 1–18.

Pooja, Nandwal, A. S., Chand, M., Kumar, A., Rani, B., Kumari, A., & Kulshrestha, N. (2017). Comparative Evaluation of Changes in Protein Profile of Sugarcane Varieties under Different Soil Moisture Regimes. *International Journal of Current Microbiology and Applied Sciences, 6*(10), 1203–1210.

Pooja, Nandwal, A. S., Chand, M., Kumari, A., Rani, B., Goel, V., & Singh, S. (2019). Genotypic differences in growth behavior and quality parameters of sugarcane (*Saccharum officinarum*) varieties under moisture stress conditions. *Indian Journal of Agricultural Sciences*, *89*(1), 65–72.

Pooja, Nandwal, A. S., Chand, M., Pal, A., Kumari, A., Rani, B., Goel, V., & Kulshreshtha, N. (2020). Soil moisture deficit induced changes in antioxidative defense mechanism of sugarcane (*Saccharum officinarum*) varieties differing in maturity. *Indian Journal of Agricultural Sciences*, *90*(3), 507–512.

Pooja, Nandwal, A. S., Chand, M., Singh, K., Mishra, A., Kumar, A., Kumari, A., & Rani, B. (2019). *Varietal variation in physiological and biochemical attributes of sugarcane varieties under different soil moisture regimes. Indian Journal of Experimental Biology*, *57*(10), 721–732.

Pote, C. L., Chougule, P. S., Shirsat, D. V., Kale, A. A., Naik, R. M., Jadhav, A. S., Garkar, R. M., & Nimbalkar, C. A. (2019). Antioxidant enzymes respond to osmotic stress in the leaves of sugarcane (*Saccharum officinarum* L.). *International Journal of Current Microbiology and Applied Sciences*, *8*(12), 1489–1495.

Rai, M. K., Kalia, R. K., Singh, R., Gangola, M. P., & Dhawan, A. (2011). Developing stress tolerant plants through in vitro selection—an overview of the recent progress. *Environmental and Experimental Botany*, *71*, 89–98.

Ramesh, P. (2000). Effect of different levels of drought during the formative phase on growth parameters and its relationship with dry matter accumulation in sugarcane. *Journal of Agronomy and Crop Science*, *185*, 83–89.

Rao, S., Vikram, T., Singh, P., Purba, M. K., & Tiwari, N. K. (2021). Bio-molecular analysis of the effects of salinity stress on sugarcane genotypes at elongation stage. *Annals of the Romanian Society for Cell Biology, 25*(6), 7339–7347.

Rao, V. P., Sengar, R. S., Singh, S., & Sharma, V. (2015). Molecular and metabolic perspectives of sugarcane under salinity stress pressure. *Progress in Agriculture, 15*, 77–84.

Raven, J. A. (1985). Regulation of pH and generation of osmolarity in vascular plants: A cost benefit analysis in relation to efficiency of use of energy, nitrogen and water. *New Phytologist, 101*, 25–77.

Raymond, M. J., & Smirnoff, N. (2002). Proline metabolism and transport in maize seedlings at low water potential. *Annals of Botany, 89*, 813–823.

Ribeiro, R. V., Machado, R. S., Machado, E. C., Machado, D. F. S. P., Magalhaes, J. R., & Landell, M. G. A. (2013). Revealing drought resistance and productive patterns in sugarcane genotypes by evaluating both physiological responses and stalk yield. *Experimental Agriculture, 49*, 212–224.

Sage, R. F., Peixoto, M. M., & Sage, T. L. (2013). Sugarcane: Physiology, biochemistry, and functional biology: Photosynthesis in sugarcane. In John Wiley & Sons Ltd (Ed.), *Photosynthesis in Sugarcane* (pp. 102–127).

Sánchez-Ken, J. G., & Clark, L. G. (2010). Phylogeny and a new tribal classification of the Panicoideae s.l. (Poaceae) based on plastid and nuclear sequence data and structural data. *American Journal of Botany, 97*, 1732–1748.

Saxena, P., Srivastava, R. P., & Sharma, M. L. (2010). Studies on salinity stress tolerance in sugarcane varieties. *Sugar Tech, 12*, 59–63.

Scandalios, J. G., Acevedo, A., & Ruzsa, S. (2000). Catalase gene expression in response to chronic high temperature stress in maize. *Plant Science, 156*, 103–110.

Seki, M., Kamei, A., Yamaguchi-Shinozaki, K., & Shinozaki, K. (2003). Molecular responses to drought, salinity and frost: Common and different paths for plant protection. *Current Opinion in Biotechnology, 14*, 194–199.

Sengar, K., Sengar, R. S., & Singh, A. (2013). Biotechnological and genomic analysis for salinity tolerance in sugarcane. *International Journal of Biotechnology and Bioengineering Research, 4*(5), 407–414.

Serrano, R., Marquez, J. A., & Rios, G. (1997). Crucial factors in salt stress tolerance. In S. Hohmann & W. H. Mager (Eds.), *Yeast Stress Responses* (pp. 147–169). RG Lande Company.

Shi, K., Li, X., Zhang, H., Zhang, G., Liu, Y., Zhou, Y., Xia, X., Chen, Z., & Yu, J. (2015). Guard cell hydrogen peroxide and nitric oxide mediate elevated CO2-induced stomatal movement in tomato. *New Phytologist, 208*, 342–353.

Silva, M. D. A., Jifon, J. L., Da-Silva, J. A. G., Dos Santos, C. M., & Sharma, V. (2014). Relationships between physiological traits and productivity of sugarcane in response to water deficit. *Journal of Agricultural Science, 152*, 104–118.

Silva, M. D. A., Jifon, J. L., Dos Santos, C. M., Jadoski, C. J., & Da Silva, J. A. G. (2013). Photosynthetic capacity and water use efficiency in sugarcane genotypes subject to water deficit during the early growth phase. *Brazilian Archives of Biology and Technology, 56*, 735–748.

Silva, M., Jifon, J., Da Silva, J., & Sharma, V. (2007). Use of physiological parameters as fast tools to screen for drought tolerance in sugarcane. *Brazilian Journal of Plant Physiology, 19*, 193–201.

Simoes, W. L., Calgaro, M., Coelho, D. S., Santos, D. B., & Souza, M. A. (2016). Growth of sugarcane varieties under salinity. *Revista Ceres, 63*(2), 265–271.

Singh, R., & Sengar, R. S. (2020a). Effect of salinity stress on morphological and yield attributes of sugarcane (*Saccharum officinarum* L) genotypes. *International Journal of Chemical Studies, 8*(5), 2312–2316.

Singh, R., & Sengar, R. S. (2020b). Impact on proline content of sugarcane (*Saccharum officinarum* L) under salinity stress. *International Journal of Current Microbiology and Applied Sciences, 9*(10), 3599–3605.

Snyman, S. J., Mhlanga, P., & Watt, M. P. (2016). Rapid screening of sugarcane plantlets for in vitro mannitol-induced stress. *Sugar Tech, 18*, 437–440.

Song, X. P., Mo, F., Yan, M., Zhang, X., Zhang, B., Huang, X., Huang, D., Pan, Y., Verma, K. K., & Li, Y. R. (2022). Effect of smut infection on the photosynthetic physiological characteristics and related defense enzymes of sugarcane. *Life, 12*, 1201. https://doi.org/10.3390/life12081201.

Song, X. P., Verma, K. K., Tian, D. D., Zhang, X. Q., Liang, Y. J., Huang, X., Li, C. N., & Li, Y. R. (2021). Exploration of silicon functions to integrate with biotic stress tolerance and crop improvement. *Biological Research, 54*, 19. https://doi.org/10.1186/s40659-021-00344-4

Song, X., Zhou, G., Xu, Z., Lv, X., & Wang, Y. (2016). Detection of photosynthetic performance of *Stipa bungeana* seedlings under climatic change using chlorophyll fluorescence imaging. *Frontiers in Plant Science, 6*, 1254.

Taiz, L., & Zeiger, E. (2002). *Plant Physiology* (3rd ed.). Sunderland, Massachusetts, USA: Sinauer Associates.

Taratima, W., Ritmaha, T., Jongrungklang, N., & Maneerattanarungroj, P. (2020). Effect of stress on the leaf anatomy of sugarcane cultivars with different drought tolerance (*Saccharum officinarum*). *Revista de Biología Tropical, 68*(4), 1159–1170.

Teixeira, F. K., Menezes-Benavente, L., Margis, R., & Margis-Pinheiro, M. (2004). Analysis of the molecular evolutionary history of the ascorbate peroxidase gene family: Inferences from the rice genome. *Journal of Molecular Evolution, 59*, 761–770.

Tsaniklidis, G., Chatzistathis, T., Fanourakis, D., Nikoloudakis, N., Kotsiras, A., Delis, C., & Tzortzakakis, E. A. (2021). Leaf antioxidant machinery stimulation by *Meloidogyne javanica* infestation: A case study on *Cucumis melo* seedlings. *Journal of Plant Stress. 1*, 100002. Advance online publication. https://doi.org/10.1016/j.stress.2021.100002.

Vasconcelos, A. C. M., Casagrande, A. A., Perecin, D., Jorge, L. A. C., & Landell, M. G. A. (2003). Evaluation of the sugar cane root system with different methods. *Revista Brasileira de Ciência do Solo, 27*, 849–858.

Verma, K. K., Anas, M., Chen, Z., Rajput, V. D., Malviya, M. K., Verma, C. L., Singh, R. K., Singh, P., Song, X. P., & Li, Y. R. (2020b). Silicon supply improves leaf gas exchange, antioxidant defense system, and growth in sugarcane responsive to water limitation. *Plants, 9*, 1032. https://doi.org/10.3390/plants9081032.

Verma, K. K., Liu, X. H., Wu, K. C., Singh, R. K., Song, Q. Q., Malviya, M. K., Song, X. P., Singh, P., Verma, C. L., & Li, Y. R. (2019). The impact of silicon on photosynthetic and biochemical responses of sugarcane under different soil moisture levels. *Silicon, 12*, 1355–1367.

Verma, K. K., Singh, P., Song, X. P., Malviya, M. K., Singh, R. K., Chen, G. L., Solomon, S., & Li, Y. R. (2020c). Mitigating climate change for sugarcane improvement: Role of silicon in alleviating abiotic stresses. *Sugar Tech, 22*(5), 741–749. https://doi.org/10.1007/s12355-020-00831-0.

Verma, K. K., Song, X. P., Li, D. M., Singh, M., Wu, J. M., Singh, R. K., Sharma, A., Zhang, B. Q., & Li, Y. R. (2022a). Silicon and soil microorganisms improve rhizospheric soil health with bacterial community, plant growth, performance, and yield. *Plant Signaling & Behavior, 17*(1), e2104004.

Verma, K. K., Song, X. P., Lin, B., Guo, D. J., Singh, M., Rajput, V. D., Singh, R. K., Singh, P., Sharma, A., Malviya, M. K., Chen, G. L., & Li, Y. R. (2021a). Silicon-induced drought tolerance in crop plants: Physiological adaptation strategies. *Silicon, 14*, 2473–2487. https://doi.org/10.1007/s12633-021-01071-x.

Verma, K. K., Song, X. P., Rajput, V. D., Boldyreva, V., Zhang, B. Q., Minkina, T., & Li, Y. R. (2022b). Morpho-physiological, biochemical, and ultrastructural modifications on sugarcane to prolonged water deficit. In K. K. Verma et al. (Eds.), Agro-industrial perspectives on sugarcane production under environmental stress. https://doi.org/10.1007/978-981-19-3955-6_11.

Verma, K. K., Song, X. P., Rajput, V. D., Solomon, S., Li, Y. R., & Rao, G. P. (Eds.). (2023). Agro-industrial perspectives on sugarcane production under environmental stress. Springer, Singapore. https://doi.org/10.1007/978-981-19-3955-6_1.

Verma, K. K., Song, X. P., Verma, C. L., Malviya, M. K., Guo, D. J., Rajput, V. D., Sharma, A., Wei, K. J., Chen, G. L., Solomon, S., & Li, Y. R. (2021c). Prediction of photosynthetic leaf gas exchange of sugarcane (*Saccharum spp.*) leaves in response to leaf positions to foliar spray of potassium salt of active phosphorus under limited water irrigation. *ACS Omega, 6*, 2396–2409.

Verma, K. K., Song, X. P., Zeng, Y., Guo, D. J., Sing, M., Rajput, V. D., Malviya, M. K., Wei, K. J., Sharma, A., Li, D. M., Chen, G. L., & Li, Y. R. (2021b). Foliar application of silicon boosts growth, photosynthetic leaf gas exchange, antioxidative response and resistance to limited water irrigation in sugarcane (*Saccharum officinarum* L.). *Plant Physiology and Biochemistry, 166*, 582–592.

Verma, K. K., Song, X. P., Zeng, Y., Li, D. M., Guo, D. J., Rajput, V. D., Chen, G. L., Barakhov, A., Minkina, T. M., & Li, Y. R. (2020a). Characteristics and correlation of leaf stomata and its relationship with photosynthesis on *Saccharum spp.* under different irrigation and silicon application. *ACS Omega, 5*, 24145–24153.

Wahid, A., & Shabbir, A. (2005). Induction of heat stress tolerance in barley seedlings by pre-sowing seed treatment with glycinebetaine. *Plant Growth Regulation, 46*, 133–141.

Waraich, E. A., Ahmad, R., Saifullah, A., & Ehsanullah, P. (2011). Role of mineral nutrition in the alleviation of drought stress in plants. *Australian Journal of Crop Science, 5*, 765–777.

Watanabe, K., Takaragawa, H., Ueno, M., & Kawamitsu, Y. (2020). Changes in agronomic and physiological traits of sugarcane grown with saline irrigation water. *Agronomy, 10*(5), 722.

Willadino, L., de Oliveira Filho, R. A., da Silva Junior, E. A., Neto, A. G., & Camara, T. R. (2011). Salinity stress in two varieties of sugarcane: Enzymes of the antioxidant system and chlorophyll fluorescence. *Revista Ciência Agronômica, 42*, 417–422.

Yamaguchi, J., & Blumwald, E. (2005). Developing salt-tolerant crop plants: Challenges and opportunities. *Trends in Plant Science, 10*, 615–620.

Yamaguchi-Shinozaki, K., & Shinozaki, K. (2006). Transcriptional regulatory networks in cellular responses and tolerance to dehydration and cold stresses. *Annual Review of Plant Biology, 57*, 781–803.

Yancey, P. H., Clark, M. B., Hands, S. C., Bowlus, R. D., & Somero, G. N. (1982). Living with water stress: Evaluation of osmolytes systems. *Science, 217*, 1214–1222.

Yeo, A. R. (1998). Molecular biology of salt tolerance in the context of the whole-plant physiology. *Journal of Experimental Botany, 49*, 915–929.

Yunita, R., Hartati, R. S., Suhesti, S., & Syafaruddin. (2020). Response of bululawang sugarcane variety to salt stress. *In 1st International Conference on Sustainable Plantation* (pp. 012–060). IOP Conference Series: Earth and Environmental Science.

Zhao, D., Zhu, K., Momotaz, A., & Gao, X. (2020). Sugarcane plant growth and physiological responses to soil salinity during tillering and stalk elongation. *Agriculture*, *10*(12), 608.

Zhu, J. K. (2001). Plant salt tolerance. *Trends in Plant Science, 6*, 66–71.

Zingaretti, S. M., Rodrigues, F. A., Da Graça, J. P., de Matos Pereira, L., & Lourenço, M. V. (2012). Sugarcane responses at water deficit conditions. In: I. M. M. Rahman & H. Hasegawa (Eds.), *Water Stress* (pp. 255–276). Intech Open Limited.

Zivcak, M., Kalaji, H. M., Shao, H. B., Olsovska, K., & Brestic, M. (2014). Photosynthetic proton and electron transport in wheat leaves under prolonged moderate drought stress. *Journal of Photochemistry and Photobiology B: Biology, 137*, 107–115.

CHAPTER 5

Environmental Stresses and Their Impact on Cane Development and Production

EID. M. MEHAREB

Sugar Crops Research Institute, ARC, Giza, Egypt

ABSTRACT

Sugarcane (*Saccharum* spp. hybrids) is necessary for the world's sugar production. Sugarcane is widely grown in several tropical and subtropical countries, with over 26.47 million hectares spread through 103 countries. Considered a long-duration crop (it takes about 12–18 months to mature), sugarcane can face many climatic changes, affecting plant growth and production. Studies have indicated that 25% of the variation in sugarcane yield was due to climate changes. Climate changes are expected to increase the intensity and frequency of abiotic stresses. Cane growth and yield are limited due to multiple abiotic stresses, such as water availability stress (drought and waterlogging), mineral stress (salinity), extreme temperatures (heat, cold, and frost), and nutritional imbalance. Abiotic stresses are the major reason for persistent suffering in global agriculture as they badly affect the germination, cane production, and sugar yield in sugarcane crops. Water or drought stress is the major abiotic stress affecting the growth phases, tillering, yield, and mineral stress (soil salinity) which decreases cane productivity. Correspondingly, under saturated conditions, sugarcane experiences a greater rate of stalk mortality and reduced growth and sugarcane yield. Lack of nutrients is the main cause of reduced sugarcane growth under soil flooding conditions.

Sugarcane Cultivation and Management: Challenges and Opportunities. Krishan K. Verma, Xiu-Peng Song, Munna Singh, Renato de Mello Prado, Jian-Ming Wu, and Yang-Rui Li (Eds.)

5.1 INTRODUCTION

Abiotic stresses, especially temperature, heavy metals, drought, water, and salt stress, induce toxicity in agronomic crops. Exposure to different abiotic stresses severely affects crop growth, morphology, productivity, and performance throughout the growing period. Further, the increased generation of reactive oxygen species under stressful conditions might be linked to the decreased yield and development of agronomic crops (Song et al., 2019; Bali & Sidhu, 2020; Verma et al., 2022a, b, c, 2023). The main determinants of sugarcane output worldwide, particularly in many developing nations, are weather and climate-related phenomena, such as temperature, precipitation, and other extreme weather (Zhao & Li, 2015; Oliveira et al., 2022). The main abiotic restrictions that lower agricultural yield include salinity, drought, nitrogen shortage, waterlogging, cold, and heat (Xue et al., 2017; Kumar, 2013; Cui et al., 2018; Verma et al., 2021a, b; Deng et al., 2022). It is common knowledge that a completely ripened sugarcane crop passes through four stages: germination, tillering, grand growth, and ripening. Tillering and grand growth have been reported to have higher water uptake rates (Ramesh, 2000). Additionally, Moore & Frederik (2014) observed germination (up to 45 days after planting—DAP), tillering (45–120 DAP), grand growth or cane elongation (120–270 DAP), and ripening or maturity (270–360 DAP) are the four distinct phases of sugarcane growth, each with a unique climatic requirement. Cane growth and yield are limited due to multiple abiotic stresses, such as limited water availability, temperatures (low and excess), saline/alkaline soil, and nutritional imbalance.

5.2 LIMITED WATER SUPPLY

Despite difficulties with biotic and abiotic stress, the sugarcane industry also faces water competition due to rising industrial demand and the approaching general population (dos Santos & de Almeida Silva, 2015). The single biggest abiotic factor impacting sugarcane yield is a lack of water. Hence, all major sugarcane-producing nations must urgently develop water-efficient and drought-tolerant cultivars (Wang et al., 2003; Rampino et al., 2006; Silva et al., 2014; Jangpromma et al., 2010a; Ferreira et al., 2017; Verma et al., 2019a, 2021c, d). Egypt has grown sugarcane since 641 AD (Mehareb et al., 2022a). However, the lack of water poses a serious problem for planting sugarcane in Egypt (Mehareb et al., 2022b). Limited photosynthesis, a result of water stress, exposes plants to excess energy from photosynthesis, which, if not safely dissipated, may cause overexcitation of the reaction centers of photosystems II and I, increasing the production of Reactive oxygen species (ROS) in chloroplasts and having a phytotoxic effect on plants (Carvalho, 2008; Verma et al., 2020b, c). Proline buildup was triggered by drought stress (Unyayar et al., 2004). Zhang & RU-KAI (1998) found that as moisture stress increased, the proline content of sugarcane varieties' leaves increased. Proline accumulation in plants helps control osmotic pressure, maintain a healthy cell structure, and endure severe environmental challenges when exposed to cold and drought stresses (Bai et al., 2003; Verma et al., 2019) and plant stress tolerance (Hamdia & Shaddad, 2010; Ashraf & Fooland, 2005). Treatments for water stress resulted in a considerable rise in proline content (Patil, 2008). The synthesis of primary metabolites, such as free proline, glycine betaine, and soluble sugars, as well as secondary metabolites, such as carotenoids, soluble phenolics, and anthocyanins, were demonstrated in sugarcane as a heat tolerance strategy. Enzymatic antioxidants also protect sugarcane from oxidative damage brought on by heat stress (Jain et al., 2007). Proline metabolism improved cell signaling processes that favored turgidity in cells and tissues, enhancing the tolerant cultivar's ability to endure drought for an extended period (Vantini et al., 2016). Proline, total phenolic content (TP), antioxidant enzyme activity (SOD and POX), lipid peroxidation (LP), membrane injury index (MII), and soluble sugar content, on the other hand, increased with increased temperature in all sugarcane clones (Kohila & Gomathi, 2018) (Figure 5.1).

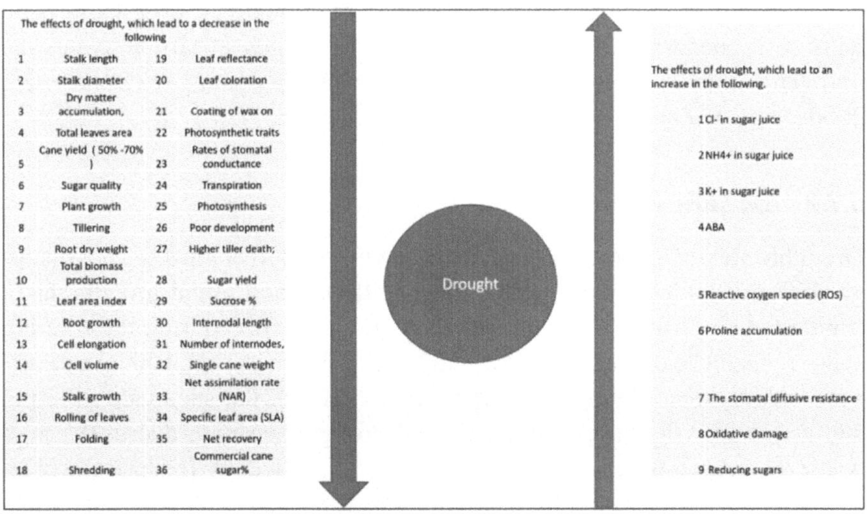

FIGURE 5.1 The effects of the drought lead to an increase and decrease in some traits.

5.2.1 GERMINATION

One of the most damaging and constricting abiotic stresses is drought, which hinders plant growth and development (Apel & Hirt, 2004; Saranga et al., 2002). Any water imbalance hampers sugarcane's growth and productivity in the plant tissues (Singh & Singh, 1974; Srivastava et al., 1996). According to earlier studies, crop roots' morphology and structural characteristics have undergone environmental adaptations that help them withstand adversity during environmental stress (Lynch, 1995). These adaptations manifest diverse forms in different genotypes (Chassot & Richner, 2002). The primary plant organ, the root, transports water and nutrients between plant channels in addition to absorbing them (Li, 2010). The crop demonstrated substantial differences between the treatments for various morphological, physiological, biophysical, and biochemical alterations due to modification in the watering interval (Patil, 2008). Reactive oxygen species are key regulators of numerous processes, comprising development, growth, and response to both abiotic and biotic stimuli besides programmed cell passing away (Luis, 2015; Baxter et al., 2014; Ferreira et al., 2017). The reduction in turgor pressure, the cessation of water passage from the xylem to the surrounding elongation cells, and a slowdown of the growth process may be caused by drought stress, which also lowered root growth in quantity and mass. In particular, in terms of reduction in cell volume and elongation (Nonami, 1998; Larcher, 2003; Jaiphong et al., 2016). Additionally, several authors have hypothesized that H_2O_2 plays a crucial intermediary role in ABA signal transduction and the induction of antioxidant gene expression because it encourages the production of H_2O_2, which in turn activates the mitogen-activated protein kinases (MAPK) signaling pathway (Guan et al., 2000; Jiang & Zhang, 2002).

5.2.2 TILLERING

Drought-tolerant cane varieties are known to be recognized and distinguished using these two crucial phases (tillering and grand growth stage) (Endres et al., 2010). Additionally, early drought stress during the tillering stage reduced leaf area index, millable stalk biomass, total biomass, and cane output. According to Robertson et al. (1999), the sugarcane tillering and grand growth phases are considered drought-sensitive due to the high water requirements for growth. Water deficit significantly reduced the plant length (Smirnoff, 1993; Ramesh & Mahadevswamy, 2000; Zhao et al.,

2010; Jangpromma et al., 2012; Lakshmanan & Robinson, 2014; Barbosa et al., 2015; Dinh et al., 2018; Srivastava & Kumar, 2020). In some areas of the subtropical region of India, more tiller production was observed when drought struck during the early stages of sugarcane growth, suggesting that canes with thinner stalks may fare better than those with thicker ones (Lal et al., 1968). Under drought conditions, modifications in leaf morphology have been noted in the reduction of tillering, rolling, folding, shredding, and reflectivity of leaves, the color of leaves, coating of wax on sugarcane stalks, reduction in leaf area, and narrower leaves (Srivastava & Srivastava, 2006). These are consistent with the reduced stalk height and width caused by the impact of drought stress in the field (Zhao et al., 2010; Begum & Islam, 2012). Early growth stage and mid-season drought stress inhibit plant growth, which manifests as stunting, and limitation of tillering, which results in unoccupied and low millable stalks, and ultimately causes yield loss in both cane and sugar yield (Dinh et al., 2018). It decreased by 11.7 to 19.2% due to stress applied at the tillering and grand growth stages (Subramanian et al., 1992).

5.2.3 GRAND GROWTH OR CANE ELONGATION

The imposed water stress during the two stages (growth and elongation) is emerging as a main constraint in realizing great cane productivity in the area (Verma et al., 2019). These damages depend on variety, the intensity and duration of the drought, and the growth stage at which drought incursions occur (Shao et al., 2009; Anjum et al., 2011). Under water stress conditions, stomata opening and closing directly affect the amount of water lost by transpiration. In both tropical and subtropical settings, moisture stress brought on by high dry temperatures results in poor development and higher tiller death; the yield loss has been estimated to be between 50% and 70% depending on the degree of water stress. There has been very little research on sugarcane's capacity to withstand drought and its physiological means of coping with water stress. Understanding the mechanisms at play in this situation aids in effectively finding solutions to the related issues (Patil, 2008).

Moreover, Barbosa et al. (2015) discovered a similar impact of early water stress on the number of leaves, the area of green leaves, the length and diameter of the stalk, the number of tillers, and likewise, the biomass produced by the stalks. Regrettably, from tillering until the early grand development phase, nitrogen is frequently employed to encourage sugarcane growth. Due to the plant's limited ability to absorb and distribute nitrogen, the water deficit that often occurs during this stage may result in an insufficient need for nitrogen supplements. Therefore, the lack of nitrogen sources caused by water stress could lead to a decline in yield and sugar quality. When irrigation is conducted normally (Dinh et al., 2018), the C_4 plant's photosynthesis is very sensitive to water shortage (Ghannoum, 2009). When plants are under drought stress, their photosynthetic properties are reduced (Du et al., 1996; Jangpromma et al., 2010b; Grac et al., 2010; Barbosa et al., 2015). While moisture stress during the grand growth stage did not significantly affect sugarcane responses to nitrogen fertilization, different irrigation levels induced throughout the growing season greatly impacted sugarcane yield and quality responses to nitrogen application (Wiedenfeld, 2000). Stressed plants lowered stalk height in early, mid-season, and late-season stages, only late season impacted cane and sugar yield (Ethan et al., 2016). Biophysical parameters like diffusive resistance and transpiration rate showed significant differences between the treatments (Figure 5.1). Maximum values for diffusive resistance were found in the stressed treatment, while the normal irrigation recorded minimum values (Patil, 2008). According to Li et al. (2016), sugarcane exposed to an increasing water deficit imposed over 9 days experienced a steady decrease in stomatal conductance and transpiration rate with a concurrent increase in ABA content. Additionally, since the plant's reaction to a water shortage is ABA-dependent, some differentially expressed genes are shared by both water stress and exogenous ABA therapy (Shinozaki & Yamaguchi-Shinozaki, 2007; Himmelbach et al., 2003; Ferreira et al., 2017). ROS outburst is caused by modifications in cellular homeostasis that occur when plants are stressed by prolonged drought. According to several studies (Miller et al., 2010; Gill & Tuteja, 2010; Ferreira et al., 2017), damage to cell membranes, DNA, and proteins results when ROS production outpaces antioxidant processes. Cell death may then occur as a result of this damage. How a plant reacts to drought will depend on the species' innate "strategy" as well as the length and intensity of the drought. Due to the excessive creation of reactive oxygen species, drought stress will certainly cause oxidative damage if it persists beyond a certain point (Smirnoff, 1993).

5.2.4 RIPENING OR MATURITY

One of the most damaging and constricting abiotic stresses is drought, which lowers productivity by resulting in significant losses (Apel & Hirt, 2004; Saranga et al., 2002). Water is one of the strongest determining influences for sugarcane yield (Gadallah & Mehareb, 2020). Likewise, drought stress is a significant factor in sugarcane yield losses, and due to climate change, it is anticipated to occur more frequently in several places. Developing sugarcane for drought tolerance remains a significant problem despite developments in traditional breeding and transformation technology. This is caused by several factors, including the sugarcane genome's complexity, the plant's responses to water deprivation, and the challenge of identifying physiological and morphological traits that can be used to select drought-tolerant varieties suitable for commercial production (Philani Justice Dlamini, 2021). Drought stress in the field decreased total dry matter and cane yield (Basnayake et al., 2012; Begum & Islam, 2012). Drought stress significantly reduced stalk diameter, root dry weight, and overall biomass production (Figure 5.1), according to Jangpromma et al. (2012). During the drought, there was a progressive decrease in the length of the stalks, as well as other characteristics, including internodal length, the number of internodes, the weight of a single cane, and finally, the proportion of commercial cane sugar (Ramesh & Mahadevswamy, 2000). Also, under conditions of soil moisture stress, Singh & Reddy (1980) noticed a larger fall in the percentage of sucrose in juice and an increase in reducing sugars. The sugarcane plants reported a 15–45% productivity drop due to waterlogging (Gomathi et al., 2015). Correspondingly, drought stress triggered proline buildup (Unyayar et al., 2004). Zhang (1998) found that as moisture stress increased, the proline content of sugarcane varieties' leaves increased. The main cause of the low yield and poor sugar quality is drought stress. Due to the high water need for sugarcane's sustainable growth, the early development and mid-season were important times for the plant to be sensitive to drought stress (Zingaretti et al., 2012). Kingston (1994) states that sugarcane yields 8–12 tonnes of cane per ML of irrigation water. A water shortage can result in up to 60% production losses (Basnayake et al., 2012; Gentile et al., 2015; Ferreira et al., 2017). Cane yield decreased with increasing soil moisture stress (Kumar et al., 1995). This reduction in output was attributed to the crop's poor performance under severe moisture stress, regardless of stage. Stomatal diffusive resistance was dramatically boosted by soil moisture stress during the crop's formative stage, which reduced transpiration rate, sugar percentage in juice, and cane yield (Singh et al., 1997). The crop demonstrated substantial differences between the treatments for various cane yields and quality attributes due to modification in the watering interval (Patil, 2008). This issue affects several cane-growing nations and can reduce cane yield by up to 60%. Due to dry conditions, cane production has suffered a 40% reduction (Misra, 2022).

5.3 SOIL FLOODING

It is stated that in conditions of waterlogging, significant morphological, anatomical, physiological, and biochemical changes occur in plants for adaptation and survival (Barcly & Crawford, 1982; Gomathi & Chandran, 2009), including the poor root system of sugarcane (Singh, 1990). When anaerobic protein genes detect a lack of oxygen, they control upstream genes at the transcriptional or post-transcriptional stages. ANPs are necessary for the acclimatization process (Jackson et al., 2003). Studies conducted at SBI in Coimbatore revealed that waterlogging resulted in a 22.4% decrease in NMC, 45.6% decrease in the weight of a single cane, a 30.0% decrease in cane length, a 15.9% decrease in internodal length, a 17.8% decrease in cane thickness, and a 40.1% decrease in cane yield (Gomathi et al., 2010). The aerenchymatous cells aid in the movement of gases, while auxins and ethylene work together to increase adventitious root formation, and ethylene works alone to encourage aerenchyma development in these circumstances (Gomathi et al., 2015). The interplay between irrigation frequency and sugarcane genotype had a considerable impact on millable cane length.

5.3.1 GERMINATION

The amount of water in the soil greatly impacts seed cane germination. For seed cane germination, soil moisture should be between 20% and 30%, with 25% being ideal. Germination will be impeded if the soil water content is over 40%, and the seed cane setts may even rot under prolonged conditions of floods or waterlogging. Consequently, proper field drainage is crucial for the growth of sugarcane (Li, 2010; Li & Yang, 2014). The germination phase lasts from planting until the seed cane setts' buds and root points have emerged successfully. The time for germination under the field circumstances of commercial production varies widely from 7 to 110 days, depending on the ambient temperature. When it comes to sugarcane, germination refers to the vegetative bud and latent roots on the node being activated and then sprouting. Both internal and external variables can influence germination. The sett's nutrient quality, bud health and hydration, lowering sugar content, and bud health are internal factors. The soil's aeration, temperature, and moisture are external influences (Li, 2010; Verma et al., 2020). Fresh seed cane often contains more than 70% water, which is more than enough to support young plants' early germination and growth. The germination rate considerably reduces as the water content falls to 50%; when it reaches 40%

or less, the buds lose their ability to germinate and cannot be revived even after soaking the seed cane in water (Li, 2010; Verma et al., 2019).

5.3.2 TILLERING

One of the abiotic stresses reducing crop plants' potential yield is waterlogging or excess water; yearly, waterlogging affects crops worldwide, placing significant stress on roots and lowering (Jackson, 2004). Studies show that the stress of standing water changes the direction in which shoots extend as well as inhibits the growth of tillers, leaves, and stalks. In addition, a regular waterlogging phenomenon severely hinders sugarcane growth and survival, causing a 15–45% decrease in cane production (Gomathi et al., 2015).

5.3.3 GRAND GROWTH OR CANE ELONGATION

Even though sugarcane is tolerant of high water levels, a higher water table during the active development period has a detrimental impact on stalk weight and plant population, which reduces production at a rate of around one ton per acre for every inch rise in extra water (Carter & Floyed, 1974; Roach & Mullins, 1985; Deren et al., 1993). Increased water table during the active development stage negatively impacts stalk weight and plant population, which reduces production by around one ton per acre for every inch of additional water (Carter, 1976).

5.3.4 RIPENING OR MATURITY

Waterlogging, a frequent occurrence accompanied by heavy rain, affects crops, lowers productivity and yield (Jackson, 2004), complicates the problem, and reduces plant productivity by 80%. In terms of stress caused by waterlogging, sensitive plants exhibit greater production losses than tolerant plants. Different growth stages of the crop will react to waterlogging stress differently (Misra, 2022). The drop in yield is attributed to low levels of moisture and nitrogen in the tissue during the main growth phase (Pandey, 1964) and the decrease in nitrogen uptake by the plant (Robertson et al., 2019). Additionally, in flood circumstances, the oxygen concentration rapidly drops, resulting in hypoxic conditions (Lee et al., 2011; Vashisht et al., 2011). According to Armstrong (1979), oxygen deprivation brought on by waterlogging circumstances is caused by a decrease in oxygen solubility and diffusion rate. Every year, waterlogging affects crops all around the world. Although less established cane looks far more susceptible to floods, it is claimed that it can survive a few months into the growing season (Deren et al., 1991b). For each day the water table was within 50 cm of the soil surface, sugarcane lost 0.5 t/ha of yield (Rudd & Chardon, 1977). Due to the high yield and productivity of the sugarcane crop, which consumes a lot of water (Misra et al., 2016), sugarcane even needs to adapt to waterlogging stress (Avivi et al., 2020). According to the affected crop stage and the flooding length, the average yield loss due to waterlogging is estimated to be between 15% and 25%, although it can approach 40% (Gomathi & Chandran, 2009). Hence inversion of sucrose is a limiting factor for achieving high sugar under waterlogging conditions (Gomathi & Chandran, 2013).

5.4 TEMPERATURES

The most damaging of them are drought, salt, and temperature stress because they impede the metabolic processes in plants (Akula & Gokare, 2011; Fujita et al., 2011; Vishwakarma et al., 2017; Kumari et al., 2019; Rastogi et al., 2019). Throughout the 21st century, temperature increases may decrease yield by 2.5 to 10% for various agronomic species (Hatfield et al., 2011). Changes in temperature and rainfall can directly and indirectly affect sugarcane. Given that it is a crop with a long growing season (12 to 15 months) and varying sensitivity to weather at different growth stages, even a small divergence from the long-term average of weather patterns during a year runs the risk of significantly reducing sugarcane yield and sugar production (Mali et al., 2014; Zhao & Li, 2015; Vandana et al., 2020). Heat stress disrupts mitochondrial functioning, causing oxidative damage that manifests as lipid peroxidation and is identified by the presence of malondialdehyde (MDA) (Vacca et al., 2004). Lipid peroxidation, which can lead to membrane damage, protein deterioration, and enzyme deactivation, is a result of oxidative stress (Meriga et al., 2004). Heat stress increases the activity of enzymes that scavenge reactive oxygen species, such as ascorbic peroxidase, catalase, guaiacol peroxidase, and superoxide dismutase (Chaitanya et al., 2002; Gomathi & Kohila, 2016). However, ROS has a role in relaying stress signals at low levels to activate defense and acclimatization pathways (Mittler et al., 2004). By contrast, levels of soluble sugar, malondialdehyde (MDA), relative electrical conductivity in plant cells, soluble protein, and proline, as well as the activity of the enzymes superoxide dismutase (SOD), catalase (CAT), and peroxidase (POD), may have increased in different genotypes during cold stress (Li et al., 2010).

Additionally, reducing sugars and acid invertase levels were low in stubble buds compared to normal temperature, but IAA, total phenols, and proline contents were high (Jain et al., 2007). The two most significant organic osmolytes (Figure 5.2) that build up in a range of plant species in response to environmental challenges such as high temperatures, drought, salinity, UV radiation, and heavy metals are glycine betaine (GB) and proline. Numerous studies and papers have demonstrated a beneficial relationship between glycine betaine and proline accumulation. Secondary metabolites, including phenolics, participate in abiotic stress reactions typically linked to heat tolerance (Wahid, 2007). Sugarcane's mechanisms for heat tolerance are still poorly known. A molecular foundation for enhancing heat tolerance in crops will be provided by discovering novel genes and examining their expression patterns in response to heat stress (Yang et al., 2006; Qin et al., 2008). Secondary metabolites, including phenolics, participate in abiotic stress reactions typically linked to heat tolerance (Wahid, 2007) (Figure 5.2).

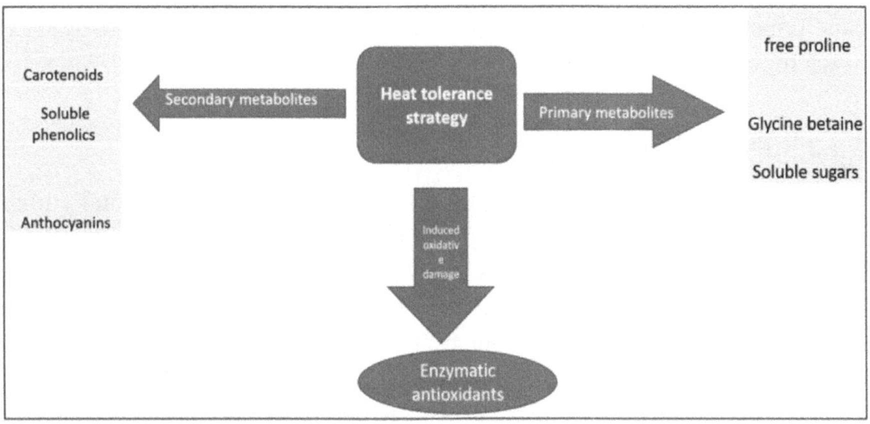

FIGURE 5.2 Responses of heat tolerance strategy.

5.4.1 GERMINATION

Heat stress significantly impacts early seedling growth and sugarcane germination percentage (Wahid & Shabbir, 2005; Wahid et al., 2010). Poor stubble bud sprouting at low temperatures seems to be caused by a decreased supply of reducing sugars and decreased acid invertase activity. An elevated amount of IAA may cause the persistence of dormancy in stubble buds at low temperatures. Complicating the problem is an in-situ buildup of toxicity brought on by an accumulation of total phenols (Jain et al., 2007). According to Russell (1978), the root is more susceptible to cold than other plant parts like the stalk and leaves. Proline builds up to act as an osmoprotectant under low-temperature stress, whereas higher catalase and peroxidase enzyme activities may protect stubble buds from oxidative damage (Jain et al., 2007). Therefore, the sugarcane crop's numerous growth processes and activities are affected differently by weather variability. When the ideal temperature is between 20°C and 32°C in the months of February and March, germination is favorable (Samui et al., 2003). The ideal temperature range for sugarcane growth is between 28°C and 32°C. Temperatures below 20°C may impede the development of the crop, but higher temperatures above 45°C can hinder its growth (Vandana et al., 2020). High temperatures also have effects on sugarcane flowering (Moore & Berding, 2014; Mehareb et al., 2016). Despite abundant water supply to roots, water potential, and its constituents are adversely impacted in sugarcane leaves under heat stress (Wahid & Close, 2007). The plant's reduced growth and water interactions demonstrate the sensitivity of sugarcane to heat, with a threshold of 36°C (Wahid et al., 2010). It has been shown that preserving dormancy in sugarcane underground buds may be caused by increased peroxidase activity at low temperatures (Kapur & Kanwar, 1985).

5.4.2 TILLERING

In both tropical and subtropical settings, moisture stress brought on by high dry temperatures results in poor development and higher tiller death (Patil, 2008). When the relative humidity is around 51% and the air temperature is between 30°C and 32°C, the tillering phase, which occurs between March and June, is more favorable. About 12% of the cane's weight and 15% of its height are acquired during this phase. The active growth and water demand at this time is supplied by regular crop irrigation using surface water and groundwater resources, which results in very high evapotranspirational demand (Verma et al., 2019). While there is now little competition for water from other crops in the area, concerns are being expressed about the groundwater's insufficient recharge as a result of the increased frequency of poor rains in recent years (Shrivastava et al., 2016), but higher temperatures above 45C reduce tillering (Vandana et al., 2020).

5.4.3 GRAND GROWTH OR CANE ELONGATION

Heat tolerance is directly correlated with the capacity of leaves to maintain gas exchange [net photosynthesis (Pn), stomatal conductance (gs), transpiration rate, and CO_2 assimilation rates] under heat stress (Hall, 1992). The sugarcane's cane elongation phase, also known as the grand growth period, lasts from July to September (Verma et al., 2019). This is the greatest active growth duration in sugarcane, where about 80% of the stalk weight is achieved. The research that is now available demonstrated that canopy temperature plays a significant role in sugarcane development and the production of new leaves (Robertson et al., 1998; Hatfield et al., 2011). The results of the study showed that, in all genotypes and species of sugarcane clones, heat stress reduced the levels of chlorophyll, the chlorophyll stability index (CSI), the SPAD value, the maximum quantum efficiency of PSII photochemistry (Fv/Fm ratio), the leaf gas exchange parameters, the relative water content (RWC), and the activities of nitrate reductase (NR), sucrose-metabolizing enzymes (Kohila & Gomathi, 2018). Canopy temperature plays a significant role in sugarcane development and the production of new leaves (Robertson et al., 1998; Hatfield et al., 2011). According to Gomathi et al. (2017), sugarcane showed an average drop in SS and SPS activity at elevated temperatures (42°C). In contrast, tolerant varieties showed increased SS and SPS activity at rising temperatures. According to Kaushal et al. (2016), unfavorable temperatures may have a major impact on the photosynthesis, respiration, water balance, and membrane stability of leaves.

5.4.4 RIPENING OR MATURITY

After October, cane growth slows down, and when the relative humidity is still moderate (61%), ripening starts when the temperature drops below 19.4°C. Beginning in October and continuing through December, sugar quickly accumulates. The ideal temperature for sucrose accumulation is between 12°C and 14°C, whereas temperatures below 19°C are conducive to sucrose accumulation (Fageria et al., 2010; Verma et al., 2019). High temperatures injure plants, which results in a substantial production loss of between 10% and 12.5% (Kaushal et al., 2016).

5.5 SALINITY AND NUTRITIONAL IMBALANCE

It is generally known that proline builds up in plants as they adjust to various environmental challenges (Figure 5.3), including drought, extreme heat, nutrient deprivation, and heavy metal exposure (Oncel et al., 2000). However, sugarcane plant age and length of heat stress exposure are crucial factors (Wahid et al., 2007). Salinity and water stress will thus be the main growing abiotic stress concerns (Ram & Karuppaiyan, 2018). Salinity stress is one of the most harmful constraints that severely reduces agricultural yields, especially in the world's arid, semi-arid, and coastal regions. At the same time, it also occurs in humid and sub-humid environments (Munns, 2002; Munns & Tester, 2008; Fageria et al., 2011; Deinlein et al., 2014; Bernstein, 2019; Cai & Gao, 2020). The most damaging of them are drought, salt, and temperature stress because they impede the metabolic processes in plants (Akula & Gokare, 2011; Fujita et al., 2011; Vishwakarma et al., 2017; Kumari et al., 2019; Rastogi et al., 2019). The Poaceae family includes sugarcane, which has a high tolerance to saline stress at various stages of crop development (Verma et al., 2021). The threshold value for sugarcane's susceptibility to salinity is 1.4 dS/m (Maas, 1986). Through the abscisic acid signaling pathway, salinity is frequently related and may result in membrane damage or macromolecule breakdown through the generation of ROS (Wang et al., 2003). Salt stress impacts sugarcane's physiological and biochemical responses related to salt tolerance (Gandonou et al., 2006; Patade & Suprasanna, 2009). Salt stress is a significant abiotic element that affects the viability of agriculture, particularly for crops like sugarcane that require a lot of water to grow across a large portion of the globe. According to estimates, high salinity affects 20% of all cultivated areas and 33% of irrigated arable land globally (Tanji, 1990) (Figure 5.3).

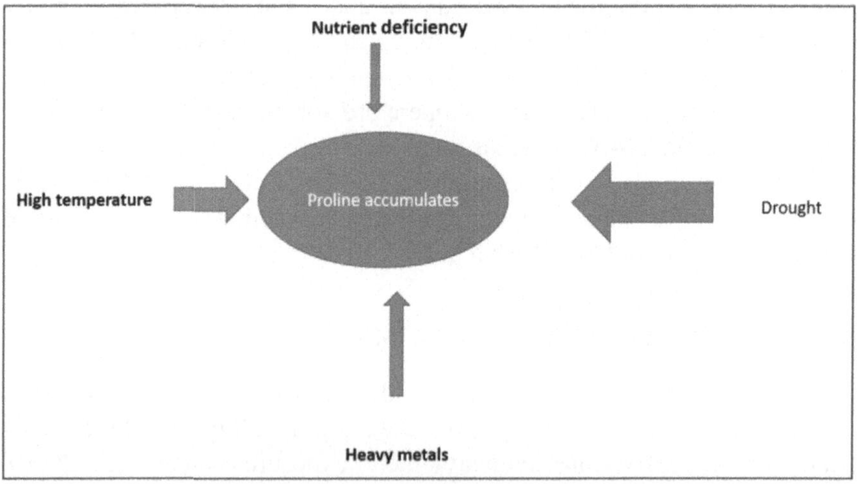

FIGURE 5.3 Factors that lead to the accumulation of proline.

5.5.1 GERMINATION

Salinity and germination had an inversely correlated connection. Salinity hinders germination, seedling growth, and enzyme activity, which are detrimental to plant growth and development (Seckin et al., 2009). The differential response of sugarcane varieties for soil salinity and acidity has been observed. At the early crop stage, germination and early growth stages of the crop become more sensitive than the later stages. Reduced cane germination and cane height (slowed crop growth) are salinity's effects on plant phenotype, as are smaller leaves (Meena et al., 2022). Salinity impacts plant growth and has a higher impact on the sugarcane root zone (Rozeff, 1995; Lingle & Weigand, 1996). Through ion toxicity, nutritional imbalances, osmotic effects, and oxidative stress, salinity prevents plant growth (Chinnusamy et al., 2005).

5.5.2 TILLERING

The salinity of irrigation water has been tested. It has been discovered to negatively affect plant growth and yields by impeding crops' ability to absorb water, obstructing transpiration and photosynthetic processes occurring on the surface of leaves (Sahab, 2020; Safdar, 2019; Kaitlyn Barnes, 2021). Abiotic stress brought on by climate change damages agronomic land and hinders the development and growth of crops (FAO, 2009; Grayson, 2013; Meena et al., 2017; Porter et al., 2019). According to the degree of salt stress and genotype makeup, sugarcane exhibits stunted development in salty environments (Patade et al., 2011; Chaum et al., 2012). Sugarcane is susceptible to high salt concentrations, and symptoms include diminished leaf margins, decreased pigment levels, and reduced chlorophyll contents. These symptoms can have fatal consequences, including wilting, chlorosis, necrosis, leaf drying, and senescence (Devi et al., 2017).

5.5.3 GRAND GROWTH OR CANE ELONGATION

The methods by which plants tolerate salt are complex and comprise numerous biochemical and developmental alterations. By priming planting materials, a plant's resistance to salinity stress can be increased (Rasheed et al., 2011). Multiple stresses, such as depleted soil minerals (iron, zinc, and copper), water stress (drought or flooding), salinity, and heavy metals (cadmium, lead, and nickel), have a significant impact on crop development and output (Verma et al., 2021b). It's possible that physiological drought and ion toxicity, which cause metabolic toxicity, membrane disorganization, and the production of reactive oxygen species (ROS), are the causes of soil salinity's effects on sugarcane development, stalk yield, and quality (Hasegawa et al., 2000).

5.5.4 RIPENING OR MATURITY

Salinity is likely the most prevalent limitation that significantly reduces crop yields, and its threat is growing every day (Schwabe et al., 2006). Reduced stalk population and weight cause a considerable drop in cane yield in irrigation water or saline soil (Lingle & Wiegand, 1997). Changes in photosynthetic and morphological properties brought about by salinity stress were critical in determining the cane yield (Plaut et al., 2000; Verma et al., 2019). Because sugarcane is a typical glycophyte, it grows slowly or doesn't develop in salty environments. Its yield drops to 50% of its potential or even less (Patade et al., 2011; Suprasanna et al., 2011; Chaum et al., 2012). Sugarcane production is impacted by salinity, which lowers sucrose yield by affecting both biomass and juice quality (Rozeff, 1995; Lingle & Wiegand, 1996). Due to its sensitivity to high salt levels, sugarcane produces less and accumulates more sucrose (Patade et al., 2011).

5.6 SOME TOLERANCE MECHANISMS

- One of the practical methods for producing sugarcane in abiotic zones is the breeding and cultivation of abiotic-tolerant and high-yielding varieties.
- Ion homeostasis, osmoregulation, antioxidant biosynthesis, and accumulation of plant growth regulators are just a few of the defense mechanisms that plants have evolved to deal with biotic and abiotic stresses, allowing halophytic plant species to adapt and defend themselves in salt and metal excess (Azevedo et al., 2011; Chaum et al., 2012).
- By applying IBA and Stimulate® treatments, higher growth rates and biomass accumulation are made possible in the early phases of sugarcane vegetative growth. In contrast to Control-treated plants, Ubyfol® and Stimulate® treatments are given to plants, allowing for improved gas exchange in the absence of drought stress and higher photosystem II yield even while plants are under drought stress. In plants under modest stress, the application of the Stimulate® therapy

promotes higher rates of stomatal conductance, transpiration, and photosynthesis (Endres et al., 2016).

- Ranking the genotypes for salinity tolerance and choosing salt-tolerant lines could be accomplished by *in vitro* and *in vivo* screening procedures that use proteins as biochemical markers (Azevedo et al., 2011; Chaum et al., 2012).
- According to Naidu & Srinivasan (1982), potassium administration boosted the cane's moisture content, aided drought tolerance, improved maturity, and increased sugar content.
- Spraying potassium sulfate throughout the hot summer at a concentration of 2.5% reduced drought and considerably enhanced cane output to 16.7% over control (Jayabal & Chockalingam, 1990).
- Proline content was highest when urea + KCl (3%) was applied topically (Patil, 2008).
- Therefore, increased nitrogen deposited into the leaf during the recovery phase of the drought stress treatment would enable higher leaf SPAD and photosynthetic rate (Dinh et al., 2018).
- Abscisic acid (ABA) is a crucial modulator in how plants perceive environmental stress signals and react to that stress at the cellular level (Himmelbach et al., 2003).
- Applying potash before the onset of moisture stress may have increased proline buildup in the plant system, which may have served as an adaptation mechanism for drought resistance (Prumal & Pasupathy, 1984).
- H_2O_2 plays a crucial intermediary role in ABA signal transduction and the induction of antioxidant gene expression because it encourages the production of H_2O_2, which activates the mitogen-activated protein kinases (MAPK) signaling pathway (Guan et al., 2000; Jiang & Zhang, 2002).
- Using K on sugarcane can reduce drought and boost cane output significantly compared to not applying K (Wood & Schroeder, 2004).

5.7 CONCLUSION

Cane growth and yield are limited due to multiple abiotic stresses, such as temperature, heavy metals, drought, water, and salt stress, inducing toxicity in sugarcane crops. The single biggest abiotic factor impacting sugarcane yield is drought, followed by salt and temperature stress because they impede the metabolic processes in plants.

KEYWORDS

- **abiotic stresses**
- **drought**
- **environmental stresses**
- **heavy metals**
- **productivity**
- **salinity**
- **soil flooding**
- **sugarcane**
- **temperature stress**
- **water deficit**

REFERENCES

Anjum, S. A., Ashraf, U., Tanveer, M., Khan, I., Hussain, S., Zohaib, A., Abbas, F., Saleem, M. F., & Wang, L. (2017). Drought tolerance in three maize cultivars is related to differential osmolyte accumulation, antioxidant defense system, and oxidative damage. *Frontiers in Plant Science, 8*, 69.

Anjum, S. A., Xie, X. Y., Wang, L. C., Saleem, M. F., Man, C., & Lei, W. (2011). Morphological, physiological and biochemical responses of plants to drought stress. *African Journal of Agricultural Research, 6*(9), 2026–2032.

Apel, K., & Hirt, H. (2004). Reactive oxygen species: Metabolism, oxidative stress, and signal transduction. *Annual Review of Plant Biology, 55*, 373–399.

Armstrong, W. (1979). Aeration in higher plants. In H. W. Woolhouse (Ed.), *Advances in Botanical Research* (pp. 225–232). Academic Press.

Augustine, S. M., Narayan, J. A., Syamaladevi, D. P., Appunu, C., Chakravarthi, M., Ravichandran, V., & Subramonian, N. (2015). Erianthus arundinaceus HSP70 (EaHSP70) overexpression increases drought and salinity tolerance in sugarcane (*Saccharum spp. hybrid*). *Plant Science, 232*, 23–34.

Avivi, S., Arnin, S. F. M., Soeparjono, S., Restanto, D. P., Fanata, W. I. D., & Widjaya, K. A. (2020). Tolerance screening of sugarcane varieties toward waterlogging stress. *E3S Web of Conferences, 142*(03007), 1–6.

Azevedo, R. A., Carvalho, R. F., Cia, M. C., & Gratao, P. L. (2011). Sugarcane under pressure: An overview of biochemical and physiological studies of abiotic stress. *Tropical Plant Biology, 4*(1), 42–51.

Bai, B. Z., Ren, Y. X., & Bai, S. (2003). Plant physiology and biochemistry (pp. 257–279).

Bali, S., & Gagan, P. S. S. (2020). Growth and morphological changes of agronomic crops under abiotic stress. In M. Hasanuzzaman (Ed.), *Agronomic Crops, Vol. 3: Stress Responses and Tolerance* (pp. 1–667).

Barbosa, A. M., Guidorizi, K. A., Catuchi, T. A., Marques, T. A., Ribeiro, R. V., & Souza, G. M. (2015). Biomass and bioenergy partitioning of sugarcane plants under water deficit. *Acta Physiologiae Plantarum, 37*, 142. https://doi.org/10.1007/s11738-015-1887-7.

Barcly, A. M., & Crawford, R. M. M. (1982). Plant growth and survival under strict anaerobiosis. *Journal of Experimental Botany, 33*, 541–549.

Basnayake, J., Jackson, P. A., Inman-Bamber, N. G., & Lakshmanan, P. (2012). Sugarcane for water-limited environments: Genetic variation in cane yield and sugar content in response to water stress. *Journal of Experimental Botany, 63*(16), 6023–6033.

Baxter, A., Mittler, R., & Suzuki, N. (2014). ROS as key players in plant stress signalling. *Journal of Experimental Botany, 65*, 1229–1240.

Begum, M. K., & Islam, M. S. (2012). Effect of drought stress on yield and yield components of sugarcane. *Journal of Agroforestry and Environment, 6*(1), 105–109.

Beijing: China Agriculture Press. Chassot, A., & Richner, W. (2002). Root characteristics and phosphorus uptake of maize seedlings in a bilayered soil. *Agronomy Journal, 94*(1), 118–127.

Cai, Z. Q., & Gao, Q. (2020). Comparative physiological and biochemical mechanisms of salt tolerance in five contrasting highland quinoa cultivars. *BMC Plant Biology, 20*(1), 1−15.

Carter, C. E. (1976). Excess water decreases cane and sugar yields. *Proceedings of the American Society of Sugar Cane Technology, 6*, 44–51.

Carter, C. E., & Floyed, J. M. (1974). Inhibition of sugarcane yields by high water level during dormant season. *Proceedings of the American Society of Sugar Cane Technology, 4*, 14–18.

Carvalho, M. H. C. (2008). Drought stress and reactive oxygen species: Production, scavenging and signaling. *Plant Signaling & Behavior, 3*, 156–165.

Chaitanya, K. V., Sundar, D., Masilamani, S., & Reddy, A. (2002). Variation in heat stress-induced antioxidant enzyme activities among three mulberry cultivars. *Plant Growth Regulation, 36*, 175–180.

Chaum, S., Chuencharoen, S., Mongkolsiriwatana, C., Ashraf, M., & Kirdmanee, C. (2012). Screening sugarcane (Saccharum sp.) genotypes for salt tolerance using multivariate cluster analysis. *Plant Cell, Tissue and Organ Culture, 110*(1), 23–33.

Clements, H. F. (1980). Sugarcane crop logging and crop control. University Press, Hawaii.

Cui, F., Sui, N., Duan, G., Liu, Y., Han, Y., Liu, S., Wan, S., & Li, G. (2018). Identification of metabolites and transcripts involved in salt stress and recovery in peanut. *Frontiers in Plant Science, 9*, 217.

De Oliveira, L. A. R., Cardoso, M. N., de Oliveira, A. C. A., Machado, C. D. A., Cardoso, B. T., Muniz, A. D. S., & Ledo, A. D. S. (2018). Effects of in vitro drought stress on growth, proline accumulation, and antioxidant defense in sugarcane. *Journal of Agricultural Science, 10*(5), 135–149.

Deinlein, U., Stephan, A. B., Horie, T., Luo, W., Xu, G., & Schroeder, J. I. (2014). Plant salt tolerance mechanisms. *Trends in Plant Science, 19*(6), 371–379. http://dx.doi.org/10.1016/j.tplants.2014.02.001.

Deng, Z. N., Huang, H. R., Huang, C. M., Verma, K. K., Luo, H. B., Jiang, S. L., Li, Y. J., Cao, H. Q., Wu, K. C., & Xu, L. (2022). Short-term limited water irrigation influences the photosynthetic pigments, enzymatic and non-enzymatic activities in *Saccharum spontaneum* L., at vegetative stage. *Sugar Tech. 25*, 110–118. https://doi.org/10.1007/s12355-022-01175-7

Deren, C. W., Cherry, R. H., & Snyder, G. H. (1993). Effects of flooding on selected sugarcane clones and soil insect pests. *Journal of the American Society of Sugarcane Technology, 13*, 22–27.

Dinh, H. T., Kenta, W. H. T., & Yoshinobu, K. (2018). Effects of drought stress at early growth stage on response of sugarcane to different nitrogen application. *Sugar Tech, 20*(4), 420–430.

dos Santos, C. M., & Silva, M. de A. (2015). Physiological and biochemical responses of sugarcane to oxidative stress induced by water deficit and paraquat. *Acta Physiologiae Plantarum, 37*, 1–14.

Du, Y. C., Kawamitsu, Y., Nose, A., Hiyane, S., Murayama, S., Wasano, K., & Uchida, Y. (1996). Effects of water stress on carbon exchange rate and activities of photosynthetic enzymes in leaves of sugarcane (Saccharum sp.). *Australian Journal of Plant Physiology, 23*, 719–726.

Endres, L., Filho, H. C. d. L. W., Clemente, P. R. A., Bezerra, B. K. L., & Ferreira, V. M. (2016). Root-promoting substances enhance sugarcane's drought tolerance. *African Journal of Agricultural Research, 11*(52), 5162–5170.

Endres, L., Silva, J. V., Ferreira, V. M., & Barbosa, G. V. S. (2010). Photosynthesis and water relations in Brazilian sugarcane. *Open Agriculture Journal, 4*(1), 31–37.

Ethan, S., Olagoke, O., & Yunusa, A. (2016). Effect of deficit irrigation on growth and yield of sugarcane. *Direct Research Journal of Agriculture and Food Science, 4*(6), 122–126.

Fageria, N. K., Moreira, A., & Coelho, A. M. (2011). Yield and yield components of upland rice as influenced by nitrogen sources. *Journal of Plant Nutrition, 34*, 361–370.

Fageria, N. K., Virupax, C., Baligar, & Jones, C. A. (2010). Growth and mineral nutrition of field crops, 3rd Ed. CRC Press, pp. 437–456.

Ferreira, T. H. S., Tsunada, M. S., Bassi, D., Araújo, P., Mattiello, L., Guidelli, G. V., Righetto, G. L., Gonçalves, V. R., Lakshmanan, P., & Menossi, M. (2017). Sugarcane water stress tolerance mechanisms and its implications on developing biotechnology solutions. *Frontiers in Plant Science, 8*, 1077. doi: 10.3389/fpls.2017.01077.

Gadallah, A. F. I., & Mehareb, E. M. (2020). Yield and quality of some sugarcane varieties as affected by irrigation number. *SVU-International Journal of Agricultural Sciences, 2*(2), 144–165.

Gandonou, C. B., Errabii, T., Abrini, J., Idaomar, M., & Senhaji, N. S. (2006). Selection of callus cultures of sugarcane (Saccharum sp.) tolerant to NaCl and their response to salt tolerance. Plant Cell, Tissue and Organ Culture, 87, 9–16.

Gentile, A., Dias, L. I., Mattos, R. S., Ferreira, T. H., & Menossi, M. (2015). MicroRNAs and drought responses in sugarcane. *Frontiers in Plant Science, 6*, 58. doi: 10.3389/fpls.2015.00058.

Ghannoum, O. (2009). C4 photosynthesis and water stress. *Annals of Botany, 103*(4), 635–644.

Gill, S. S., & Tuteja, N. (2010). Reactive oxygen species and antioxidant machinery in abiotic stress tolerance in crop plants. *Plant Physiology and Biochemistry, 48*, 909–930.

Gomathi, R., & Chandran, K. (2009). Effect of waterlogging on growth and yield of sugarcane clones. Sugarcane Breeding Institute (SBI-ICAR). *Quarterly Newsletter, 29*(4), 1–2.

Gomathi, R., & Chandran, K. (2010). Physiological and growth response of sugarcane clones to waterlogging. *In Proceedings of ISPP Zonal Conference on Recent Trends in Plant Physiology and Crop Improvement* (p. 17). VIT University, Vellore, Tamil Nadu.

Gomathi, R., & Chandran, K. (2013). Juice quality as influenced by water-logging stress in sugarcane. *In Proceedings in National Conference of Plant Physiology on Current Trends in Plant Biology Research* (pp. 410–411). Directorate of Groundnut Research (DGR), Junagath, Gujarat.

Gomathi, R., & Kohila, S. (2016). Impact of temperature extremes on sugarcane growth and development. *In Biotechnological and conventional tools for biotic and abiotic stresses management in sugarcane* (pp. 162–170). Edited & Compiled by Dr. Bakshi Ram, P. Govindaraj, A. Ramesh Sundar & A. Bhaskaran. Published by Dr. Bakshi Ram, The Director, Sugarcane Breeding Institute.

Gomathi, R., Gururaja Rao, P. N., Chandran, K., & Selvi, A. (2015). Adaptive responses of sugarcane to waterlogging stress: An overview. *Sugar Tech, 17*, 325–338. https://doi.org/10.1007/s12355-014-0319-0.

Gomathi, R., Vasantha, S., Kohila, S., & Lakshmi, K. (2017). Sucrose synthesis and accumulation in sugarcane in response to elevated temperature. *In International symposium on sugarcane research since Co 205: 100 years and beyond* (pp. 667–668). Coimbatore.

Graca, J. P., Rodrigues, F. A., Farias, J. R. B., Oliveira, M. C. N., Hoffmann-Campo, C. B., & Zingaretti, S. M. (2010). Physiological parameters in sugarcane cultivars submitted to water deficit. *Brazilian Journal of Plant Physiology, 22*(3), 189–197.

Grayson, M. (2013). Agriculture and drought. *Nature, 501*, S1. doi: 10.1038/501S1a.

Grichko, V. P., & Glick, B. R. (2001). Amelioration of flooding stress by ACC deaminase-containing plant growth-promoting bacteria. *Canadian Journal of Microbiology, 47*, 77–80.

Guan, L., Zhao, J., & Scandalios, J. G. (2000). Cis-elements and trans-factors that regulate expression of the maize Cat1 antioxidant gene in response to ABA and osmotic stress: H2O2 is the likely intermediary signaling molecule for the response. *The Plant Journal, 22*, 87–95. doi: 10.1046/j.1365-313x.2000.00723.x.

Hall, A. E. (1992). Breeding for heat tolerance. *Plant Breeding Reviews, 10*, 129–168.

Hamdia, M. A., & Shaddad, M. A. K. (2010). Salt tolerance of crop plants. *Journal of Stress Physiology & Biochemistry, 6*(3), 64–90.

Hasegawa, P. M., Bressan, R. A., Zhu, J. K., & Bohnert, H. J. (2000). Plant cellular and molecular responses to high salinity. *Annual Review of Plant Biology, 51*(1), 463–499.

Hatfield, J. L., Boote, K. J., Kimball, B. A., Ziska, L. H., Izaurralde, R. C., Ort, D., & Thomson, A. M. (2011). Climate impacts on agriculture: Implications for crop production. *Agronomy Journal, 103*, 351–370.

Himmelbach, A., Yang, Y., & Grill, E. (2003). Relay and control of abscisic acid signaling. *Current Opinion in Plant Biology, 6*, 470–479. doi: 10.1016/S1369-5266(03)00090-6.

Jackson, M. B. (2004). The impact of flooding stress on plants and crops. Retrieved from http://www.plantstress.com/Articles/waterlogging_i/waterlog_i.htm (accessed on 27 May 2024).

Jackson, M. B., Saker, L. R., Crisp, C. M., Else, M. A., & Janowiak, F. (2003). Ionic and pH signalling from roots to shoots of flooded tomato plants in relation to stomatal closure. *Plant and Soil, 253*, 103–113.

Jain, R., Shrivastava, A. K., Solomon, S., & Yadav, R. L. (2007). Low temperature stress-induced biochemical changes affect stubble bud sprouting in sugarcane (Saccharum spp. hybrid). *Plant Growth Regulation, 53*, 17–23. DOI 10.1007/s10725-007-9199-6.

Jaiphong, T., Tominaga, J., Watanabe, K., Nakabaru, M., Takaragawa, H., Suwa, R., Ueno, M., & Kawamitsu, Y. (2016). Effects of duration and combination of drought and flood conditions on leaf photosynthesis, growth and sugar content in sugarcane. *Plant Production Science, 19*, 427–437.

Jangpromma, N., Songsri, P., Thammasirirak, S., & Jaisil, P. (2010a). Rapid assessment of chlorophyll content in sugarcane using a SPAD chlorophyll meter across different water stress conditions. *Asian Journal of Plant Sciences, 9*, 368–374.

Jangpromma, N., Songsri, P., Thammasirirak, S., & Jaisil, P. (2010b). Rapid assessment of chlorophyll content in sugarcane using a SPAD chlorophyll meter across different water stress conditions. *Asian Journal of Plant Sciences, 9*(6), 368–374.

Jangpromma, N., Thammasirirak, S., Jaisil, P., & Songsri, P. (2012). Effects of drought and recovery from drought stress on above-ground and root growth, and water use efficiency in sugarcane (Saccharum officinarum L.). *Australian Journal of Crop Science, 6*(8), 1298–1304.

Jaybal, V., & Chokalingam, S. (1990). Studies on drought management in sugarcane. *Cooperative Sugar, 21*(8), 571–574.

Jiang, M., & Zhang, J. (2002). Water stress-induced abscisic acid accumulation triggers the increased generation of reactive oxygen species and up-regulates the activities of antioxidant enzymes in maize leaves. *Journal of Experimental Botany, 53*, 2401–2410.

Kapur, A., & Kanwar, R. S. (1985). Biochemical aspects of subterranean buds under low-temperature conditions in a sugarcane ratoon crop. *Sugarcane, 6*, 10–12.

Kaushal, N., Bhandari, K., Siddique, K. H. M., & Nayyar, H. (2016). Food crops face rising temperatures: An overview of responses, adaptive mechanisms, and approaches to improve heat tolerance. *Cogent Food & Agriculture, 2*, 1–42.

Kazuo, S., & Kazuko, Y. S. (2007). Gene networks involved in drought stress response and tolerance. *Journal of Experimental Botany, 58*, 221–227.

Kingston, G. (1994). Benchmarking yield of sugarcane from estimates of crop water use. *In Proceedings of the Australian Society for Sugar Cane Technologists (Bundaberg)* (pp. 201–209).

Kohila, S., & Gomathi, R. (2018). Adaptive physiological and biochemical response of sugarcane genotypes to high-temperature stress. *Indian Journal of Plant Physiology, 23*, 245–260. https://doi.org/10.1007/s40502-018-0363-y.

Kumar, M. (2013). Crop plants and abiotic stresses. *Journal of Biomolecular Research & Therapeutics, 3*, e125. doi: 10.4172/2167-7956.1000e125.

Lakshmanan, P., & Robinson, N. (2014). Stress physiology: Abiotic stress. In Sugarcane: Physiology, Biochemistry and Functional Biology (Eds. P. Moore & F. Botha). Hoboken: Wiley.

Lal, K. N., Mehrotra, D. N., & Tandon, J. N. (1968). Growth behaviour, root extension and juice characters of sugarcane in relation to nutrient deficiency and drought resistance. *Indian Journal of Agricultural Sciences, 38*(5), 790–804.

Larcher, W. (2003). Plants under stress. *In Physiological Plant Ecology* (4th ed., Ed. W. Larcher) (pp. 401–416). Stuttgart: Verlag Eugen Ulmer.

Lee, S. C., Mustroph, A., Sasidharan, R., Vashisht, D., Pedersen, O., Oosumi, T., Voesenek, L. A. C. J., & Bailey-Serres, J. (2011). Molecular characterization of the submergence response of the Arabidopsis thaliana ecotype Columbia. *New Phytologist, 190*, 457–471.

Li, C., Nong, Q., Solanki, M. K., Liang, Q., Xie, J., Liu, X., Li, Y., Wang, W., Yang, L., & Li, Y. (2016). Differential expression profiles and pathways of genes in sugarcane leaf at elongation stage in response to drought stress. *Scientific Reports, 6*, 25698. doi: 10.1038/srep25698

Li, H., Zhang, X. X., Peng, L. X., & Yan, G. R. (2010). Effects of low-temperature stress on antioxidant enzyme activities and membrane lipid peroxidation in gourd seedlings. *Journal of Anhui Agricultural Sciences, 38*(26), 14258–14259.

Li, Y. R. (2010). Modern sugarcane science. Beijing: China Agriculture Press.

Li, Y. R. (2010). Sugarcane Science. Beijing, China: China Agriculture Press.

Li, Y. R., & Yang, L. T. (2014). Research and development priorities for sugar industry of China: Recent research highlights. *Sugar Tech, 17*, 9–12.

Lingle, S. E., & Weigand, C. L. (1996). Growth and yield responses of sugarcane to saline soil: II. Sucrose biochemistry in individual internodes. *Proceedings of the International Society of Sugar Cane Seminars* (pp. 93–102).

Lingle, S. E., & Wiegand, C. L. (1997). Soil salinity and sugarcane juice quality. *Field Crop Research, 54*, 259–268.

Luis, A. (2015). ROS and RNS in plant physiology: An overview. *Journal of Experimental Botany, 66*, 2827–2837.

Lynch, J. P. (1995). Root architecture and plant productivity. *Plant Physiology, 109*(1), 7–13.

Maas, E. V. (1986). Salt tolerance in plants. *Applied Agricultural Research, 1*, 12–26.

Madhuri, K. V. N., Kumar, M. H., & Sarala, N. V. (2011). Influence of higher doses of nitrogen on yield and quality of early maturing sugarcane varieties. *Sugar Tech, 13*(1), 96–98.

Mali, S. C., Shrivastava, P. K., & Thakare, H. S. (2014). Impact of weather changes on sugarcane production. *Research in Environment and Life Sciences, 7*(4), 243–246.

Meena, K. K., Sorty, A. M., Bitla, U. M., Choudhary, K., Gupta, P., Pareek, A., Singh, D. P., Prabha, R., Sahu, P. K., Gupta, V. K., Singh, H. B., Krishnanani, K. K., & Minhas, P. S. (2017). Abiotic stress responses and microbe-mediated mitigation in plants: The omics strategies. *Frontiers in Plant Science, 8*, 172. doi: 10.3389/fpls.2017.00172.

Meena, M. R., Tabish, A., Appunu, C., Raja, A. K., Vasantha, S., Kumar, P. R., & Pandey, S. K. (2022). Impact of salinity stress on sugarcane yield and quality: Management approaches for higher cane sugar productivity. In K. K. Verma, X. P. Song, V. D. Rajput, S. Solomon, Y. R. Li, & G. P. Rao (Eds.), *Agro-industrial perspectives on sugarcane production under environmental stress* (pp. 1–433).

Mehareb, E. M., El-Shafai, A. M. A., & Fouz, F. M. A. (2022). History and current status of sugarcane breeding in Egypt. *Sugar Tech, 24*, 267–271. https://doi.org/10.1007/s12355-021-01010-5.

Mehareb, E. M., Osman, A. E., Attia, M. A., Bekheet, A. E., & Fouz, F. M. (2022). Stability assessment for selection of elite sugarcane clones across multi-environment based on AMMI and GGE-biplot models. *Euphytica, 218*, 115. https://doi.org/10.1007/s10681-022-03061-5.

Mehareb, E. M., Salem, E. S. R., & Ghonema, M. A. (2016). Effects of environmental factors on sugarcane flowering under natural conditions in Egypt. *Egyptian Journal of Plant Breeding, 20*(6), 1093–1105.

Meriga, B., Reddy, K., Rao, B. R., Reddy, K. A. L., & Kishor, P. B. K. (2004). Aluminium-induced production of oxygen radicals, lipid peroxidation and DNA damage in seedlings of rice (Oryza sativa). *Journal of Plant Physiology, 161*, 63–68.

Miller, G., Suzuki, N., Ciftci-Yilmaz, S., & Mittler, R. (2010). Reactive oxygen species homeostasis and signaling during drought and salinity stresses. *Plant, Cell & Environment, 33*, 453–467. doi: 10.1111/j.1365-3040.2009.02041.x.

Misra, V., Solomon, S., Mall, A. K., Abid, M., Abid Ali Khan, M., & Ansari, M. I. (2022). Drought Stress and Sustainable Sugarcane Production. In: Arora, N. K., & Bouizgarne, B. (eds) *Microbial BioTechnology for Sustainable Agriculture, Volume 1. Microorganisms for Sustainability, 33*, 353–368. Springer, Singapore. https://doi.org/10.1007/978-981-16-4843-4.

Misra, V., Solomon, S., Mall, A. K., Prajapati, C. P., Hassan, A., Allah, E. F. E., & Ansari, M. I. (2020). Morphological assessment of water stressed sugarcane: A comparison of waterlogged and drought affected crop. *Saudi Journal of Biological Sciences, 27*(5), 1228–1236.

Misra, V., Solomon, S., Singh, P., Prajapati, C. P., & Ansari, M. I. (2016). Effect of water logging on post-harvest sugarcane deterioration. *Agri., 5*(2), 119–132.

Moore, P. H., & Berding, N. (2014). Flowering. In Sugarcane: Physiology, Biochemistry, and Functional Biology.

Moore, P. H., & Frederik, C. S. (2014). Sugarcane: Physiology, Biochemistry and Functional Biology. *Wiley-Blackwell,* 1–716.

Naidu, K. M., & Srinivasan, T. R. (1982). Effect of potash on sugarcane crop subjected to moisture stress. *In Potassium and Plant Physiology: Proceedings of Group Discussion, Potash Research Institute of India* (pp. 75–80).

Nonami, H. (1998). Plant water relations and control of cell elongation at low water potentials. *Journal of Plant Research, 111*, 373–382.

Oliveira, M. A., Verma, K. K., Bhatt, R., & Oliveira, T. B. A. (2022). Impact of green and organic fertilizers on soil fertility and sugarcane productivity. In K. K. Verma et al. (Eds.), Agro-industrial perspectives on sugarcane production under environmental stress.

Patade, V. Y., & Suprasanna, P. (2009). An in vitro radiation induced mutagenesis-selection system for salinity tolerance in sugarcane. *Sugar Tech, 11*, 246–251.

Patade, V. Y., Bhargava, S., & Suprasanna, P. (2011). Salt and drought tolerance of sugarcane under iso-osmotic salt and water stress: Growth, osmolytes accumulation, and antioxidant defense. *Journal of Plant Interactions, 6*(4), 275–282.

Patil, R. P. (2008). Physiological approaches for drought tolerance in sugarcane (Saccharum officinarum L.) (Doctoral dissertation, University of Agricultural Sciences).

Philani, J. D. (2021). Drought stress tolerance mechanisms and breeding effort in sugarcane: A review of progress and constraints in South Africa. *Plant Stress, 2*, 100027. https://doi.org/10.1016/j.stress.2021.100027.

Plaut, Z., Meinzer, F. C., & Federman, E. (2000). Leaf development, transpiration and ion uptake and distribution in sugarcane cultivars grown under salinity. *Plant and Soil, 218*(1–2), 59–69.

Qin, D., Wu, H., Peng, H., Yao, Y., Ni, Z., Li, Z., Zhou, C., & Sun, Q. (2008). Heat stress-responsive transcriptome analysis in heat susceptible and tolerant wheat (Triticum aestivum L.) by using wheat genome array. *BMC Genomics, 9*, 432.

Rahman, A. A. A., Shalaby, A. F., & El Monayeri, M. O. (1971). Effect of moisture stress on metabolic products and ions accumulation. *Plant and Soil, 34*, 65–90.

Ram, B., & Karuppaiyan, R. (2018). Genetic improvement in sugarcane for biotic and abiotic stresses. In *Status Paper. Annual Convention (2018) and Asian Conclave of Sugar Millers Vision-2022* (pp. 63–76). North Indian Sugarcane and Sugar Technologists' Association, Lucknow.

Ramesh, P. (2000). Sugarcane Breeding Institute, Coimbatore, India effect of different levels of drought during the formative phase on growth parameters and its relationship with dry matteraccumulation in sugarcane. *Journal of Agronomy and Crop Science, 185*(2), 83–89.

Ramesh, P., & Mahadevswamy, M. (2000). Effect of formative phase drought on different classes of shoots, shoot mortality, cane attributes, yield and quality of four sugarcane cultivars. *Journal of Agronomy and Crop Science, 185*(4), 249–258.

Rampino, P., Pataleo, S., Gerardi, C., Mita, G., & Perrotta, C. (2006). Drought stress response in wheat: Physiological and molecular analysis of resistant and sensitive genotypes. *Plant, Cell & Environment, 29*, 2143–2152. https://doi.org/10.1111/j.1365-3040.2006.01588.x.

Rasheed, R., Wahid, A., Farooq, M., Hussain, I., & Basra, S. M. A. (2011). Role of proline and glycinebetaine pretreatments in improving heat tolerance of sprouting sugarcane (Saccharum sp.) buds. *Plant Growth Regulation, 65*, 35–45.

Roach, B. T., & Mullins, R. T. (1985). Testing sugarcane for water logging tolerance. *Proceedings of the Australian Society of Sugar Cane Technologists*, pp. 95–102.

Robertson, F., Suraweera, D., McCaskill, M., Christy, B., Armstrong, R., Zollinger, R., Byron, J., Partington, D., & Clark, S. (2019). Waterlogging effects on soils and wheat crops in the high rainfall zone of Victoria. In *Proceedings of the 2019 Agronomy Australia Conference, 25–29 August 2019, Wagga Wagga, Australia.*

Robertson, M. J., Bonnett, G. D., Hughes, R. M., Muchow, R. C., & Campbell, J. A. (1998). Temperature and leaf area expansion of sugarcane: Integration of controlled-environment, field and model studies. *Australian Journal of Plant Physiology, 25*, 819–828.

Robertson, M. J., Inman-Bamber, N. G., Muchow, R. C., & Wood, A. W. (1999). Physiology and productivity of sugarcane with early and mid-season water deficit. *Field Crops Research, 64*, 211–227.

Rozeff, N. (1995). Sugarcane and salinity—a review paper. *Sugarcane, 5*, 8–19.

Rudd, A. V., & Chardon, C. W. (1977). The effect of drainage on cane yields as measured by watertable heights in the Macknade mill area. *Proceedings of the Queensland Society of Sugar Cane Technologists, 44*, 111–117.

Russell, T. E. (1978). Planting and seeding shortleaf pine: Proceedings, symposium for the management of pines of the interior south. *USDA Forest Service*, 7–13.

Samui, R. P., John, G., & Kulkarni, M. B. (2003). Impact of weather on yield of sugarcane at different growth stages. *Journal of Agricultural Physics, 3*, 119–125.

Saranga, Y., Jiang, C. X., Wright, R. J., Yakir, D., & Paterson, A. H. (2002). Genetic dissection of cotton physiological responses to arid conditions and their interrelationships with productivity. *Plant, Cell & Environment, 27*(3), 47–58.

Schwabe, K. A., Iddo, K., & Knap, K. C. (2006). Drain water management for salinity mitigation in irrigated agriculture. *American Journal of Agricultural Economics, 88*, 133–140.

Seckin, B., Sekmen, A. H., & Turkan, I. (2009). An enhancing effect of exogenous mannitol on the antioxidant enzyme activities in roots of wheat under salt stress. *Journal of Plant Growth Regulation, 28*, 12–20.

Shao, H. B., Chu, L. Y., Jaleel, C. A., Manivannan, P., Panneerselvam, R., & Shao, M. A. (2009). Understanding water deficit stress-induced changes in the basic metabolism of higher plants-biotechnologically and sustainably improving agriculture and the ecoenvironment in arid regions of the globe. *Critical Reviews in Biotechnology, 29*(2), 131–151.

Shrivastava, A. K., & Srivastava, M. K. (2006). *Abiotic stresses affecting sugarcane: Sustaining productivity.* International Book Distributing Co, Lucknow.

Shrivastava, A. K., Srivastava, T. K., Srivastava, A. K., Misra, V., Srivastava, S., Singh, V. K., & S. P. Shukla. (2016). *Climate change induced abiotic stresses affecting sugarcane and their mitigation.* ICAR-Indian Institute of Sugarcane Research, Lucknow.

Silva, M. A., John, J. L., da Silva, J. A. G., Santos, C. M., & Sharma, V. (2014). Relationships between physiological traits and productivity of sugarcane in response to water deficit. *Journal of Agricultural Science, 152*, 104–118.

Singh, K. (1990). Sugarcane cultivated in waterlogging areas. *Indian Sugar, 10*, 537.
Singh, R. G., & Singh, S. (1974). Studies on drought tolerance of sugarcane varieties. *Proceedings of the 40th Annual Convention Sugarcane Technologists Association India*, pp. 13–15.
Singh, S. R. (1980). Growth and yield in soil moisture stress. *ISSCT Proceedings, 17*, 541–555.
Smirnoff, N. (1993). The role of active oxygen in the response of plants to water deficit and desiccation. *New Phytologist, 125*, 27–58.
Song, X. P., Mo, F., Verma, K. K., Wei, J., Zhang, X., Yang, L., & Li, Y. R. (2019). Effect of sugarcane smut (*Ustilago scitaminea* Syd.) on ultrastructure and biochemical indices of sugarcane. *Biomedical Journal of Scientific & Technical Research*. MS. ID. 002950. DOI: 10.26717/ BJSTR. 2019. 17. 002950
Srivastava, K. (2020). *Abiotic Stress Responses and Tolerance Mechanisms for Sustaining Crop.* (pp. 1–667).
Srivastava, R. P., Jouhari, D., Singh, S., & Singh, G. P. (1996). Effect of moisture stress on growth, yield and juice quality of sugarcane. *Bharatiya Sugar, 21*(7), 37–44.
Subramanian, K. S., Selvakumari, G., Subramanian, P., & Sanmugasundaram, V. S. (1992). Response of potassium under constraints of moisture in sugarcane. *Cooperative Sugar, 23*(9), 605–609.
Suprasanna, P., Patade, V. Y., Desai, N. S., Devarumath, R. M., Kawar, P. G., Pagariya, M. C., Ganapathi, A., Manickavasagam, M., & Babu, K. H. (2011). Biotechnological developments in sugarcane improvement: An overview. *Sugar Technology, 13*, 322–335.
Unyayar, S., Keles, Y., & Ünal, E. (2004). Proline and ABA levels in two sunflower genotypes subjected to water stress. *Bulgarian Journal of Plant Physiology, 30*, 34–47.
Vacca, R. A., Pinto, M. C., Valenti, D., Passarella, D., Maria, S. E., & Gara, L. D. (2004). Production of reactive oxygen species, alteration of cytosolic ascorbate peroxidase, and impairment of mitochondrial metabolism are early events in heat shock-induced programmed cell death in tobacco Bright Yellow 2 cells. *Plant Physiology, 134*, 1100–1112.
Vandana, P., Singh, D., Srivastava, S., Dayal, G., & Guru, R. (2020). Effect of climate change on sugarcane crop: A review. *Journal of Pharmacognosy and Phytochemistry, SP6*, 255–261.
Vantin, J. d. S. I., Carlin, S. D., Gimenez, D. F. J., Perecin, D., Ferro, J. A., & Ferro, M. I. T. (2016). Proline accumulation in sugarcane roots subjected to drought conditions. *Científica, Jaboticabal, 44*, 592–598. http://dx.doi.org/10.15361/1984-5529.2016v44n4p592-598.
Vashisht, D., Hesselink, A., Pierik, R., Ammerlaan, J. M. H., Bailey-Serres, J., Visser, E. J. W., Pederson, O., Van Zanten, M., Vreugdenhil, D., Jamar, D. C. L., Voesenek, L. A. C. J., & Sasidharan, R. (2011). Natural variation of submergence tolerance among Arabidopsis thaliana accessions. *New Phytologist, 190*, 299–310.
Verma, K. K., Anas, M., Chen, Z., Rajput, V. D., Malviya, M. K., Verma, C. L., Singh, R. K., Singh, P., Song, X. P., & Li, Y. R. (2020a). Silicon supply improves leaf gas exchange, antioxidant defense system and growth in sugarcane responsive to water limitation. *Plants, 9*, 1032. DOI: 10.3390/plants9081032.
Verma, K. K., Liu, X. H., Wu, K. C., Singh, R. K., Song, Q. Q., Malviya, M. K., Song, X. P., Singh, P., Verma, C. L., & Li, Y. R. (2020c). The impact of silicon on photosynthetic and biochemical responses of sugarcane under different soil moisture levels. *Silicon, 12*, 1355–1367.
Verma, K. K., Singh, P., Song, X. P., Malviya, M. K., Singh, R. K., Chen, G. L., Solomon, S., & Li, Y. (2020b). Mitigating climate change for sugarcane improvement: Role of silicon in alleviating abiotic stresses. *Sugar Tech, 22*(5), 741–749. DOI: 10.1007/s12355-020-00831-0.

Verma, K. K., Singh, R. K., Song, Q. Q., Singh, P., Zhang, B. Q., Song, X. P., Chen, G. L., & Li, Y. R. (2019). Silicon alleviates drought stress of sugarcane plants by improving antioxidant responses. *Biomedical Journal of Scientific & Technical Research, 17*, 002957. https://doi.org/10.26717/BJSTR.2019.17.002957

Verma, K. K., Song, X. P., Budeguer, F., Nikpay, A., Enrique, R., Singh, M., Zhang, B. Q., Wu, J. M., & Li, Y. R. (2022b). Genetic engineering: An efficient approach to mitigating biotic and abiotic stresses in sugarcane cultivation. *Plant Signaling & Behavior, 17*(1), e2108253.

Verma, K. K., Song, X. P., Lin, B., Guo, D. J., Singh, M., Rajput, V. D., Singh, R. K., Singh, P., Sharma, A., Malviya, M. K., Chen, G. L., & Li, Y. R. (2021b). Silicon induced drought tolerance in crop plants: Physiological adaptation strategies. *Silicon 14*, 2473–2487. https://doi.org/10.1007/s12633–021–01071-x.

Verma, K. K., Song, X. P., Rajput, V. D., Boldyreva, V., Zhang, B. Q., Minkina, T., & Li, Y. R. (2022c). Morpho-physiological, biochemical and ultrastructural modifications on sugarcane to prolonged water deficit. In K. K. Verma et al. (Eds.), *Agro-industrial perspectives on sugarcane production under environmental stress*. https://doi.org/10.1007/978-981-19-3955-6_11

Verma, K. K., Song, X. P., Rajput, V. D., Solomon, S., Li, Y. R., & Rao, G. P. (Eds.) (2023). *Agro-industrial Perspectives on Sugarcane Production under Environmental Stress*. Springer, Singapore. https://doi.org/10.1007/978-981-19-3955-6_1.

Verma, K. K., Song, X. P., Rajput, V. D., Solomon, S., Li, Y. R., & Rao, G. P. (2022). *Agro-industrial Perspectives on Sugarcane Production under Environmental Stress* (pp. 1–433).

Verma, K. K., Song, X. P., Verma, C. L., Chen, Z. L., Rajput, V. D., Wu, K. C., Liao, F., Chen, G. L., & Li, Y. R. (2021a). Functional relationship between photosynthetic leaf gas exchange in response to silicon application and water stress mitigation in sugarcane. *Biological Research, 54*, 15. https://doi.org/10.1186/s40659-021-00338-2.

Verma, K. K., Song, X. P., Verma, C. L., Malviya, M. K., Guo, D. J., Rajput, V. D., Sharma, A., Wei, K. J., Chen, G. L., Solomon, S., & Li, Y. R. (2021d). Predication of photosynthetic leaf gas exchange of sugarcane (Saccharum spp.) leaves in response to leaf positions to foliar spray of potassium salt of active phosphorus under limited water irrigation. *ACS Omega, 6*, 2396–2409.

Verma, K. K., Song, X. P., Yadav, G., Degu, H. D., Parvaiz, A., Singh, M., Huang, H. R., Mustafa, G., Xu, L., & Li, Y. R. (2022a). Impact of agroclimatic variables on proteogenomics in sugarcane (Saccharum spp.) plant productivity. *ACS Omega, 7*, 22997–23008.

Verma, K. K., Song, X. P., Zeng, Y., Guo, D. J., Sing, M., Rajput, V. D., Malviya, M. K., Wei, K. J., Sharma, A., Li, D. P., Chen, G. L., & Li, Y. R. (2021c). Foliar application of silicon boosts growth, photosynthetic leaf gas exchange, antioxidative response and resistance to limited water irrigation in sugarcane (Saccharum officinarum L.). *Plant Physiology and Biochemistry, 166*, 582–592.

Verma, K. K., Song, X. P., Zeng, Y., Li, D. M., Guo, D. J., Rajput, V. D., Chen, G. L., Barakhov, A., Minkina, T. M., & Li, Y. R. (2020). Characteristics and correlation of leaf stomata and its relationship with photosynthesis on Saccharum spp. under different irrigation and silicon application. *ACS Omega, 5*, 24145–24153.

Verma, K. K., Wu, K. C., Singh, P., Malviya, M. K., Singh, R. K., Song, X. P., & Li, Y. R. (2019). The protective role of silicon in sugarcane under water stress: Photosynthesis and antioxidant enzymes. *Biomedical Journal of Scientific & Technical Research*, MS. ID. 002685. DOI: 10.26717/BJSTR.2019.15.002685.

Waclawovsky, A. J., Sato, P. M., Lembke, C. G., Moore, P. H., & Souza, G. M. (2010). Sugarcane for bioenergy production: An assessment of yield and regulation of sucrose content. *Plant Biotechnology Journal, 8*, 1–14.

Wahid, A. (2007). Physiological implications of metabolite biosynthesis for net assimilation and heat-stress tolerance of sugarcane (*Saccharum officinarum*) sprouts. *Journal of Plant Research, 120*, 219–228.

Wahid, A., Close, T. (2007). Expression of dehydrins under heat stress and their relationship with water relations of sugarcane leaves. *Biologia Plantarum, 51*, 104–109. https://doi.org/10.1007/s10535-007-0021-0.

Wahid, A., Farooq, M., Rasheed, R., Gelani, S., & Rasul, E. (2010). Sugarcane under thermal stress: Some biotechnological considerations. In A. K. Ashwani (Ed.), *Plant genetic transformation and molecular markers* (pp. 109–123). Pointer Publisher.

Wahid, A., Gelani, S., Ashraf, M., & Foolad, M. (2007). Heat tolerance in plants: An overview. *Environmental and Experimental Botany, 61*, 199–223.

Wahid, A., Shabbir, A. (2005). Induction of heat stress tolerance in barley seedlings by pre-sowing seed treatment with glycine-betaine. *Plant Growth Regulation, 46*, 133–141.

Wang, W., Vinocur, B., & Altman, A. (2003). Plant responses to drought, salinity and extreme temperatures: Towards genetic engineering for stress tolerance. *Planta, 218*, 1–14.

Wiedenfeld, R. P. (2000). Water stress during different sugarcane growth periods on yield and response to N fertilization. *Agricultural Water Management, 43*, 173–182.

Wood, A. W., Schroeder, B. L. (2004). Potassium: A critical role in sugarcane production particularly in drought conditions. *Proceedings of the Australian Society of Sugarcane Technologists, 26*, 2004.

Xue, D., Zhang, X., Lu, X., Chen, G., & Chen, Z. -H. (2017). Molecular and evolutionary mechanisms of cuticular wax for plant drought tolerance. *Frontiers in Plant Science, 8*, 621.

Yang, J. W., Huang, T. D., Hua, Y. W., & Huang, H. S. (2012). Low temperature resistance analysis of young Brazilian rubber tree. *Chinese Journal of Tropical Crops, 33*(7), 1235–1238.

Yang, K. A., Lim, G. J., Hong, J. K., Cheong, Y. H., Chung, W. S., Lee, K. O., Lee, S. Y., Cho, M. J., & Co, L. (2006). Identification of cell wall genes modified by a permissive high temperature in Chinese cabbage. *Plant Science, 171*(1), 175–182.

Zhang, M. D., Quing, C., & Ru, K. (1998). Osmotic adjustment in leaves of sugarcane (*S. officinarum*) in response to water stress. *Indian Sugar*, pp. 707–713.

Zhao, D., & Li, Y. R. (2015). Climate Change and sugarcane production: potential impact and mitigation strategies. *International Journal of Agronomy, 2015*, 1–11. https://doi.org/10.1155/2015/547386.

Zhao, D., Glaz, B., & Comstock, J. C. (2010). Sugarcane response to water-deficit stress during early growth on organic and sand soils. *American Journal of Agricultural and Biological Sciences, 5*, 403–414.

Zingaretti, S. M., Rodrigues, F. A., Graca, J. P., Pereira, L. M., & Lourenco, M. V. (2012). Sugarcane response at water deficit conditions. In P. Rhaman (Ed.), *Water Stress*. Rijeka: InTech. https://doi.org/10.5772/30986.

CHAPTER 6

Microbiome Intervention in Sugarcane Growth Against Abiotic Factors

MUKESH KUMAR MALVIYA,[1] PRACHI SHARMA,[1]
MANOJ KUMAR SOLANKI,[2] ANJALI CHANDROL SOLANKI,[3]
PRAVEEN KUMAR DIVVELA,[4] and KRISHAN K. VERMA[5]

[1]*Institute of Biological Science, SAGE University, Indore, Madhya Pradesh, India*

[2]*Plant Cytogenetics and Molecular Biology Group, Institute of Biology, Biotechnology, and Environmental Protection, Faculty of Natural Sciences, University of Silesia in Katowice, Katowice, Poland*

[3]*Department of Agriculture Science, Mansarovar Global University, Bhopal, Madhya Pradesh, India*

[4]*Contec Global Agro Limited, Abuja, Nigeria*

[5]*Sugarcane Research Institute, Guangxi Academy of Agricultural Sciences, Guangxi Key Laboratory of Crop Genetic Improvement and Biotechnology, Nanning, Guangxi, China*

ABSTRACT

This chapter explores the potential of microbiomes to boost the growth of sugarcane plants under abiotic stress conditions. Abiotic stress factors such as drought, high temperature, and salinity are major challenges in sugarcane cultivation, leading to reduced crop yield and quality. Microorganisms present in the sugarcane rhizosphere play crucial roles in plant growth, stress tolerance, and nutrient acquisition. The interaction between plants and microorganisms creates a variety of ecosystems because, occasionally, both

Sugarcane Cultivation and Management: Challenges and Opportunities. Krishan K. Verma, Xiu-Peng Song, Munna Singh, Renato de Mello Prado, Jian-Ming Wu, and Yang-Rui Li (Eds.)

partners have a mutualistic relationship. Microorganisms such as endophytes, plant growth-promoting rhizobacteria (PGPR), and arbuscular mycorrhizal fungus (AMF) are examples that are crucial in reducing abiotic stressors and thus enhancing plant growth. Therefore, understanding the interactions between plants and microorganisms can provide insights into developing microbiome-based interventions to improve sugarcane growth and stress tolerance. The study utilizes various techniques such as next-generation sequencing, metagenomics, and metatranscriptomics to characterize the sugarcane microbiome under different stress conditions. The findings show that abiotic stress variables impact the makeup and diversity of the sugarcane microbiome. Moreover, the microbiome positively correlates with plant growth and stress tolerance. Based on these findings, the study proposes microbiome interventions such as inoculation with beneficial microbes, microbial consortia, and biostimulants to enhance sugarcane growth and stress tolerance. These interventions could improve nutrient uptake, reduce water loss, and enhance photosynthesis in sugarcane plants under abiotic stress conditions. The present chapter emphasizes the possibility of microbiome-based interventions to enhance the resilience and productivity of sugarcane crops against abiotic stress factors.

6.1 INTRODUCTION

In many nations across the world, sugarcane is grown as a commercial crop. Due to its long growth cycle, sugarcane is subject to significant environmental limitations that influence its development (Kumar et al., 2019; Verma et al., 2022a, b, 2023). Abiotic stress has emerged as a persistent issue due to the sudden changes in the climatic situation. Abiotic stressors directly or indirectly influence sugarcane by changing the plant's growth, metabolism, and development. As a result of climate change, abiotic stresses on sugarcane affect the health of the soil, the crop's growth and development, chemical makeup, sugar accumulation and synthesis, availability of seed cane, and escalating other abiotic/biotic stresses that cause losses (Shrivastava et al., 2016; Verma et al., 2021a, b, c; Deng et al., 2022; Xu et al., 2022). In terms of sugarcane, which is a plant known for liking water, water stress is the main major abiotic stress. Abiotic stress is the main factor limiting agricultural yield globally among the several stress situations that impact plants. Abiotic factor strength or amount will determine how it affects the plant. For optimal growth, plants need a specific amount of abiotic environmental factors. Abiotic stress is defined as any deviation from these ideal environmental parameters that results in a chemical or physical deficiency and significantly negatively influences plant growth, development, and productivity (Jalil & Ansari, 2018; Verma et al., 2020a, b, c, 2021d, e).

The advancement of molecular biology enabled the creation of omics techniques, which have recently gained favor in studying microbial diversity and abundance (Alawiye & Babalola, 2019; Fadiji & Babalola, 2020; Verma et al., 2020d, e; Huang et al., 2021). Recent advances in molecular biology, particularly developing low-cost, high-throughput DNA sequencing technology and related data analytics, have increased culture-independent research understanding of rhizosphere microflora (Reuter et al., 2015; Wei et al., 2018). Regardless of microbial abundance, the next-generation sequencing techniques offer an effective and complete methodology to detect microbial species in the rhizosphere and root-associated microbial communities (Uroz et al., 2013; Mishra et al., 2022). As a result, high-throughput sequencing has achieved substantial advancements in the taxonomic characterization of rhizospheric and diazotrophic bacteria diversity (Dong et al., 2017; Gong et al., 2019; Malviya et al., 2022).

Brazil is the greatest producer of sugarcane, which is used to make ethanol, electricity, and sugar. Since sugarcane has a high water requirement, changes in the rainfall pattern anticipated in the upcoming years due to environmental changes could jeopardize sugarcane output. Exploring the interaction processes between this plant and the rhizosphere microorganisms and determining the function of this microbial community in promoting plant health and growth is one strategy for reducing the impacts of water stress on sugarcane crops. Understanding the relationship between sugarcane and microbes also requires research on root exudates (Mehnaz, 2013; Pieterse et al., 2016; Pereira et al., 2020).

The majority of soil microorganisms aid plants in surviving adverse conditions, even though some are thought to be hazardous to plants since they can cause disease. To grow crops, these microbes are being utilized in agriculture. Microbes are important for fixing atmospheric nitrogen, detoxifying pesticides, reducing plant disease, and producing bioactive substances, including vitamins, hormones, enzymes, growth-promoting rhizobacteria (PGPR), and plant-growth-promoting fungus (PGPF). Beneficial microbes adopt a number of techniques to reduce abiotic stress, including the synthesis of phytohormones, a decrease in ethylene oxide levels, an increase in the dehydration response, and the stimulation of genes encoding antioxidant genes. Typically, phytohormones secreted by bacteria in the plant's root reduce salinity and slow the growth of seedlings. In response to stress, indole acetic acid and siderophores are secreted by bacteria that promote plant development, with *Pseudomonas* sp. and Bacillus. Lower ethylene concentrations encourage root development, finally leading to a healthy plant (Al-Arjan et al., 2016; Munir et al., 2022).

This chapter thus emphasizes elucidating the mechanisms involved in plant-microbe interactions, highlighting these beneficial microorganisms and their ability to sustain sugarcane plants under challenging abiotic stress environmental conditions. Integrating them with biochemical and physiological data is crucial to learning more about the mechanisms behind the interactions.

6.2 BIOTIC AND ABIOTIC STRESSES

In the field of mechanics, the word "stress" has been properly defined; nevertheless, in biology, it has taken on entirely new meanings. Several of these definitions coincide in that they attribute "stress" to any environmental aspect that is "unfavorable" for the live creature under discussion, probably as a result of an expansion of the physical meaning (Levitt, 1980; Verma et al., 2023). Throughout their daily lives, plants are regularly subjected to various forms of stress. Any biotic or abiotic restriction that slows down photosynthesis and impairs a plant's capacity to convert energy into biomass is referred to as stress. More generally, stress is any outside force that negatively affects a plant. The main metrics used to assess stress are the principal absorption mechanisms, crop production, plant survival, and growth. Any physical or chemical condition that the environment may impose on the plant is referred to as abiotic stress. Water stress, reduced soil nutrient availability (or, equally, the buildup of harmful ions in salt stress), temperature extremes, or abundant light, particularly when photosynthesis is limited, are examples of abiotic stress (Akter et al., 2014; Ghasemi et al., 2018; Sytar et al., 2019; Song et al., 2022).

A biological factor (such as a pest or pathogen) to which a plant may be subjected during its lifetime is known as biotic stress. Abiotic environmental influences impact plant growth and development at the molecular, morphological, biochemical, physiological, and cellular levels. So, the way plants react to stress must take these conditions into account. Abiotic stress typically causes plants to respond by inhibiting germination, reducing growth, hastening senescence, and decreasing output (Taiz & Zeiger, 2009; Zhu, 2016). Physiological effects include reduced photosynthesis, decreased nitrogen absorption, changed transpiration and respiration rates, decreased water intake, and metabolic toxicity. Stressed plants exhibit changed gene expression, reduced activity of essential enzymes, decreased protein synthesis, disorganized membrane systems, and degradation of macromolecules at the biochemical and molecular levels (Rao et al., 2006).

Salinity, drought, floods, temperature extremes, heavy metals, radiation, etc., are examples of abiotic stressors. They play a significant role in the loss of important crop plants across the globe. Pathogen attacks from fungi, bacteria, oomycetes, nematodes, and herbivores are examples of biotic stress. Globally, diseases brought on by these pathogens are a major cause of yield loss. Sessile plants are unable to ignore these environmental stimuli since they are sessile. Successful completion of the lifetime requires expertise in handling these stressors. Because of this, plants have created a variety of defense mechanisms to adapt to these conditions and survive (Zhu et al., 2002).

6.3 ROLE OF MICROORGANISMS IN ABIOTIC STRESS

A complex biological community comprises millions of microorganisms that live in plant roots and influence plant development and productivity through metabolic processes and interactions with other plants (Schmidt et al., 2014). Modifications in the organization of plant-associated bacterial communities in the root zone, which favor the selection of assemblages adapted to abiotic stress, increase a plant's resistance to stress and its ability to tolerate drought (Cherif et al., 2015; Vurukonda et al., 2015). Generally, microorganisms have great metabolic capacities to reduce abiotic stressors. Microbial interactions with plants are considered the natural partners that regulate local and systemic mechanisms in plants to provide a defense under challenging external situations since they are a crucial component of the living ecosystem. Complex systems exist within the plant cellular system for interactions between plants and microbes. Studies in biochemistry, molecular biology, and physiology are laying the foundation for knowledge of the intricate yet interconnected cellular processes (Kang, 2017). In the following section, the following microorganisms play a significant role in alleviating abiotic stress in plants.

6.3.1 BACTERIA

Abiotic stress negatively affects plants, yet certain specific functional features are present in bacteria that aid in their recovery. The optimum isolates to target in any bioprospecting effort for bacterial stress reduction exhibit high levels of these characteristics (Selvakumar et al., 2012). Several mechanisms emphasizing the part bacteria play in reducing abiotic stress have been presented. It has been demonstrated that soil-dwelling microorganisms from the genera *Variovorax, Achromobacter, Bacillus, Azospirillum, Enterobacter, Klebsiella, Azotobacter, Aeromonas,* and *Pseudomonas* promote plant growth in poor ecological circumstances (Meena et al., 2017).

6.3.2 ACTINOMYCETES

Gram-positive filamentous bacteria called actinomycetes, which frequently associate with plants as rhizosphere dwellers and plant endophytes, are gaining interest for their possible use in stressed ecosystems. Phosphate solubilization, Indole acetic acid (IAA), siderophore synthesis, biocontrol of phytopathogens, and 1-aminocyclopropane-1-carboxylic acid (ACC) deaminase activity are only a few of the plant growth-promoting (PGP) traits that many actinomycetes possess. Moreover, they may thrive in stressful environments, including high temperatures, salinity, alkalinity, and a wide pH range. Numerous papers have shown actinomycetes to help crop plants cope with salinity and drought stress (Grover et al., 2016). *Streptomyces* spp. has been reported to alleviate stress tolerance in several crops, such as maize (Chukwuneme et al., 2020), Chickpea (Srivastava et al., 2014), and Sugarcane (Wang et al., 2019).

6.3.3 FUNGI

Every plant investigated in the natural ecosystem has been discovered to have a fungal symbiont that colonizes and lives wholly or partially in the interior tissues of the host plant. In response to host genotype and environmental conditions, fungal endophytes can display or create a variety of various lifestyles and connections with different hosts, including symbiotic, mutualistic, commensalistic, and parasitic partnerships. In a mutualistic relationship, a fungal endophyte can benefit the host plant by promoting growth, improving reproductive efficiency, and granting biotic and abiotic stress resistance. All across the world, abiotic stressors such as drought, excessive heat, soil salinity, cold, heavy metal toxicity, and oxidative stress negatively impact crop growth (Singh et al., 2011). *Aspergillus fumigatus* and *Aspergillus niger* are two examples of fungal endophytes that solubilize phosphorus, potassium, and zinc salts, respectively, and *Penicillium, Aspergillus, Curvularia, Trichoderma,* and *Mesorhizobium,* among others, increase plant metabolic activity, growth, and crop yield (Verma et al., 2022).

6.3.4 ARBUSCULAR MYCORRHIZAL FUNGI (AMF)

AMF promotes complex communication between plants and fungi, resulting in improved photosynthetic rate and increased water intake (Birhane et al., 2012). Several research studies examine how the presence of fungal symbioses makes plants more resistant to a variety of stresses, such as drought, salinity, heat, metals, and diseases (Ahanger et al., 2014; Salam et al., 2017). Bryophytes, ferns, and flowering plants are among the more than 90% of plant species that can coexist harmoniously with AMF (Ahanger et al., 2014). In addition to vesicles, arbuscules, and hyphae in roots, AMF also produces spores and hyphae in the rhizosphere. When AMF and plant roots work together to produce a hyphal network, the plant grows more effectively because the roots have better access to a large amount of soil surface (Begum et al., 2019). Mycorrhizal associations with plants under drought stress boost plant performance by promoting plant growth, nutrient accumulation, and water status. AMF colonization encourages the development of large hyphal networks and glomalin secretion throughout this process, which in turn aids in water and nutrient uptake and hence improves soil structure. Studies have shown that AMF can affect plant systems such as photosynthetic rate, root hydraulic conductivity, and root architecture to develop drought-adaptive mechanisms (Bahadur et al., 2019).

6.4 SUGARCANE-ASSOCIATED MICROORGANISMS

The surfaces of leaves, the rhizosphere of plants, and other plant tissues are all open to microbial habitation. Plant microbiomes are the collective name for these bacteria. By influencing physiology and growth, these bacteria that are related to plants promote the health of the plant. Environmental factors control how the microbial population of plants behaves and how it functions. In the agricultural industry, plant-associated microbiomes have numerous prospects to enhance plant health and productivity. By employing contemporary biological technology that uses advantageous plant microbes and their metabolites, crop productivity, nutrient availability, disease resistance, and plant stress responses can all be improved (Trivedi et al., 2017). Sugarcane plants have relationships with various microorganisms throughout their life cycle, making it possible to use helpful fungi and bacteria as vital tools to promote plant growth. They are also an excellent choice to rely on to satisfy the present expectations for increased output and sustainability.

Expanding bacterial variety results from contemporary biological techniques like high-throughput sequencing and the rise in microbial culture libraries (Hug et al., 2016). De Souza et al. (2016) discovered 11,727 fungal operational taxonomic units (OTUs) and 23,811 bacterial OTUs living in sugarcane shoots, roots, and leaves. These microbial communities invade the tissues in various sections of the plants and are primarily derived from rhizosphere soil. Dong et al. (2018) described the root endophytic bacterial communities and diazotroph diversity of four kinds of sugarcane using 16S rRNA and nifH gene sequencing. Five different species of sugarcane's rhizosphere and diazotroph bacterial communities were recently examined by high-throughput sequencing (HTS) technology. According to the HTS findings, 107 different bacterial groupings containing a total of 6,202 operational taxonomic units (OTUs) specific to bacteria were discovered. 1,099 OTUs in all were found to be diazotrophs (Malviya et al., 2022). Several reports show that numerous abiotic stress tolerance PGPR and endophytes are associated with sugarcane plants and play important roles in improving overall sugarcane plant health (Table 6.1).

TABLE 6.1 Sugarcane Associated Abiotic Stress Tolerant Microbiome and Their Mechanisms

Microorganisms	Source	Abiotic Stress Tolerant	Mechanism	References
Endophyes and rhizobacteria	Root and rhizosphere	Salt	Production of IAA, phosphate solubilization.	Leite et al. (2014)
Gluconacetobacter diazotrophicus PAL5	Rhizosphere	Drought	ABA-dependent signaling genes	Vargas et al. (2014)
Klebsiella spp., *Pantoea* spp., and *Entrobacter* spp.	Root, leaf, and stem	Drought	PGPR	Rodrigues et al. (2016)
Enterobacter cloacae, *Bacillus pumilus*, *Pseudomonas* sp.	Endophytes	Salt	Potassium and zinc for their solubilizing and antifungal activity.	Pirhadi et al. (2016)
Enterobacter cloacae R13, *Enterobacter cloacae* R33, *Paenibacillus lactis*, and *Pseudomonas* sp.	Rhizosphere	Salt	PGPR	Lamizadeh et al. (2016)
Bacillus subtilis (BSSC11) and *Bacillus megaterium* (BMSE7)	Rhizosphere	Drought	PGPR	Chandra et al. (2018)
Pichia kudriavzevii;	Sugarcane fields	Heat	Thermotolerant and ethanogenic	Pongcharoen et al. (2018)
Arthrobacter, pseudomonas, mycobacterium, and bacillus	Rhizosphere	Drought	PGPB	Pereira et al. (2019)

TABLE 6.1 *(Continued)*

Microorganisms	Source	Abiotic Stress Tolerant	Mechanism	References
Bacillus spp., *Bacillus cereus*, *Bacillus altitudinis*	Sugarcane industries	Heavy metal	Bioremediation	Elhady et al. (2019)
Streptomyces chartreusis WZS021	Rhizosphere	Drought	PGPR	Wang et al. (2019)
Enterobacter roggen-kampii ED5	Root	Drought	Nitrogen fixation and plant defense.	Gua et al. (2020)
Bacillus xiamenensis PM14	Rhizosphere	Drought, salinity, heat stress	Antioxidative enzymes, induced systemic resistance	Amna et al. (2020)
Bacillus xiamenensis ASN-1	Rhizosphere	Salt	PGPR	Sharma et al. (2021)
Rhizobiales and Streptomycetales	Rhizosphere	Drought	PGPR	Liu et al. (2021)
Acinetobacter sp.	Rhizosphere	Salt	PGPR	Patel et al. (2021)
Actinobacteria fermicutes	Rhizosphere	Drought	PGPR	Liu et al. (2021)
Trichoderma asperellum	Rhizosphere	Drought	Antioxidative enzymes, proline synthesis	Scudeletti et al. (2021)
Bacillus subtilis	Rhizosphere	Drought/ nutrient uptake	PGPR	Cássia da Fonseca et al. (2022)
82-*Bacillus* and 336-*Pseudomonas*	Rhizosphere, roots, and stems	Drought	PGPR	Michavila et al. (2022)

6.5 MECHANISM OF THE MICROBIOME TO MITIGATE ABIOTIC STRESSES

Globally, abiotic stress is becoming a bigger issue for agricultural productivity. The most common abiotic stresses include temperature, drought, salinity, and nutrient deficiency (Verma et al., 2022, 2023). Plant-microbe interactions activate a number of pathways and cause abiotic stress (Figure 6.1). These pressures may alter metabolomics and transcriptomics, changing exudates from the roots and leaves and impacting the plant microbial population (Liu et al., 2020; Kamran et al., 2022). A vital component of the soil ecosystem, plant-associated microorganisms work in unison with plants to defend themselves against harmful environmental influences through local and systemic defense mechanisms. This section focuses on the underlying mechanisms and how interactions between plants and microbes alter responses to important abiotic stimuli like temperature, drought, and salinity. Induced systemic tolerance can greatly lessen the detrimental effects of abiotic stresses (IST). Because plants produce phytohormones like IAA, cytokinins, and abscisic acid (ABA), for example, they can endure environmental obstacles. Additionally, the creation of antioxidants like superoxide dismutase (SOD), glutathione reductase (GR), ascorbate peroxidase (APX), and catalase (CAT) helps to lessen the negative effects of abiotic stressors as well as the degradation of the enzyme that converts amino acids into ethylene, ACC deaminase (Munir et al., 2020). Reactive oxygen species (ROS) are byproducts of plant metabolic processes like respiration and photosynthesis (Ahmad et al., 2010). Their biological function depends on the quantity of molecules they make and how quickly cellular scavenging mechanisms can remove those molecules (Mittler et al., 2011). Many activities, such as cellular growth, development, response to biotic and abiotic

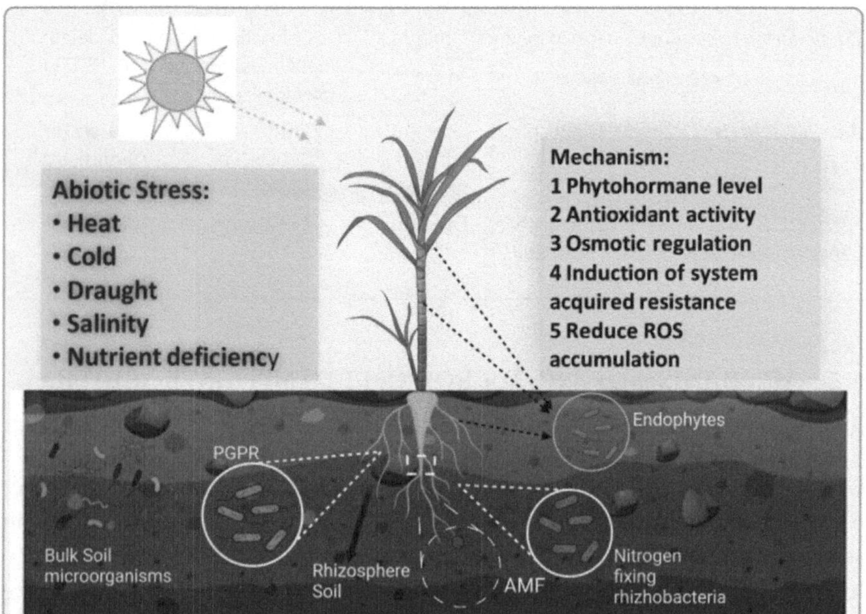

FIGURE 6.1 An overview graphic illustrating the main environmental stresses and the ways in which plant-microbe interactions affect the resilience of plants to those stresses. The key mechanisms to tolerate these challenges include induction of system-acquired resistance, increase in antioxidant defense enzymes, osmotic adjustment, and phytohormone synthesis.

stimuli, and programmed cell death, are all regulated by ROS (Baxter et al., 2014; Luis, 2015). ROS outburst is caused by modifications in cellular homeostasis that occur when plants are stressed by prolonged drought. Cell membranes, DNA, and proteins are damaged when ROS generation overwhelms antioxidant defenses, potentially leading to cell death (Gill & Tuteja, 2010; Miller et al., 2010). Nonetheless, ROS has a role in relaying stress signals to activate defense and acclimatization pathways at low levels. The increased drought resilience of plants by rhizobacteria has been explored using a variety of techniques and methods, but the precise mechanism is still unknown. Nonetheless, the literature suggests that the following probable processes could be at play in helping beneficial microbes become more drought-tolerant.

6.5.1 IMPACT OF THE MICROBIOME ON PHYTOHORMONAL LEVEL

It has been used to complement exogenous phytohormones to enhance growth and metabolism in demanding conditions. Several studies have demonstrated that phytohormones produced by microorganisms linked with roots may prove to be significant metabolic engineering targets for promoting host tolerance to abiotic stressors (Egamberdieva et al., 2017).

Plants produce phytohormones that are vital to their growth and development, including ethylene, abscisic acid, cytokinins, indole acetic acid, and gibberellins. Under drought stress, ABA has been shown to regulate root development and water content (Cutler et al., 2010). *Gluconacetobacter diazotrophicus* activates the ABA-dependent signaling genes and induces drought tolerance in sugarcane (Vargas et al., 2014). IAA and phosphate-solubilizing endophytes and rhizobacteria induce salt tolerance in sugarcane (Leite et al., 2014). Moreover, several studies indicate that rhizobacteria that produce IAA can help plants under drought stress. For instance, inoculation of wheat and maize with IAA-producing *Bacillus* and *Enterobacter* strains boosted both plant species' tolerance to drought (Jochum et al., 2019).

6.5.2 ROLE OF THE MICROBIOME IN THE PRODUCTION OF AMINOCYCLOPROPANE-1-CARBOXYLATE (ACC) DEAMINASE ACTIVITY

Reduced root and shoot growth results from ethylene's endogenous regulation of plant homeostasis during drought stress. By destroying ACC and lowering the quantity of ethylene in the plant, PGPM produces the enzyme ACC deaminase, improving drought resistance and increasing plant growth (Glick, 2005). For instance, *Pseudomonas fluorescens* biotype G (ACC-5) inoculation caused *Pisum sativum* to develop larger roots, which improved the plant's capacity to absorb water from the soil during a drought (Zahir et al., 2008). ACC-deaminase-producing PGPR promotes plant development and reduces the negative effects of drought stress on maize growth (Danish et al., 2020). *Diazotrophic Enterobacter* species produce ACC deaminase, boosting the sugarcane plant's response to salt stresses (Kruasuwan & Thamchaipene, 2018). Rhizobacteria that produce ACC deaminase include *Achromobacter* spp., *Variovorax* sp., *Ochrobactrum* sp., and *Pseudomonas* spp. These bacteria also raised finger millet leaves' overall growth metrics and nutrient concentrations by a significant amount (Chandra et al., 2020).

6.5.3 ROLE OF OSMOTIC SECRETING MICROBIOME

According to multiple studies, in reaction to dryness stress, PGPR secretes osmolytes, which work in concert with osmolytes generated by plants to promote plant development (Paul et al., 2008). By accumulating proline in maize plants under drought stress, PGPR *Pseudomonas putida* inoculation increased plant biomass, leaf water potential, and relative water content (Sandhya et al., 2010). Haseem et al. (2016) reported that *Bacillus subtilis* and arbuscular mycorrhizal fungi (AMF) induced acquired systemic resistance in *Acacia gerrardii* and modulated the effect of salt stress.

6.5.4 IMPACT OF THE MICROBIOME ON PLANT ANTIOXIDATIVE SYSTEM

It has been established that the production of ROS is the primary mechanism underlying the physiological response of plants to drought, with increased oxidative damage, restricted development, and, ultimately, cell death occurring when the ROS level exceeds a particular threshold (Asghari et al., 2020). The use of PGPRs improves plant drought tolerance by modifying antioxidant activity and lowering active oxygen species in water-scarce circumstances (Gusain et al., 2015). Plant-growing rhizobacteria respond to suppress oxidative stress and antioxidant enzyme activities in potatoes to alleviate drought stress (Batool et al., 2020). PGPR *viz. Promicromonospora* sp., *Burkholderia cepacia*, and *Acinetobacter calcoaceticus* by regulating the phytohormones and antioxidants in *Cucumis sativus*, can lessen the negative effects of salt and osmotic stress (Kang et al., 2013). *K. variicola* is a promising bioinoculant for managing salinity stress because of its prodigious production of several PGP, salt-improving, and antioxidant metabolites (Kusale et al., 2013).

6.6 CONCLUSIONS AND FUTURE PERSPECTIVE

Several microbiomes have been used in the modern age to improve agricultural output. The microbiomes of plants impact their development, production, and health. However, more in-depth investigation is needed to uncover additional pathways of plant-microbiome interactions. Plant microbiomes are commonly employed for creating various crop kinds. It's possible that the microbiomes connected to the relevant plants can reduce abiotic stress. Another study claimed that transgenic plants might be created by introducing helpful genes from microorganisms into the target plant. Likewise, transgenic plants were given the ACC deaminase gene from bacteria. Although the application of PGPR increases the stress tolerance of crops, other crops have exhibited detrimental effects from PGPR or no effects at all. Also, the microbial consortium has been used in the field to modify the rhizobiome for treating both biotic and abiotic stress. Therefore, more research must be done to improve the effectiveness of microbial formulations for the reduction of ecological stressors. In addition to sugarcane-associated microbiome techniques, microbial technologies would suggest new and more environmentally friendly ways to boost sugarcane productivity. We may implement sustainable sugarcane production by using plant-microbiome interaction as a beneficial technology. Experts are interested in this interaction's contribution and intend to conduct additional research employing beneficial microorganisms to apprehend this goal.

ACKNOWLEDGMENT

The authors are thankful to the Head of the Department, Institute of Biological Science, SAGE University Indore, for his constant support.

KEYWORDS

- **abiotic stress**
- **aminocyclopropane-1-carboxylate**
- **arbuscular mycorrhizal fungi**
- **deaminase activity**
- **microbiome**
- **PGPR**
- **rhizosphere**
- **sugarcane**

REFERENCES

Ahanger, M. A., Tyagi, S. R., Wani, M. R., & Ahmad, P. (2014). Drought tolerance: role of organic osmolytes, growth regulators, and mineral nutrients. In P. Ahmad & M. R. Wani (Eds.), *Physiological mechanisms and adaptation strategies in plants under changing environment* (Vol. 1, pp. 25–55). New York, NY: Springer. doi: 10.1007/978-1-4614-8591-9_2.

Akter, N., Rafiqul, I. M., & Abdul, K. M. (2014). Alleviation of drought stress in maize by exogenous application of gibberellic and cytokinin. *Journal of Crop Science and Biotechnology, 4*(17), 41–48.

Al-Arjani, A.-B. F., Hashem, A., & Abd Allah, E. F. (2020). Arbuscular mycorrhizal fungi modulates dynamics tolerance expression to mitigate drought stress in *Ephedra foliata* boiss. *Saudi Journal of Biological Sciences, 27*, 380–394.

Alawiye, T. T., & Babalola, O. O. (2019). Bacterial diversity and community structure in typical plant rhizosphere. *Diversity, 11*(10), 179.

Amaral, J., Lamelas, L., Valledor, L., Castillejo, M. Á., Alves, A., & Pinto, G. (2021). Comparative proteomics of *Pinus–Fusarium circinatum* interactions reveal metabolic clues to biotic stress resistance. *Physiologia Plantarum, 173*, 2142–2154.

Asghari, B., Khademian, R., & Sedaghati, B. (2020). Plant growth promoting rhizobacteria (PGPR) confer drought resistance and stimulate biosynthesis of secondary metabolites in pennyroyal (*Mentha pulegium* L.) under water shortage condition. *Scientia Horticulturae, 263*, 109132. doi: 10.1016/j.scienta.2019.109132.

Atkinson, N. J., Lilley, C. J., & Urwin, P. E. (2013). Identification of genes involved in the response of *Arabidopsis* to simultaneous biotic and abiotic stresses. *Plant Physiology, 162*, 2028–2041.

Bahadur Ali, Batool, A., Nasir, F., Jiang, S., Mingsen, Q., Zhang, Q., Pan, J., Liu, Y., & Feng, H. (2019). Mechanistic Insights into Arbuscular Mycorrhizal Fungi-Mediated Drought Stress Tolerance in Plants. *International Journal of Molecular Sciences, 20*, 4199. doi:10.3390/ijms20174199.

Begum, N., Qin, C., Ahanger, M. A., Raza, S., Khan, M. I., Ashraf, M., Ahmed, N., & Zhang, L. (2019). Role of Arbuscular Mycorrhizal Fungi in Plant Growth Regulation: Implications in Abiotic Stress Tolerance. *Frontiers in Plant Science, 10*, 1068.

Bhattacharya, A. (2022). Effect of low temperature stress on photosynthesis and allied traits: A Review. In *Physiological Processes in Plants Under Low Temperature Stress*. Springer, Singapore.

Birhane, E., Sterck, F., Fetene, M., Bongers, F., & Kuyper, T. (2012). Arbuscular mycorrhizal fungi enhance photosynthesis, water use efficiency, and growth of frankincense seedlings under pulsed water availability conditions. *Oecologia, 169*, 895–904. doi: 10.1007/s00442-012-2258-3.

Bohnert, H. J., Gong, Q., Li, P., & Ma, S. (2006). Unraveling abiotic stress tolerance mechanisms- getting genomics going. *Current Opinion in Plant Biology, 9*, 180–188.

Chandra, P., Tripathi, P., & Chandra, A. (2018). Isolation and molecular characterization of plant growth-promoting *Bacillus* spp. and their impact on sugarcane (*Saccharum spp.* hybrids) growth and tolerance towards drought stress. *Acta Physiologiae Plantarum, 40*, 199. https://doi.org/10.1007/s11738-018-2770-0.

Chen, P., Yu, S., Zhan, Y., & Kang, X. L. (2006). Effects of jasmonate acid on thermotolerance of grape seedlings. *Journal of Shihezi University (Natural Science), 24*(1), 87–91.

Conrath, U., Beckers, G. J., Flors, V., Garcia-Agustin, P., Jakab, G., Mauch, F., Mauch, F., Newman, M. A., Pieterse, C. M. J., Poinssot, B., Pozo, M. J., Pugin, A., Schaffrath, U., Ton, J., Wendehenne, D., Zimmerli, L., & Mauch-Mani, B. (2006). Priming: Getting ready for battle. *Molecular Plant-Microbe Interactions, 19*, 1062–1071.

Dar, M. Z., Masood, A., Mughal, A. H., Asif, M., & Malik, M. A. (2018). Review on Drought Tolerance in Plants Induced by Plant Growth Promoting Rhizobacteria. *International Journal of Current Microbiology, 7*(5), 412–422.

Deng, Z. N., Huang, H. R., Huang, C. M., Verma, K. K., Luo, H. B., Jiang, S. L., Li, Y. J., Cao, H. Q., Wu, K. C., & Xu, L. (2022). Short-term limited water irrigation influences the photosynthetic pigments, enzymatic and non-enzymatic activities in *Saccharum spontaneum* L. at vegetative stage. *Sugar Tech. 25*, 110–118. https://doi.org/10.1007/s12355-022-01175-7.

Dong, L., Xu, J., Zhang, L., Yang, J., Liao, B., Li, X., & Chen, S. (2017). High-throughput sequencing technology reveals that continuous cropping of American ginseng results in changes in the microbial community in arable soil. *Chinese Medicine, 12*, 18. doi: 10.1186/s13020-017-0139-8.

Egamberdieva, D., Wirth, S. J., Alqarawi, A. A., Abd_Allah, E. F., & Hashem, A. (2017). Phytohormones and Beneficial Microbes: Essential Components for Plants to Balance Stress and Fitness. *Frontiers in Microbiology, 7*(8), 2104.

Fadiji, A. E., & Babalola, O. O. (2020). Metagenomics methods for the study of plant-associated microbial communities: A review. *Journal of Microbiological Methods, 170*, 105860. https://doi.org/10.1016/j.mimet.2020.105860.

Glick, B. R. (2005). Modulation of plant ethylene levels by the bacterial enzyme ACC deaminase. *FEMS Microbiology Letters, 251*, 1–7.

Glick, B. R., & Pasternak, J. J. (2003). *Molecular Biotechnology: Principles and Application Recombinant Dna Technology* (3rd ed.). ASM Press, Washington.

Gong, B., Cao, H., Peng, C., Perculija, V., Tong, G., & Fang, H. (2019). High-throughput sequencing and analysis of microbial communities in the mangrove swamps along the coast of Beibu Gulf in Guangxi, China. *Scientific Reports, 9*, 9377. doi: 10.1038/s41598-019-45804-w.

Hashem, A., Abd_Allah, E. F., Alqarawi, A. A., Al-Huqail, A. A., & Shah, M. A. (2016). Induction of Osmoregulation and Modulation of Salt Stress in *Acacia gerrardii* Benth. by Arbuscular Mycorrhizal Fungi and *Bacillus subtilis* (BERA 71) 6294098, 11.

Huang, H. R., Gao, Y. J., Malviya, M. K., Verma, K. K., Solanki, M. K., Huang, Y. X., Li, X., Deng, Y., Yan, J., Tang, S. Y., Wang, L. W., & Xu, L. (2021). Genetic diversity analysis of sugarcane (*Saccharum spp.* Hybrids) among high-sucrose clones of GT series and commonly used parents by using microsatellite markers in Guangxi, China. *Sugar Tech. 24*, 397–407. https://doi.org/10.1007/s12355-021-01033-y.

Jalil, S. U., & Ansari, M. I. (2018). Plant microbiome and its functional mechanism in response to environmental stress. *International Journal of Green Pharmacy, 12*(1), S81.

Jochum, M. D., McWilliams, K. L., Pierson, E. A., & Jo, Y. K. (2019). Host-mediated microbiome engineering (HMME) of drought tolerance in the wheat rhizosphere. *PLoS One, 14*, e0225933.

Kruasuwan, W., & Thamchaipene, A. (2018). 1-Aminocyclopropane-1-carboxylate (ACC) Deaminase-Producing Endophytic *Diazotrophic Enterobacter* sp. EN-21 Modulates Salt–Stress Response in Sugarcane. *Journal of Plant Growth Regulation, 37*, 849–858.

Kusale, S. P., Attar, Y. C., Sayyed, R. J., Malek, R. A., Ilyas, N., Suriani, N. L., Khan, N., & El-Enshasy, H. A. (2021). Production of Plant Beneficial and Antioxidants Metabolites by *Klebsiella variicola* under Salinity Stress. *Molecules, 26*, 1894. https://doi.org/10.3390/molecules26071894.

Lamizadeh, E., Enayatizamir, N., & Motamedi, H. (2016). Identification of Plant Growth-Promoting Rhizobacteria (PGPR) from the Rhizosphere of Sugarcane in Saline and Non-Saline Soil. *International Journal of Current Microbiology and Applied Sciences, 5*(10), 1072–1083.

Leite, M. C. B. S., de Farias, A. R. B., Freire, F. J., Andreote, F. D., Kuklinsky-Sobral, J., & Freire, M. B. G. S. (2014). Isolation, bioprospecting and diversity of salt-tolerant bacteria associated with sugarcane in soils of Pernambuco, Brazil. *Revista Brasileira de Engenharia Agrícola e Ambiental, 18*, S73–S79. https://doi.org/10.1590/1807-1929/agriambi.v18nsuppS73-S79.

Liu, Q., Xie, S., Zhao, X., Liu, Y., Xing, Y., Dao, J., Wei, B., Peng, Y., Duan, W., & Wang, Z. (2021). Drought Sensitivity of Sugarcane Cultivars Shapes Rhizosphere Bacterial Community Patterns in Response to Water Stress. *Frontiers in Microbiology, 12*, 732989.

Malviya, M. K., Li, C. N., Lakshmanan, P., Solanki, M. K., Wang, Z., Solanki, A. C., Nong, Q., Verma, K. K., Singh, R. K., Singh, P., Sharma, A., Guo, D. J., Dessoky, E. S., Song, X. P., & Li, Y. R. (2022). High-Throughput Sequencing-Based Analysis of Rhizosphere and Diazotrophic Bacterial Diversity Among Wild Progenitor and Closely Related Species of Sugarcane (*Saccharum spp.* Inter-Specific Hybrids). *Frontiers in Plant Science, 13*, 829337. https://doi.org/10.3389/fpls.2022.829337.

Mehnaz, S. (2013). Microbes–friends and foes of sugarcane. *Journal of Basic Microbiology, 53*, 954–971. https://doi.org/10.1002/jobm.201200299.

Michavila, G., Alibrandi, P., Cinà, P., Welin, B., Castagnaro, A. P., Chalfoun, N. R., Noguera, A. S., Puglia, A. M., Ciaccio, M., & Racedo, J. (2022). Plant growth-promoting bacteria isolated from sugarcane improve the survival of micropropagated plants during acclimatization. *Italian Journal of Agronomy, 17*(2), 1–9. https://doi.org/10.4081/ija.2022.2006.

Mishra, A. K., Sudalaimuthuasari, N., Hazzouri, K. M., Saeed, E. E., Shah, I., & Amiri, K. M. A. (2022). Tapping into Plant–Microbiome Interactions through the Lens of Multi-Omics Techniques. *Cells, 11*, 3254. https://doi.org/10.3390/cells11203254.

Munir, N., Hanif, M., Abideen, Z., Sohail, M., El-Keblawy, A., Radicetti, E., Mancinelli, R., & Haider, G. (2022). Mechanisms and Strategies of Plant Microbiome Interactions to Mitigate Abiotic Stresses. *Agronomy, 12*, 2069.

Pereira, L. B., de Oliveira Gambarini, V. M., de Menezes, A. B., Ottoboni, L. M. M., & Vicentini, R. (2020). Responses of the sugarcane rhizosphere microbiota to different levels of water stress. *Applied Soil Ecology. 159*, 103817. https://doi.org/10.1016/j.apsoil.2020.103817.

Pieterse, C. M., de Jonge, R., & Berendsen, R. L. (2016). The soil-borne supremacy. *Trends in Plant Science, 21*, 171–173. https://doi.org/10.1016/j.tplants.2016.01.018.

Pritchard, S. G. (2011). Soil organisms and global climate change. *Plant Pathology, 60*, 82–99. https://doi.org/10.1111/j.1365-3059.2010.02405.x.

Raton, T. D. L. M. O., Yano, R., Gamez, O. R., Floh, E. I. S., Dıaz, M. D. J. S., & Barbosa, H. R. (2012). Isolation and characterization of aerobic endospore forming *Bacilli* from sugarcane rhizosphere for the selection of strains with agriculture potentialities. *World Journal of Microbiology and Biotechnology, 28*, 1593–1603.

Reuter, J. A., Spacek, D. V., & Snyder, M. P. (2015). High-throughput sequencing technologies. *Molecular Cell, 58*, 586–597. doi: 10.1016/j.molcel.2015.05.004.

Salam, E. A., Alatar, A., & El-Sheikh, M. A. (2017). Inoculation with arbuscular mycorrhizal fungi alleviates harmful effects of drought stress on damask rose. *Saudi Journal of Biological Sciences, 25*(8), 1772–1780. doi: 10.1016/j.sjbs.2017.10.015.

Scudeletti, D., Crusciol, C. A. C., Bossolani, J. W., Moretti, L. G., Momesso, L., Servaz Tubaña, B., de Castro, S. G. Q., De Oliveira, E. F., & Hungria, M. (2021). Trichoderma asperellum Inoculation as a Tool for Attenuating Drought Stress in Sugarcane. *Frontiers in Plant Science, 12*, 570.

Selvakumar, G., Panneerselvam, P., & Ganeshamurthy, A. N. (2012). Bacterial Mediated Alleviation of Abiotic Stress in Crops. In D. K. Maheshwari (Ed.), *Bacteria in Agrobiology: Stress Management (pp.* 205–224*)*. Springer-Verlag Berlin Heidelberg. https://doi.org/10.1007/978-3-642-23465-1_10.

Singh, L. P., Gill, S. S., & Tuteja, N. (2011). Unraveling the role of fungal symbionts in plant abiotic stress tolerance. *Plant Signaling & Behavior, 6*(2), 175–191.

Song, X. P., Mo, F., Yan, M., Zhang, X., Zhang, B., Huang, X., Huang, D., Pan, Y., Verma, K. K., & Li, Y. R. (2022). Effect of smut infection on the photosynthetic physiological characteristics and related defense enzymes of sugarcane. *Life, 12*, 1201. https://doi.org/10.3390/life12081201.

Spaepen, S., Vanderleyden, J., & Remans, R. (2007). Indole-3-acetic acid in microbial and microorganism-plant signaling. *FEMS Microbiology Reviews, 31*, 425–448.

Uroz, S., Ioannidis, P., Lengelle, J., Cébron, A., Morin, E., Buée, M., & Martin, F. (2013). Functional assays and metagenomic analyses reveal differences between the microbial communities inhabiting the soil horizons of a Norway spruce plantation. *PLoS One, 8*, e55929. https://doi.org/10.1371/journal.pone.0055929.

Vargas, L., Brıgida, A. B. S., Filho, J. P. M., de Carvalho, T. G., Rojas, C. A., Vaneechoutte, D., Bel, M. V., Farrinelli, L., Ferreira, P. C. G., Vandepoele, K., & Hemerly, A. S. (2014). Drought Tolerance Conferred to Sugarcane by Association with *Gluconacetobacter diazotrophicus*: A Transcriptomic View of Hormone Pathways. *PLOS ONE, 9*(12), e114744.

Verma, A., Shameem, N., Jatav, H. S., Sathyanarayana, E., Parray, J. A., Poczai, P., & Sayyed, R. Z. (2022). Fungal Endophytes to combat biotic and abiotic stresses for climate-smart and sustainable agriculture. *Frontiers in Plant Science, 13*, 953836. https://doi.org/10.3389/fpls.2022.953836.

Verma, K. K., Anas, M., Chen, Z., Rajput, V. D., Malviya, M. K., Verma, C. L., Singh, R. K., Singh, P., Song, X. P., & Li, Y. R. (2020d). Silicon supply improves leaf gas exchange, antioxidant defense system and growth in sugarcane responsive to water limitation. *Plants, 9*, 1032. https://doi.org/10.3390/plants9081032.

Verma, K. K., Li, D. M., Singh, M., Rajput, V. D., Malviya, M. K., Minkina, T., Singh, R. K., Singh, P., Song, X. P., & Li, Y. R. (2020b). Interactive role of silicon and plant-rhizobacteria mitigating abiotic stresses: A new approach for sustainable agriculture and climate change. *Plants, 9*, 1055. https://doi.org/10.3390/plants9091055.

Verma, K. K., Singh, P., Song, X. P., Malviya, M. K., Singh, R. K., Chen, G. L., Solomon, S., & Li, Y. R. (2020e). Mitigating climate change for sugarcane improvement: role of silicon in alleviating abiotic stresses. *Sugar Tech, 22*(5), 741–749. https://doi.org/10.1007/s12355-020-00831-0.

Verma, K. K., Song, X. P., Budeguer, F., Nikpay, A., Enrique, R., Singh, M., Zhang, B., Wu, J. M., & Li, Y. R. (2022a). Genetic engineering: an efficient approach to mitigating biotic and abiotic stresses in sugarcane cultivation. *Plant Signaling & Behavior, 17*(1), e2108253.

Verma, K. K., Song, X. P., Li, D. M., Singh, M., Wu, J. M., Singh, R. K., Sharma, A., Zhang, B. Q., & Li, Y. R. (2022b). Silicon and soil microorganisms improve rhizospheric soil health with bacterial community, plant growth, performance, and yield. *Plant Signaling & Behavior, 17*(1), e2104004.

Verma, K. K., Song, X. P., Lin, B., Guo, D. J., Singh, M., Rajput, V. D., Singh, R. K., Singh, P., Sharma, A., Malviya, M. K., Chen, G. L., & Li, Y. R. (2021c). Silicon-induced drought tolerance in crop plants: physiological adaptation strategies. *Silicon, 14*, 2473–2487. https://doi.org/10.1007/s12633-021-01071-x.

Verma, K. K., Song, X. P., Tian, D. D., Singh, M., Verma, C. L., Rajput, V. D., Singh, R. K., Sharma, A., Singh, P., Malviya, M. K., & Li, Y. R. (2021a). Investigation of the defensive role of silicon during drought stress induced by irrigation capacity in sugarcane: physiological and biochemical characteristics. *ACS Omega, 6*, 19811–19821.

Verma, K. K., Song, X. P., Verma, C. L., Chen, Z. L., Rajput, V. D., Wu, K. C., Liao, F., Chen, G. L., & Li, Y. R. (2021b). Functional relationship between photosynthetic leaf gas exchange in response to silicon application and water stress mitigation in sugarcane. *Biological Research, 54*, 15. https://doi.org/10.1186/s40659-021-00338-2.

Verma, K. K., Song, X. P., Verma, C. L., Malviya, M. K., Guo, D. J., Rajput, V. D., Sharma, A., Wei, K. J., Chen, G. L., Solomon, S., & Li, Y. R. (2021e). Prediction of photosynthetic leaf gas exchange of sugarcane (Saccharum spp.) leaves in response to leaf positions to foliar spray of potassium salt of active phosphorus under limited water irrigation. *ACS Omega, 6*, 2396–2409.

Verma, K. K., Song, X. P., Zeng, Y., Guo, D. J., Sing, M., Rajput, V. D., Malviya, M. K., Wei, K. J., Sharma, A., Li, D. P., Chen, G. L., & Li, Y. R. (2021d). Foliar application of silicon boosts growth, photosynthetic leaf gas exchange, antioxidative response, and resistance to limited water irrigation in sugarcane (Saccharum officinarum L.). *Plant Physiology and Biochemistry, 166*, 582–592.

Verma, K. K., Song, X. P., Zeng, Y., Li, D. M., Guo, D. J., Rajput, V. D., Chen, G. L., Barakhov, A., Minkina, T. M., & Li, Y. R. (2020c). Characteristics and correlation of leaf stomata and its relationship with photosynthesis in *Saccharum spp.* under different irrigation and silicon application. *ACS Omega, 5*, 24145–24153.

Verma, K. K., Wu, K. C., Verma, C. L., Li, D. M., Malviya, M. K., Singh, R. K., Singh, P., Chen, G. L., Song, X. P., & Li, Y. R. (2020a). Developing mathematical models for diurnal dynamics of photosynthetic responses in sugarcane responsive to different irrigation and silicon application. *PeerJ, 8*, e10154. https://doi.org/10.7717/peerj.10154.

Wang, Z., Solanki, M. K., Yu, Z. X., Yang, L. T., An, Q. L., Dong, D. F., & Li, Y. R. (2019). Draft genome analysis offers insights into the mechanism by which Streptomyces chartreusis wzs021 increases drought tolerance in sugarcane. *Frontiers in Microbiology, 9*, 3262. doi: 10.3389/fmicb.2018.03262.

Wei, Y. J., Wu, Y., Yan, Y. Z., Zou, W., Xue, J., Ma, W. R., Wang, W., & Wang, L. (2018). High-throughput sequencing of microbial community diversity in soil, grapes, leaves, grape juice, and wine of grapevine from China. *PloS One, 13*, e0193097. doi: 10.1371/journal.pone.0193097.

Xu, L., Deng, Z. N., Wu, K. C., Malviya, M. K., Solanki, M. K., Verma, K. K., Pang, T., Li, Y. J., Liu, X. Y., Kashyap, B. K., Dessoky, E. S., Wang, W. Z., & Huang, H. R. (2022). Transcriptome analysis reveals a gene expression pattern that contributes to sugarcane bud propagation induced by Indole-3-Butyric Acid. *Frontiers in Plant Science, 13*, 852886. doi: 10.3389/fpls.2022.852886.

Zahir, Z. A., Munir, A., Asghar, H. N., Shahroona, B., & Arshad, M. (2008). Effectiveness of rhizobacteria containing ACC deaminase for growth promotion of peas (*Pisum sativum*) under drought conditions. *Journal of Microbiology and Biotechnology, 18*, 958–963.

Zhu, J. K. (2016). Abiotic Stress Signaling and Responses in Plants. *Cell, 167*(2), 313–324. doi: 10.1016/j.cell.2016.08.029.

CHAPTER 7

Exploring the Physiology of Silicon and Its Impact on Sugarcane Growth, Development, Yield, and Juice Quality

MILTON GARCIA COSTA,[1] RENATO DE MELLO PRADO,[1]
GELZA CARLIANE MARQUES TEIXEIRA,[1] KAMILLA SILVA OLIVEIRA,[1]
GILMARA PEREIRA DA SILVA,[1] and
JONAS PEREIRA DE SOUZA JÚNIOR[2]

[1]São Paulo State University (UNESP), School of Agricultural and Veterinarian Sciences, Jaboticabal, Brazil

[2]Citrus Research and Education Center, University of Florida, Lake Alfred, Brazil

ABSTRACT

The use of silicon (Si) in agriculture is a sustainable management strategy for both stressed and unstressed sugarcane. The benefits of Si are multiple, including modifying the biochemical, physiological, and nutritional processes of sugarcane plants, attributing greater tolerance to stress, and extending the longevity of the sugarcane field. These Si benefits correlate with different antioxidant defense mechanisms, from the improvement of photosynthetic machinery and photosynthetic efficiency of plants, directly impacting plant performance. In this scenario, this study will address the role of Si in the physiology and nutritional balance of plants and the impact of this beneficial element in mitigating nutritional disorders and water deficit to reduce yield losses and improve the quality of sugarcane juice.

Sugarcane Cultivation and Management: Challenges and Opportunities. Krishan K. Verma, Xiu-Peng Song, Munna Singh, Renato de Mello Prado, Jian-Ming Wu, and Yang-Rui Li (Eds.)

7.1 INTRODUCTION

Sugarcane plants are silicon (Si) accumulators due to their high capacity to absorb this element, with a concentration above 1% of the shoot dry matter (Majumdar & Prakash, 2020; Verma et al., 2022a, 2023). Plants predominantly absorb silicon in the form of orthosilicic acid (H_4SiO_4) present in soil solution at a concentration of 0.1 to 0.6 mM of Si (Imtiaz et al., 2016). However, silica and silicates are the predominant forms of silicon in the soil, and the element occurs mainly as $SiO_2.nH_2O$, which is not available for plant uptake (Zhu et al., 2019; Song et al., 2021; Verma et al., 2022b, c). There are currently three models of Si uptake: active, passive, and exclusive, each associated with different mechanisms of how plants accumulate this beneficial element (Cornelis et al., 2011; Wu et al., 2013; Verma et al., 2020a, b, 2021a, b). After Si uptake, interactions occur with hydrophilic compounds or silica gel. After hydrolysis, precipitation occurs (Farooq & Dietz, 2015; Tubaña & Heckman, 2015). Sugarcane plants have specific Si transporters, such as LSi2 and LSi6. These transporters ensure active uptake and facilitate the uptake process (Souri et al., 2020; Tubaña & Heckman, 2015).

Silicon is one of the most important stress mitigators. This is relevant due to the pressure from climate change in recent decades, causing different stresses such as thermal, water deficit, flooding, saline and nutritional deficiencies, which limit growth and yield. Silicon is a potential alternative to mitigate stress, modifying biochemical, physiological, and nutritional processes in plants (Bezerra et al., 2019; Costa et al., 2022, 2023; Oliveira et al., 2021; Verma et al., 2020b, 2021c, d, e, f). Moreover, it promotes crop growth and yield (Camargo et al., 2017; Li, 2019a), directly enhancing the quality of sugarcane juice.

Recent findings have demonstrated the role of Si in nutritional homeostasis, modifying stoichiometric C:N:P ratios under water deficit, altering C use efficiency, and directly impacting sugarcane growth and yield (Verma et al., 2020c, d; Costa et al., 2022; Oliveira Filho et al., 2021d, c; Teixeira et al., 2020b, 2022b, Costa et al., 2024). Research has also shown that the impact of Si on nutritional homeostasis is not limited to stressful conditions, indicating improvement in crop growth and yield in the absence of stress (Frazão et al., 2020; Souza Júnior et al., 2022).

This study will address the main advances made by the scientific community regarding Si. The study will also point out the challenges to be addressed in the coming decades, based on advancements in biochemical, physiological, and nutritional processes and their interaction with stressful conditions, highlighting their impacts on plant growth, health, yield, and the quality of sugarcane juice.

7.2 IMPORTANCE OF NUTRITIONAL BALANCE IN SUGARCANE YIELD

The assessment of the nutritional status of crops highlights the importance of nutrient balance in agriculture. It is crucial to prioritize which nutritional combinations should be examined first, particularly those relevant to interest in specific regions. To optimize the nutrient use efficiency, improper nutrient composition and antagonistic (negative) interactions should be minimized, while synergistic (positive) nutrient interactions must be maximize. These actions require adequate knowledge of possible beneficial interactions between nutrients (Bouain et al., 2019). An imbalance of nutrient composition can lead to increased nutrient absorption, which can disturb or at least affect regulatory processes, consequently impacting the absorption and metabolism of other nutrients. This information must be extensively studied to maximize crop yield (Kumar et al., 2021; Verma et al., 2022b, 2023).

Nutrient balance is an emerging and important research topic necessary for optimal plant nutrition. However, there is concern that sugarcane fertilization is conducted based solely on soil chemistry data without considering the interactions between nutrients in plant tissue, which can limit fertilization efficiency. One way to evaluate the nutritional balance of a crop is through the conventional methods of soil chemical analysis and leaf chemical analysis. Leaf chemical analysis is a direct method for optimal leaf diagnosis, and it can be performed using univariate interpretation methods from the critical level or adequate range. More recently, the bivariate methods like Integrated Diagnosis and Recommendation System (DRIS) and the multivariate Nutritional Composition Diagnosis (CND) have been developed.

The DRIS and CND methods are advantageous in identifying nutritional limitations. They allow for the rapid updating of norms, and can pinpoint which nutrients may be responsible for nutritional imbalance and, consequently, low yield. Nutritional diagnoses using these methods are indicated by negative indices for deficiencies, positive incides for excesses, and near-zero indices for nutritional balance (Nowaki et al., 2017). The DRIS method allows for identifying the order of nutritional limitation, grouping nutrients from the most deficient to the most excessive (Beaufils, 1973). The CND method, in turn, relies on log-transformed relationships between each nutrient and the geometric mean of the nutritional composition of the sampled tissue (Parent & Dafir, 1992). This multivariate method can be interpreted similarly to the DRIS indices, depending on the order of nutritional limitation.

One study evaluated growth limitations due to nutritional deficiency and excess in sugarcane using the DRIS and CND methods (Silva & Chiaia,

2021). The study was carried out in the field across three Brazilian municipalities (Caione et al., 2015; Moda et al., 2015). Information from the chemical analysis of leaf contents and yield in sugarcane crops in each agricultural year enabled the creation of a database addressing a reference population that presented yield above 172 t ha^{-1} (average yield). The authors calculated the CND index (Parent & Dafir, 1992) and the DRIS index (Beaufils, 1973). They found the following decreasing order of nutrient limitation using the DRIS method in the cane-plant cycle: Cu (excess) > B (excess) > S (excess) > Zn (excess) > Ca (excess) > P (excess) > K (deficiency) > N (deficiency) > Mg (excess) > Mn (deficiency) > Fe (excess) (Figure 7.1).

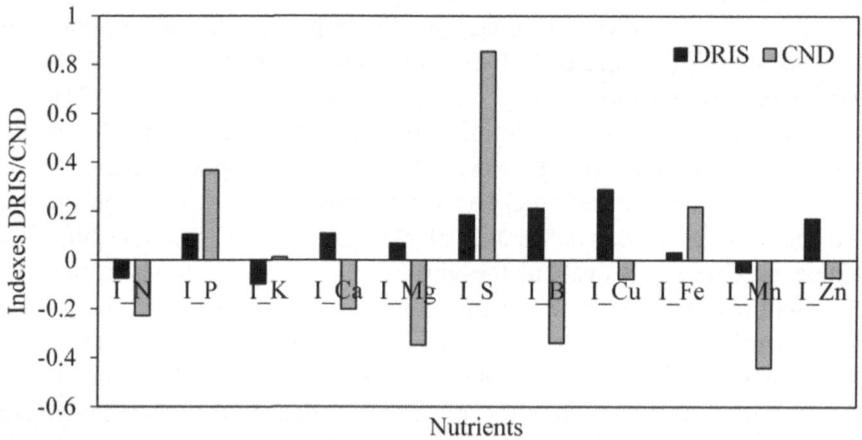

FIGURE 7.1 Order of limitation for the nutrients N, P, K, Ca, Mg, S, B, Cu, Fe, Mn, and Zn from the high productivity population, using the DRIS and CND methodology in cane plants in the year 2013. Negative, positive, and close-to-zero indices recognized nutrient deficiencies, excesses, and balance, respectively.

Source: Silva & Chiaia (2021).

Using the CND methodology in sugarcane plants, the descending order of nutrient limitation was: S (excess) > Mn (deficiency) > P (excess) > Mg (deficiency) > B (deficiency) > N (deficiency) > Fe (excess) > Ca (deficiency) > Cu (deficiency) > Zn (deficiency) > K (balanced) (Figure 7.1). The most limiting nutrients in the first ratoon were Cu (excess), according to the DRIS methodology, and Mg (deficiency), according to the CND methodology. Studies with sugarcane cultivars have found differences in the extraction of certain nutrients for the different production cycles. In Red Latosol, nutrient accumulation in the shoots of sugarcane cultivar SP81-3250 presented the descending order of K > N > Ca > S > Mg > P at harvest (Franco et al.,

2007). In ratoon cane, the descending order was: K > N > P > Mg > S > Ca. However, in the present research, the order of limitation due to nutritional deficiency and nutritional excess using the DRIS and CND methods in the sugarcane plant and ratoon cycles (Figure 7.1) is different from the order of nutrient extraction by the crop. This can indicate valuable results in nutrient use by the crop and increase fertilization efficiency (Silva & Chiaia 2021).

The variability in the order of limitation due to the lack and excess of nutrients in the present study (Figure 7.1) can be explained by the phenological cycles of sugarcane growth. These cycles, known as plant and ratoon cane, have different periods of vegetative development. For example, the sugarcane-plant cycle (one-and-a-half-year-old cane) has an average vegetative growth of 16 months. In ratoon cane, the average period of vegetative development decreases to approximately 12 months (Segato et al., 2006). In this scenario, the importance of nutritional balance for increasing sugarcane yield and the complete understanding of the interaction mechanisms, either by synergism or by antagonism between nutrients, are essential to optimize plant growth and ensure sustainable agricultural production around the world.

Foliar diagnosis is essential for obtaining scientifically-based information to identify potential nutrient deficiencies and excess in cultivated areas. Nutritional disorders causing nutritional stress in crop can be mitigated with the use of Si, and this type of research is important for ensuring the sustainability of sugarcane crop, as sugarcane plants can accumulate significant amounts of Si in their organs (Camargo & Keeping, 2021). Thus, studies with Si have promoted the nutritional balance of sugarcane in relation to changes in C:N:P stoichiometry (Costa et al., 2022; Frazão et al., 2020; Oliveira Filho et al., 2021d, c; Souza Júnior et al., 2022; Tombeur et al., 2020).

Plants absorb Si in the form of monosilicic acid, which accumulates in cell walls as amorphous silica (Neethirajan et al., 2009; Verma et al., 2020c, e, f, 2021d, e) and in organic complexes with cellulose (Tombeur et al., 2020), hemicelluloses (He et al., 2015), and other cell wall components. It suggests that Si enhances plant growth (Isa et al., 2010; Verma et al., 2022a). Once taken up by plants, Si can partially replace carbon (C) in organic compounds (Klotzbücher et al., 2018) and modify C:N:P stoichiometry in plant tissues (Frazão et al., 2020; Hao et al., 2020; Oliveira Filho et al., 2021c; Teixeira et al., 2020b). This may indicate that Si uptake can have a profound impact on plant nutrient use (Li et al., 2018; Song et al., 2014).

Replacement of structural carbon by phytoliths (organic silica compounds) in plant structures (Neu et al., 2017) requires less metabolic energy for their formation and incorporation (Schaller et al., 2016), and plants can use excess energy to synthesize other organic compounds (Klotzbücher et al., 2018). Another benefit of Si is the increase in PPase and ATPase activity, favoring the transport of nutrients (Liang et al., 2006). Research evidence shows that Si supply increases N and P accumulation in sugarcane (Frazão et al., 2020). These nutrients play a vital role as components of compounds such as chlorophyll and adenosine triphosphate (ATP), which are essential for plant biological processes and increase sugarcane production (Silva et al., 2018). Furthermore, Si may play a promising role in influencing the biogeochemical cycles of C, N, and P in terrestrial ecosystems that are dominated by Si-accumulating plants, allowing plants to govern nutrient balance and alleviate nutritional deficiencies (Hao et al., 2020). Therefore, a complete understanding of the mechanisms of interaction between nutrients is essential to optimize plant growth and ensure sustainable agricultural production worldwide.

7.3 IMPORTANCE OF PHYSIOLOGY IN SUGARCANE GROWTH AND YIELD

The use of Si in sugarcane crops has improved crop sustainability, ensuring greater yield. This enhancement is achievable because of the improvements in the concentration of photosynthetic pigments such as chlorophyll a, chlorophyll b, and carotenoids (Camargo et al., 2019; Verma et al., 2020c, d, 2021c, 2022a). Stress usually affects pigment content, impairing the photosynthetic machinery from the reduction in the quantum efficiency of photosystem II (Teixeira et al., 2021b; Verma et al., 2021e). The Si-induced increase in carotenoid content reduces reactive oxygen species, which are responsible for the degradation of lipids in cell membranes (Fu & Huang, 2001). Carotenoids are powerful antioxidants responsible for consuming reactive oxygen species, absorbing the thermal energy of the singlet oxygen and, consequently, releasing it by polyene vibration (Maoka, 2020). This function of carotenoids is provided by their ability to conjugate double bonds with more than 11 bonds responsible for quenching singlet oxygen (Maoka, 2020). Additionally, the increased concentration of photosynthetic pigments improves the transfer of light energy from carotenoids and chlorophylls, consequently inhibiting the excess formation of reactive oxygen species (Havaux, 2014).

Silicon can also improve water absorption by plants, modulating the activities of aquaporins (Chen et al., 2016; Song et al., 2021). This occurs because Si positively regulates genes, such as *SbPIP1* and *SbPiP2*, responsible for coding the plasma membrane protein (Liu et al., 2015; Manivannan & Ahn, 2017). The benefit of Si in improving water content prevents the exposure of chloroplasts to excess light energy, consequently inhibiting the formation of reactive oxygen species (Smirnoff, 1993). The Si-induced improvement in the photosynthetic machinery increases the quantum efficiency of photosystem II, ensuring that photons are efficient in using light energy to promote excitation and proper functioning of the photochemical phase of photosynthesis (Teixeira et al., 2021b). Improved photosynthetic machinery also reflects on gas exchange, improving C assimilation, stomatal conductance, and transpiration (Frazão et al., 2020). The reduction of stomatal conductance correlates directly with Si deposition, affecting cell wall properties and decreasing the loss of guard cell turgor (Zhu & Gong, 2014). Si can reduce intracellular carbon concentration in sugarcane plants due to the greater activity of electrons available for fixation (Frazão et al., 2020; Verma et al., 2021a, 2022b).

The beneficial roles of Si in plants are countless, including improving the concentration of photosynthetic pigments, photosystem II quantum efficiency, and plant gas exchange, ensuring better plant performance, and reducing yield losses under stress conditions. However, there is evidence that the role of Si in sugarcane plants is not limited to stress conditions, indicating that Si also improves plant performance in the absence of stress, increasing crop yield (Frazão et al., 2020; Souza Júnior et al., 2022). Different mechanisms correlate directly or indirectly with Si, promoting stress tolerance in cultivated plants. These include regulation of the antioxidant defense system and ion homeostasis (Alamri et al., 2020; Parveen et al., 2019), induction of the biosynthesis of osmolytes and hydrolytic enzymes (Biju et al., 2017), improved nutrient uptake, suberization, and lignification of roots (Chen et al., 2011; Fleck et al., 2011), and regulation of the proteins involved in photosynthesis (Zahedi et al., 2020).

In this scenario, the beneficial role of Si in sugarcane cultivation is indisputable (Figure 7.2), showing positive responses in physiological and biochemical processes and ensuring the reduction of losses under stress conditions. Studies have proven these findings in conditions of drought (Li, 2019a, b; Verma et al., 2020a, c), salinity (Asraf et al., 2010), and nutritional disorder (Oliveira et al., 2021, 2022). In recent decades, new advances have been made in understanding the role of Si in biochemical and physiological processes. However, further advances are still needed, mainly to understand the impact of different stresses on the physiology of sugarcane plants in the face of constant climate change and limited resources.

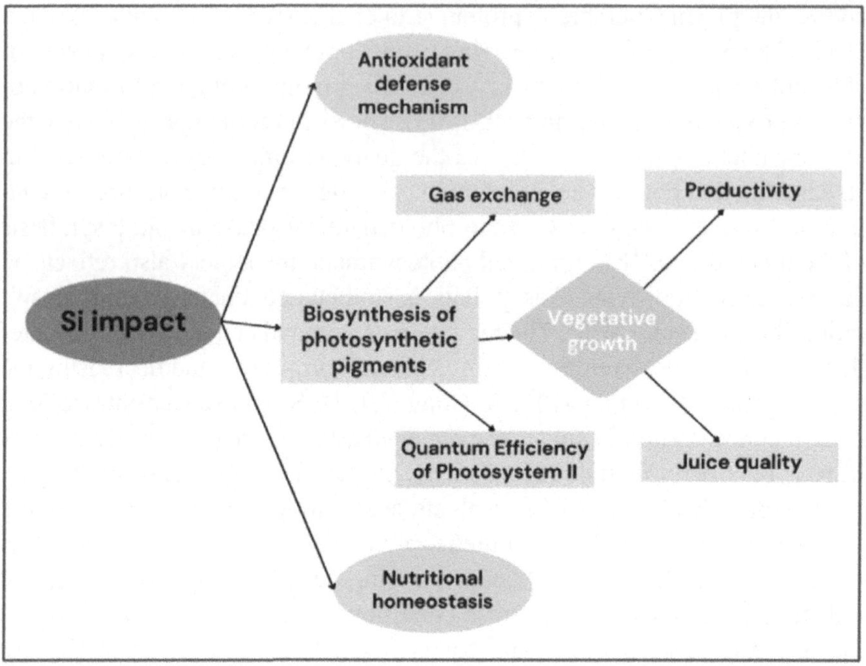

FIGURE 7.2 The benefit of Si on antioxidant defense mechanisms, photosynthetic machinery, and nutritional homeostasis and its impact on growth, yield, and juice quality in sugarcane.

7.4 SILICON IN THE ATTENUATION OF NUTRITIONAL DISORDERS (DEFICIENCY AND TOXICITY) IN SUGARCANE

Silicon is one of the most studied elements in the attenuation of plant stress. Research has shown its beneficial effects in several pathways in the metabolism and morphology of several species in recent decades. Nutritional stress (deficiency and toxicity) is also the focus of research involving the attenuating and mitigating role of Si. Nevertheless, research in this area is more recent, mainly in large species (such as sugarcane) and in cases involving nutrient deficiency. Researchers have made efforts in this area, obtaining advances that reveal biochemical, physiological, and growth benefits. These studies indicate that Si can affect the processes of absorption, transport, availability, and even nutrient use efficiency in plants, but this depends on the species, nutrients, and environment (Ali et al., 2020). In this sense, it can act beneficially in nutritional deficiency or toxicity. In addition, Si triggers potentiated responses to combat oxidative stress generated by nutritional disorders, acting on the antioxidant metabolism of plants (Kim et al., 2017).

Ali et al. (2020) made a broad review summarizing some mechanisms of Si action in nutritional deficiency and toxicity in several plant species. The findings of that study are as follows: (i) in nutritional deficiency, Si increases nutrient absorption and translocation in the roots as well as nutrient transport to the shoots; influences the defense system of plants in the shoots, reducing stress; improves physiological and growth parameters; and (ii) in elemental toxicity, Si binds to the element by forming a complex in the root cell wall, and reduces the translocation of the element to the shoots; in the leaves, it induces compartmentalization of the toxic element in the vacuole; moreover, it improves physiological and growth aspects as a primary response to the presence of the beneficial element in plants or as a secondary effect of reducing oxidative stress.

Research on the effects of Si in addressing nutritional disorders in sugarcane reveals a vast field for further investigation, particularly regarding the interaction of Si with various elements of plant nutrition. However, existing research has already demonstrated some positive effects in both nutrient deficiency and toxicity scenario. For example, Si increases N and Mn nutrition in sugarcane (Borges et al., 2016; Oliveira et al., 2021, 2022). In this sense, the supply of Si to the plants can improve the responses of other nutrients, with greater potential when sugarcane plants grow in low soil nutrient content conditions.

Nutrient toxicity studies indicate different relationships between Si nutrition and the absorption of essential elements. However, these studies are in greater numbers in the literature about nutritional deficiency studies. A huge amount of initial studies in several species indicate that plants tend to decrease the absorption of toxic elements in the presence of Si (Etesami & Jeong, 2018), with several proposed and proven mechanisms. This induced the tendency of this topic for decades; notwithstanding, for sugarcane, there are still many gaps to be filled in relation to nutrient excess and the effects of Si.

Reports of P deficiency in sugarcane are more common than reports of P toxicity in this crop. In tropical and weathered soils, conditions are favorable for P deficiency in sugarcane crops, with a great response to fertilization (Silva et al., 2020). This is due to the high adsorption of P to soil colloids (Cherubin et al., 2016), which reduces the efficiency of phosphate fertilization. Research on sugarcane has shown that Si increased phosphorylation, having a marked effect on the formation of organic phosphates and the performance of kinases (Cheong & Chan, 1973). Despite the scarce reports for this nutrient in sugarcane, the presence of Si in crops with wheat plants (another grass with high absorption of the element) under conditions of low P has increased the exudation of citrate and malate by the roots, increasing P acquisition (Kostic et al., 2017). Moreover, silicon modulates carbohydrate metabolism, increasing P use efficiency (Silva et al., 2021; Ulloa et al., 2021) and improving antioxidant compounds in sorghum plants as well as electron transport and growth. On the contrary, P toxicity is not perceived under normal conditions of sugarcane cultivation in tropical soils. This is because the specificities that favor the above-mentioned deficiency conditions reduce available P and prevent the concentrations absorbed by the plants from being sufficient to cause toxicity.

Silicon can affect both N deficiency and toxicity in sugarcane. In this sense, Borges et al. (2016) indicate a strong interaction between silicate and N doses applied to sugarcane. The benefit of Si in sugarcane stumps occurs mainly in the nutritional disorder of N (deficiency and excess). Although there are no reports of Si in N deficiency in sugarcane in rice plants, the change in C and Si stoichiometry and lignin content was due to the benefits of this element (Deus et al., 2020). In forage grasses, the beneficial effects of Si involve increasing N use efficiency and improving antioxidant defense compounds (phenols) and physiological aspects such as quantum efficiency of PSII and green color index (Araújo et al., 2022).

Sugarcane plants are sensitive to excess NH_4^+, with damage to stomatal conductance and carboxylation efficiency, which reduces photosynthesis and growth (Pissolato et al., 2019). Therefore, although studies on the interaction of excess N and Si in sugarcane are still incipient, there are indications of improvement in the metabolism of other plant species (Campos et al., 2020). In horticultural plants, silicon application increased chlorophyll and carotenoid contents (Barreto et al., 2016), the physical integrity of cell membranes, and water use efficiency (Campos et al., 2020; Ferreira Barreto et al., 2017), in addition to reducing damage to transpiration and stomatal conductance, with an increase in biomass (Barreto et al., 2016; Viciedo et al., 2019). Furthermore, Si can protect cells under NH_4^+ excess from its ability to accumulate in plant cell walls (Campos et al., 2020). The fact that sugarcane is an efficient species in absorbing Si may lead to a positive influence of Si in reducing the effects of N toxicity.

In situations of Mn deficiency, current studies reveal that Si supplied in nutrient solution has attenuated the effects of deficiency in energy cane cultivated at low Mn. Silicon increased Mn absorption during the seedling period, increasing Mn use efficiency and inducing a higher concentration of pigments, also improving the efficiency of PSII and the concentration of nonenzymatic antioxidants, reducing oxidative stress (Oliveira et al., 2021).

In a later study, the same authors indicated that in more fibrous energy cane, Si increased Mn absorption efficiency in plants cultivated under low Mn concentrations by inducing greater activity of plant defense system enzymes (SOD and GPOX). The authors mentioned a reduced reactive oxygen species and oxidative damage reduction, which improved physiological performance and plant growth (Oliveira et al., 2022). On the other hand, a study of varying doses of Mn and Si showed that sugarcane reduced Mn content in tissues when grown in a solution with Si (Samuels & Alexander, 1969), being a typical feature of combating toxicity effects.

A recent study involving the nutrient Ni in sugarcane indicated that its soil application improved N metabolism and the antioxidant system, enhancing amino acids and total sugars. This demonstrates that its deficiency in plants can reduce product yield and quality (Rampazzo et al., 2022). On the other hand, some authors mention the effects of Ni toxicity on plants (Hassan et al., 2019). In sugarcane, the element reduces chlorophyll content and plant biomass, leading to chlorosis symptoms (Misra et al., 2010). In this sense, although there are no effects of Si on Ni deficiency, in corn plants under Ni toxicity, the element attenuated damage by increasing chlorophylls and improving the enzymatic and nonenzymatic defense system, in addition to reducing damage to membrane stability and promoting plant growth (Fiala et al., 2021). Other authors verified antioxidant, cell stability, and root anatomy benefits in corn roots under Ni toxicity but treated with Si (Vaculík et al., 2021). These effects of Si are thus likely to occur in sugarcane plants, but the responses require investigation and may be the focus of future research.

Aluminum (Al) toxicity often occurs in tropical soils, and researchers have identified Si as a tool to combat the damage caused by this toxicity in sugarcane. A study revealed that in this condition, the use of Si attenuated Al toxicity by improving the elemental stoichiometry of C, N, and P in plants (Sousa Junior et al., 2022). In addition, another recent study with two sugarcane cultivars indicated that the presence of Si in an Al-toxic growth medium decreased Al uptake and translocation to the shoots while improving the mineral composition of plants, nutrient use efficiency, lignin content, and biomass (Sousa Junior et al., 2021).

Researchers have also explored sodium chloride (NaCl) toxicity in sugarcane. It is important to analyze this toxicity and the Si response-inducing effects, as excess NaCl reduces the absorption of K^+ and Ca^{2+} and, therefore, can induce a deficiency of these nutrients in sugarcane plants (Asraf et al., 2010). In this sense, the authors verified that Si decreased Na+ absorption by the plants while increasing the absorption of Ca and K. A recent study on cabbage plants showed that the presence of Si in a Ca-deficient nutrient solution increased the concentration of the latter nutrient in the shoots of plants (Silva et al., 2021). In K-deficient maize and bean plants, the supplied Si increased water use efficiency, pigment contents, and dry matter production (Santos Sarah et al., 2021, 2022).

7.5 SILICON IN THE ATTENUATION OF WATER DEFICIT IN SUGARCANE

Water is a resource of unquestionable importance for agricultural production and the proper functioning of plant mechanisms. Roots absorb about 500 g of water to produce 1 g of organic matter (Mengel et al., 2001). In the photosynthetic process, the amount of water needed is small, about 0.01% of the total water used by the plant. However, water is the main solvent for inorganic salts, sugars, and organic anions, in addition to being the medium in which all biochemical reactions occur, also maintaining cell turgor (Mengel et al., 2001). Thus, plants grown under conditions that favor dehydration may present several physiological alterations that limit plant growth and reduce production (Verma et al., 2020c, 2021a, c, 2022a, b).

Sugarcane plants have a C4 metabolism and great adaptability to different environments. However, the agronomic management used for this species, such as planting in all seasons of the year and various types of soil with contrasting water storage capacities, without irrigation, can expose sugarcane to periods of water deficit (Basnayake et al., 2012), especially in drought-sensitive cultivars (Camargo et al., 2022). The phenological phase is also a factor that can influence the damage caused by water deficit in sugarcane (Camargo et al., 2022). In the early stages of growth, the damage is evidenced by the limitation of root growth and post-transplantation setting, causing seedling death and a decrease in plant population in the field (Jain et al., 2015). In phases of intensive relative plant growth (e.g., tillering), damage from lack of water can compromise plant growth, especially affecting stem elongation (Dingre et al., 2021).

Water deficit is one of the environmental stresses that cause major obstacles to sugar and alcohol production, occurring with greater frequency, duration, and intensity (Camargo et al., 2022). These events occur mainly in regions with sandy soils and low water retention capacity (Schaller et al., 2020). Physiological changes caused by water deficit in sugarcane plants decrease photosynthetic efficiency and biomass production. These limitations initially occur due to dehydration of root cells (main site of absorption), decreasing cell water, osmotic potential, and hormonal signaling, with increased synthesis of abscisic acid (ABA), which induces total or partial closure of stomata (Verma et al., 2021c, 2022b). This causes an emergency protective effect on the plant to reduce leaf transpiration rate, but with impairment of CO_2 influx and synthesis of photosynthetic assimilates.

The decrease in the relative content of cellular water results in secondary damage because of the limitation of the adequate functioning of the photosynthetic apparatus. The smallest influx of CO_2 generates an electron space that can form reactive oxygen species (ROS), leading to oxidative stress. These species (ROS) have the potential to oxidize plant cell compounds; that is, lipid peroxidation occurs (Teixeira et al., 2022a), with increased extravasation of electrolytes (Camargo et al., 2022). Degradation of cell membrane lipids compromises the proper functioning of plant systems, including the defense mechanism, from the denaturation of both enzymatic antioxidant agents (superoxide dismutase (SOD); catalase (CAT); ascorbate peroxidase (APX); glutathione reductase (GSR)) and nonenzymatic antioxidant agents (carotenoids and tocopherols) (Bezerra et al., 2019; Teixeira et al., 2020a).

Long-term water deficit induces secondary stress due to nutritional deficiency, as nutrient absorption requires water (Rizwan et al., 2015). The transport of water and nutrients via the xylem thus also decreases (Ibrahim et al., 2020). Potassium nutritional deficiency, for example, is common in plants under water deficit. This deficiency aggravates biological damage in the plant, as this nutrient correlates with leaf water status and water use efficiency (Martineau et al., 2017). Damage caused by water deficit also causes an ionic imbalance in plants, inducing a decrease in C, N, and P contents in sugarcane seedlings subjected to short-term water deficit (30 days) (Teixeira et al., 2020a) and in sugarcane plants in the tillering phase with longer water deficit (160 days) (Teixeira et al., 2022b). In both conditions, alterations in the contents of these nutrients affect stoichiometric homeostasis (Si:C:N:P), reducing nutrient use efficiency.

In this context, to maintain the production of sugar, energy, and ethanol from sugarcane, further research is required to assess the responses of plants under the growing environments that may occur in regions intended for the cultivation of this species. Advances in the knowledge of improvements for agricultural production in the current scenario require a concern with sustainability. At the same time, high production is required, which reinforces the need for greater efficiency in the use of water and fertilizers (Firouzabadi et al., 2021). This will even enable the expansion of sugarcane planting in new cultivation areas with restricted or irregular rainfall.

The literature indicates Si as an efficient, beneficial element in mitigating stress damage in crops. In the soil, Si predominates in the form of silicon dioxide (SiO_2), representing about 50–70% of the soil mass. However, this element has low availability in solution in the form assimilable by plants (H_4SiO_4) (<1 mM) (Epstein, 1994). In mitigating the damage caused by water deficit, silicon increases the efficiency of the water present in cellular tissues; that is, it stimulates water use efficiency. For this, after being absorbed by the plant, Si is transported via the xylem (Mitani et al., 2005) to the leaves. As transpiration occurs, silicon concentrates in the leaves, specifically in the epidermal cells, in the cell wall. The increased Si concentration in leaf tissues favors polymerization to form amorphous or biogenic silica (Yoshida, 1965). This exerts a physical effect by reducing water loss by transpiration, a frequent impact in sugarcane plants (Pei et al., 2010).

Silicon can also exert a secondary physical effect by transforming into Si-cellulose structures in cell walls. After polymerization, the Si-cellulose structures formed in the cell wall can replace high-energy cost structural compounds formed by long C chains, such as lignin (Schaller et al., 2012). This replacement contributes to the maintenance of elemental stoichiometric homeostasis (C:N:P), favoring the efficient use of C (Oliveira Filho et al., 2021c; Teixeira et al., 2020b, 2022b) and other nutrients with a structural function. Researchers have already reported these effects in sugarcane and energy cane plants (Teixeira et al., 2022b). The Si-induced increase in nutritional efficiency in plants under water deficit is an interesting, beneficial effect, as it attenuates the secondary damage caused by water deficit by decreasing nutrient absorption. This occurs due to the limitation in the processes of diffusion and mass flow. Silicon applied via fertigation with an adequate soluble source induced a beneficial effect that increased the efficiency of N, P, K, Ca, Mg, and S use in maize plants under severe water deficit (Teixeira et al., 2022a). However, physical effects are more common in plants with high Si absorption rates (Si accumulators), such as sugarcane, since 90% of the absorbed Si is directed to these effects (Yoshida, 1965).

Silicon can also exert physiological and biochemical mitigating effects, which are more frequent in plants that do not accumulate large amounts of Si. This is because Si expresses these effects at the nanometer level contained in intercellular structures (↓1%) (Woesz et al., 2006). Physiologically, the Si-induced increase in water use efficiency starts from the stimulation of osmotic adjustment by increasing proline levels and sugars (Avila et al., 2021). Osmotic adjustment favors balance in the hydraulic conductivity of roots, increasing the absorption of water (Teixeira et al., 2020a) and, therefore, nutrients (Ibrahim et al., 2020). In addition, Si stimulates the expression of enzymes involved in water absorption, such as aquaporins (Abbas et al., 2015). These combined effects prevent cellular dehydration of plants under low water availability.

The maintenance of cellular water content helps to reduce the damage caused to photosystem II reaction centers, increasing the use of solar radiation in the photochemical phase of photosynthesis (Verma et al., 2021) and preventing excessive ROS formation (Gratão et al., 2005). Less formation of degrading agents maintains cell membrane integrity, preservation of photosynthetic pigments (Rocha et al., 2021), and the activity of agents of the secondary defense system. These agents may be enzymatic such as SOD, CAT, APX, and GSR (Teixeira et al., 2021a), and/or non-enzymatic: carotenoids and phenols (Teixeira et al., 2021b), increasing photosynthetic rates and dry matter production.

The inclusion of Si in agricultural management can collaborate with the rational use of water for irrigation (Cardozo et al., 2018). In this context, future scenarios may include the worst occurrences of more intense drought events that may demand the frequent use of Si in agriculture to guarantee food security. To increase the viability of Si used in agriculture, it is necessary to increase the efficiency of its applications. Fertigation is a strategy for Si application in commercial crops, and this strategy can reduce Si polymerization rates in the soil, allowing the use of low doses below 20 kg ha^{-1}. Researchers evidenced the efficiency of Si application via fertigation in increasing Si absorption using low doses per area and with high-frequency applications in the initial phase of plant growth in sugarcane either in the seedling phase (Teixeira et al., 2020a), during the tillering phase (Teixeira et al., 2021a), or in sugarcane ratoon (Oliveira Filho et al., 2021d). These beneficial effects also occurred in other plants of the family Poaceae, including forage (Rocha et al., 2021) and maize (Teixeira et al., 2022a). Therefore, Si emerges as a sustainable alternative to enhance sugarcane production and minimize the water demand for cultivation without any environmental risk.

7.6 SILICON IN THE QUALITY OF SUGARCANE JUICE

Silicon favors the development of sugarcane plants, especially under stress conditions (Costa et al., 2023; Oliveira Filho et al., 2021b, a; Teixeira et al., 2022b). Still, reports also mention the benefit of silicate fertilization in plants grown without apparent stress (Frazão et al., 2020; Souza Júnior et al., 2022). For sugarcane, in addition to the increase in yield, it is also important to highlight the increments of silicate fertilization in juice quality.

In this sense, studies indicate that Si application in sugarcane plants can increase the juice °Brix and polymerization value (Pol) (Ashraf et al., 2009; Jain et al., 2016, 2018; Shen et al., 2019). The Brix is measured on a scale of degrees and corresponds to the concentration (in g of solute per 100 g of solution) of pure sucrose in water, having the same density as the solution at the same temperature (Nawi et al., 2014). Sugarcane juices generally have between 12 and 30 °Brix (Nawi et al., 2014). The Pol indicates the concentration (in g of solute per 100 g of solution) of a solution of pure sucrose in water having the same rotational optics at the same temperature, representing the sucrose content expressed in percentage (Nawi et al., 2014). Silicate fertilization can increase °Brix and Pol in sugarcane under stress. In this regard, some authors observed an increase in °Brix and Pol in different sugarcane genotypes sensitive to salinity and grown under saline conditions (Jain et al., 2018).

Salinity can reduce the quality of sugarcane juice due to greater sodium retention. This can reduce the activity of sucrose synthase and starch phosphorylase, in addition to increasing the activity of acid and neutral invertases (Dubey & Singh, 1999). Silicon can regulate the activity of invertases, peroxidases, polyphenol oxidases, adenosine triphosphatases, and phosphatases. The reduction in phosphatase activity can provide the essential high-energy supplement necessary for sugar production (Savant et al., 1999). Studies with cultivation without induced stress also report an increase in °Brix content in sugarcane juice with Si application. The authors attribute this effect to the benefits of the element in increasing the photosynthetic rate, which correlates directly with the °Brix in the juice (Jain et al., 2016).

Silicate fertilization also increases the recoverable sugar content in sugarcane juice (Camargo et al., 2017). This occurs because the Si supply induces sugar production in the leaves, which is then transported to sugarcane stems (Alexander, 1969; Ayres, 1966; Samuels & Alexander, 1969). In the stem, the enzymatic action of Si delays the oxidation and inversion reactions of sugars, decreasing the production of reducing sugars and increasing the production of sucrose (Alexander, 1969). Additionally, the silicon deposited in the cell wall can regulate the transport of soluble sugars through physiological or physical mechanisms not detailed by the authors (Zhu et al., 2016). Fiber content and juice purity are qualitative variables less influenced by silicate fertilization, with no reports of significant results (Shen et al., 2019). Nevertheless, the Si supply also enhances the quality of sugarcane juice by reducing the production of reducing sugars (Shen et al., 2019).

7.7 FINAL CONSIDERATIONS AND FUTURE PERSPECTIVES

The beneficial role of Si in sugarcane cultivation is indisputable, positively impacting biochemical, physiological, and nutritional mechanisms, reflecting productivity and juice quality. The latest evidence indicates that the use of Si in sugarcane cultivation is the main strategy for reducing yield loss in the face of different stressful conditions, ensuring sustainable crop management. The impact of Si on the interaction between nutrients is evident through mechanisms that are still not fully understood. Further studies are thus needed, mainly on the role of Si in nutrient absorption and the homeostatic balance of C:N:P and nutrient use efficiency. Understanding the interaction of Si and nutrients is vital, especially to understand the implications of changes in nutrient dynamics on the biochemical and physiological processes of sugarcane plants.

Research also mentions the impact of Si on activating or reinforcing antioxidant defense mechanisms under different stress conditions, evidencing the greater tolerance provided by Si to sugarcane plants. However, whether antioxidant defense mechanisms change under various stress conditions is still not fully understood. Also, few studies report the benefit of Si on gene expression, requiring a further understanding of the role of Si in the genetics of cultivated plants, mainly with regard to nutritional genomics and proteomics. In recent decades, advances in understanding the role of Si in plant life have allowed this beneficial element in agriculture to become an efficient strategy in the face of stressful conditions. The use of Si in sugarcane cultivation is strategic to reduce yield losses and ensure crop sustainability. In conditions of absence of stress, the element becomes a strategy to improve yield and juice quality in sugarcane plants.

ACKNOWLEDGMENTS

We acknowledge the Coordination of Superior Level Staff Improvement (CAPES), National Council for Scientific and Technological Development (CNPq), and São Paulo State University (UNESP) for granting scholarships to the first, third, and fourth authors, respectively.

KEYWORDS

- **beneficial element**
- **nutritional balance**
- **plant stress**
- **Saccharum officinarum L.**
- **sustainable management**
- **water deficit**

REFERENCES

Abbas, T., Balal, R. M., Shahid, M. A., Pervez, M. A., Ayyub, C. M., Aqueel, M. A., & Javaid, M. M. (2015). Silicon-induced alleviation of NaCl toxicity in okra (*Abelmoschus esculentus*) is associated with enhanced photosynthesis, osmoprotectants, and antioxidant metabolism. *Acta Physiologiae Plantarum, 37*, 1–15. https://doi.org/10.1007/s11738-014-1768-5.

Alamri, S., Hu, Y., Mukherjee, S., Aftab, T., Fahad, S., Raza, A., Ahmad, M., & Siddiqui, M. H. (2020). Silicon-induced postponement of leaf senescence is accompanied by modulation of antioxidative defense and ion homeostasis in mustard (*Brassica juncea*) seedlings exposed to salinity and drought stress. *Plant Physiology and Biochemistry, 157*, 47–59. https://doi.org/10.1016/j.plaphy.2020.09.038.

Alexander, A. G. (1969). Enzyme-sugar relationships in immature sugarcane treated with ascorbic acid, cysteine, hydroxylamine, cyanide, silicon, and iron. *The Journal of Agriculture of the University of Puerto Rico, 51*, 46–54. https://doi.org/10.46429/jaupr.v51i1.11218.

Ali, N., Réthoré, E., Yvin, J. C., & Hosseini, S. A. (2020). The regulatory role of silicon in mitigating plant nutritional stresses. *Plants, 9*(12), 1779. https://doi.org/10.3390/plants9121779.

Araújo, W. B. S., Teixeira, G. C. M., de Mello Prado, R., & Rocha, A. M. S. (2022). Silicon mitigates nutritional stress of nitrogen, phosphorus, and calcium deficiency in two forages plants. *Scientific Reports, 12*(1), 6611. https://doi.org/10.1038/s41598-022-10615-z.

Ashraf, M., Rahmatullah, Ahmad, R., Afzal, M., Tahir, M. A., Kanwal, S., & Maqsood, M. A. (2009). Potassium and silicon improve yield and juice quality in sugarcane (*Saccharum officinarum* L.) under salt stress. *Journal of Agronomy and Crop Science, 195*(4), 284–291. https://doi.org/10.1111/j.1439-037X.2009.00364.x.

Ashraf, M., Rahmatullah, Ahmad, R., Bhatti, A. S., Afzal, M., Sarwar, A., Maqsood, M. A., & Kanwal, S. (2010). Amelioration of salt stress in sugarcane (*Saccharum officinarum* L.) by supplying potassium and silicon in hydroponics. *Pedosphere, 20*(2), 153–162. https://doi.org/10.1016/S1002-0160(10)60003-3.

Avila, R. G., Magalhães, P. C., da Silva, E. M., de Souza, K. R. D., Campos, C. N., de Alvarenga, A. A., & de Souza, T. C. (2021). Application of silicon to irrigated and water deficit sorghum plants increases yield via the regulation of primary, antioxidant, and osmoregulatory metabolism. *Agricultural Water Management, 255*, 107004. https://doi.org/10.1016/j.agwat.2021.107004.

Ayres, A. S. (1966). Calcium silicate slag as a growth stimulant for sugarcane on low-soils. *Soil Science, 103*(3), 216–227.

Barreto, R. F., Prado, R. M., Leal, A. J. F., Troleis, M. J. B., Junior, G. B. S., Monteiro, C. C., Santos, L. C. N., & Carvalho, R. F. (2016). Mitigation of ammonium toxicity by silicon in tomato depends on the ammonium concentration. *Acta Agriculturae Scandinavica, Section B — Soil & Plant Science, 66*(5), 483–488. https://doi.org/10.1080/09064710.2016.1178324.

Basnayake, J., Jackson, P. A., Inman-Bamber, N. G., & Lakshmanan, P. (2012). Sugarcane for water-limited environments: Genetic variation in cane yield and sugar content in response to water stress. *Journal of Experimental Botany, 63*(16), 6023–6033. https://doi.org/10.1093/jxb/ers251.

Beaufils, E. R. (1973). Diagnosis and recommendation integrated system (DRIS): A general scheme for experimentation and calibration based on principles developed from research in plant nutrition [Doctoral dissertation, University of Natal]. Pietermaritzburg.

Bezerra, B. K. L., Lima, G. P. P., dos Reis, A. R., Silva, M. de A., & Camargo, M. S. (2019). Physiological and biochemical impacts of silicon against water deficit in sugarcane. *Acta Physiologiae Plantarum, 41*(1), 1–12. https://doi.org/10.1007/S11738-019-2980-0/FIGURES/3.

Biju, S., Fuentes, S., & Gupta, D. (2017). Silicon improves seed germination and alleviates drought stress in lentil crops by regulating osmolytes, hydrolytic enzymes and antioxidant defense system. *Plant Physiology and Biochemistry, 119*, 250–264. https://doi.org/10.1016/j.plaphy.2017.09.001.

Borges, B. M. M. N., de Almeida, T. B. F., & de Mello Prado, R. (2016). Response of sugarcane ratoon to nitrogen without and with the application of silicon. *Journal of Plant Nutrition, 39*, 793–803. https://doi.org/10.1080/01904167.2015.1109101.

Bouain, N., Krouk, G., Lacombe, B., & Rouached, H. (2019). Getting to the root of plant mineral nutrition: Combinatorial nutrient stresses reveal emergent properties. *Trends in Plant Science, 24*, 542–552. https://doi.org/10.1016/j.tplants.2019.03.008.

Caione, G., Prado, R. de M., Campos, C. N. S., Rosatto Moda, L., de Lima Vasconcelos, R., & Pizauro Júnior, J. M. (2015). Response of sugarcane in a red ultisol to phosphorus rates, phosphorus sources, and filter cake. *The Scientific World Journal, 2015*, 1–10. https://doi.org/10.1155/2015/405970.

Camargo, M. S., & Keeping, M. G. (2021). Silicon in sugarcane: Availability in soil, fertilization, and uptake. *Silicon, 13*, 3691–3701. https://doi.org/10.1007/S12633-020-00935-Y.

Camargo, M. S., Baltieri, G. J., Santos, H. L., Carnietto, M. R. A., dos Reis, A. R., Pacheco, A. C., & de Almeida Silva, M. (2022). Silicon fertilization enhances photosynthetic activity and sugar metabolism in sugarcane cultivars under water deficit at the ripening phase. *Silicon, 15*, 3021–3033. https://doi.org/10.1007/s12633-022-02236-y.

Camargo, M. S., Bezerra, B. K. L., Holanda, L. A., Oliveira, A. L., Vitti, A. C., & Silva, M. A. (2019). Silicon fertilization improves physiological responses in sugarcane cultivars grown under water deficit. *Journal of Soil Science and Plant Nutrition, 19*, 81–91. https://doi.org/10.1007/S42729-019-0012-1.

Camargo, M. S., Bezerra, B. K. L., Vitti, A. C., Silva, M. A., & Oliveira, A. L. (2017). Silicon fertilization reduces the deleterious effects of water deficit in sugarcane. *Journal of Soil Science and Plant Nutrition, 17*, 99–111. https://doi.org/10.4067/S0718-95162017005000008.

Campos, C. N. S., Silva Júnior, G. B. da, Prado, R. de M., David, C. H. O. de, Souza Junior, J. P. de, & Teodoro, P. E. (2020). Silicon mitigates ammonium toxicity in plants. *Agronomy Journal, 112*, 635–647. https://doi.org/10.1002/agj2.20069.

Cardozo, N. P., de Oliveira Bordonal, R., & la Scala, N. (2018). Sustainable intensification of sugarcane production under irrigation systems, considering climate interactions and agricultural efficiency. *Journal of Cleaner Production, 204*, 861–871. https://doi.org/10.1016/j.jclepro.2018.09.004.

Chen, D., Cao, B., Wang, S., Liu, P., Deng, X., Yin, L., & Zhang, S. (2016). Silicon moderated the K deficiency by improving the plant-water status in sorghum. *Scientific Reports, 6*, 22882. https://doi.org/10.1038/srep22882.

Chen, W., Yao, X., Cai, K., & Chen, J. (2011). Silicon alleviates drought stress of rice plants by improving plant water status, photosynthesis and mineral nutrient absorption. *Biological Trace Element Research, 142*, 67–76. https://doi.org/10.1007/s12011-010-8742-x.

Cheong, Y. W. Y., & Chan, P. Y. (1973). Incorporation of P32 in phosphate esters of the sugar cane plant and the effect of Si and Al on the distribution of these esters. *Plant and Soil, 38*, 113–123. https://doi.org/10.1007/BF00011221.

Cherubin, M. R., Karlen, D. L., Franco, A. L. C., Tormena, C. A., Cerri, C. E. P., Davies, C. A., & Cerri, C. C. (2016). Soil physical quality response to sugarcane expansion in Brazil. *Geoderma, 267*, 156–168. https://doi.org/10.1016/j.geoderma.2016.01.004.

Cornelis, J. T., Titeux, H., Ranger, J., & Delvaux, B. (2011). Identification and distribution of the readily soluble silicon pool in a temperate forest soil below three distinct tree species. *Plant and Soil, 342*, 369–378. https://doi.org/10.1007/s11104-010-0702-x.

Costa, M. G., dos Santos, Sarah M. M., de Mello Prado, R., Palaretti, L. F., de Cássia Piccolo, M., & de Souza Júnior, J. P. (2022). Impact of Si on C, N, and P stoichiometric homeostasis favors nutrition and stem dry mass accumulation in sugarcane cultivated in tropical soils with different water regimes. *Frontiers in Plant Science, 13*, 949909. https://doi.org/10.3389/fpls.2022.949909.

Costa, M. G., Prado, R. de M., Santos, Sarah M. M., Palaretti, L. F., Piccolo, M. de C., & Souza Júnior, J. P. (2023). New approaches to the effects of Si on sugarcane ratoon under irrigation in Quartzipsamments, Eutrophic Red Oxisol, and Dystrophic Red Oxisol. *BMC Plant Biology, 23*, 51. https://doi.org/10.1186/s12870-023-04077-2.

Costa, M.G., Prado, R. M., Palaretti, L. G., Souza Júnior, J. P. (2024) The effect of abiotic stresses on plant C:N:P homeostasis and their mitigation by silicon. *The Crop Journal*, 12(2), 340-353. https://doi.org/10.1016/j.cj.2023.11.012.

Deus, A. C. F., Prado, R. de M., Alvarez, R. de CF., Oliveira, R. LL. de, & Felisberto, G. (2020). Role of silicon and salicylic acid in the mitigation of nitrogen deficiency stress in rice plants. *Silicon, 12*, 997–1005. https://doi.org/10.1007/s12633-019-00195-5.

Dingre, S. K., Gorantiwar, S. D., & Kadam, S. A. (2021). Correlating the field water balance derived crop coefficient (Kc) and canopy reflectance-based NDVI for irrigated sugarcane. *Precision Agriculture, 22*, 1134–1153. https://doi.org/10.1007/s11119-020-09774-8.

Dubey, R. S., & Singh, A. K. (1999). Salinity induces accumulation of soluble sugars and alters the activity of sugar metabolizing enzymes in rice plants. *Biologia Plantarum, 42*, 233–239. https://doi.org/10.1023/A:1002160618700.

Epstein, E. (1994). The anomaly of silicon in plant biology. *Proceedings of the National Academy of Sciences of the United States of America, 91*, 11–17. https://doi.org/10.1073/pnas.91.1.11.

Etesami, H., & Jeong, B. R. (2018). Silicon (Si): Review and future prospects on the action mechanisms in alleviating biotic and abiotic stresses in plants. *Ecotoxicology and Environmental Safety, 147*, 881–896. https://doi.org/10.1016/j.ecoenv.2017.09.063.

Farooq, M. A., & Dietz, K-J. (2015). Silicon as versatile player in plant and human biology: Overlooked and poorly understood. *Frontiers in Plant Science, 6*, 165767. https://doi.org/10.3389/fpls.2015.00994..

Ferreira Barreto, R., Schiavon Júnior, A. A., Maggio, M. A., & de Mello Prado, R. (2017). Silicon alleviates ammonium toxicity in cauliflower and in broccoli. *Scientia Horticulturae, 225*, 743–750. https://doi.org/10.1016/j.scienta.2017.08.014.

Fiala, R., Fialová, I., Vaculík, M., & Luxová, M. (2021). Effect of silicon on the young maize plants exposed to nickel stress. *Plant Physiology and Biochemistry, 166*, 645–656. https://doi.org/10.1016/j.plaphy.2021.06.026.

Firouzabadi, A. G., Baghani, J., Jovzi, M., & Albaji, M. (2021). Effects of wheat row spacing layout and drip tape spacing on yield and water productivity in sandy clay loam soil in a semi-arid region. *Agricultural Water Management, 251*, 106868. https://doi.org/10.1016/j.agwat.2021.106868.

Fleck, A. T., Nye, T., Repenning, C., Stahl, F., Zahn, M., & Schenk, M. K. (2011). Silicon enhances suberization and lignification in roots of rice (*Oryza sativa*). *Journal of Experimental Botany, 62*, 2001–2011. https://doi.org/10.1093/jxb/erq392.

Franco, H. C. J., Bologna, I. R., Faroni, C. E., Vitti, A. C., & Trivelin, P. C. O. (2007). Acúmulo de macronutrientes em cana-de-açúcar em função da adubação nitrogenada e dos resíduos culturais incorporados ao solo no plantio. *Bragantia, 66*, 669–674. https://doi.org/10.1590/S0006-87052007000400017.

Frazão, J. J., Prado, R. de M., Souza Júnior, J. P., & Rossatto, D. R. (2020). Silicon changes C:N:P stoichiometry of sugarcane and its consequences for photosynthesis, biomass partitioning and plant growth. *Scientific Reports, 10*, 1–10. https://doi.org/10.1038/s41598-020-69310-6.

Fu, J., & Huang, B. (2001). Involvement of antioxidants and lipid peroxidation in the adaptation of two cool-season grasses to localized drought stress. *Environmental and Experimental Botany, 45*, 105–114. https://doi.org/10.1016/S0098-8472(00)00084-8.

Gratão, P. L., Polle, A., Lea, P. J., & Azevedo, R. A. (2005). Making the life of heavy metal-stressed plants a little easier. *Functional Plant Biology, 32*, 481–494. https://doi.org/10.1071/FP05016.

Hao, Q., Yang, S., Song, Z., Li, Z., Ding, F., Yu, C., Hu, G., & Liu, H. (2020). Silicon affects plant stoichiometry and accumulation of C, N, and P in grasslands. *Frontiers in Plant Science, 11*, 1304. https://doi.org/10.3389/fpls.2020.01304.

Hassan, M. U., Chattha, M. U., Khan, I., Chattha, M. B., Aamer, M., Nawaz, M., Ali, A., Khan, M. A. U., & Khan, T. A. (2019). Nickel toxicity in plants: reasons, toxic effects, tolerance mechanisms, and remediation possibilities—a review. *Environmental Science and Pollution Research, 26*, 12673–12688. https://doi.org/10.1007/s11356-019-04892-x.

Havaux, M. (2014). Carotenoid oxidation products as stress signals in plants. *The Plant Journal, 79*, 597–606. https://doi.org/10.1111/tpj.12386.

He, C., Ma, J., & Wang, L. (2015). A hemicellulose-bound form of silicon with potential to improve the mechanical properties and regeneration of the cell wall of rice. *New Phytologist, 206*, 1051–1062. https://doi.org/10.1111/nph.13282.

Ibrahim, M. F. M., El-Samad, G. A., Ashour, H., El-Sawy, A. M., Hikal, M., Elkelish, A., El-Gawad, H. A., El-Yazied, A. A., Hozzein, W. N., & Farag, R. (2020). Regulation of agronomic traits, nutrient uptake, osmolytes and antioxidants of maize as influenced by exogenous potassium silicate under deficit irrigation and semiarid conditions. *Agronomy, 10*, 1212. https://doi.org/10.3390/agronomy10081212.

Imtiaz, M., Rizwan, M. S., Mushtaq, M. A., Ashraf, M., Shahzad, S. M., Yousaf, B., Saeed, D. A., Rizwan, M., Nawaz, M. A., Mehmood, S., & Tu, S. (2016). Silicon occurrence, uptake, transport and mechanisms of heavy metals, minerals and salinity enhanced tolerance in plants with future prospects: A review. *Journal of Environmental Management, 183*, 521–529. https://doi.org/10.1016/j.jenvman.2016.09.009.

Isa, M., Bai, S., Yokoyama, T., Ma, J. F., Ishibashi, Y., Yuasa, T., & Iwaya-Inoue, M. (2010). Silicon enhances growth independent of silica deposition in a low-silica rice mutant, lsi1. *Plant and Soil, 331*, 361–375. https://doi.org/10.1007/s11104-009-0258-9.

Jain, R., Chandra, A., Venugopalan, V., & Solomon, S. (2015). Physiological changes and expression of SOD and P5CS genes in response to water deficit in sugarcane. *Sugar Tech, 17*, 276–282. https://doi.org/10.1007/s12355-014-0317-2

Jain, R., Singh, A., Jain, N., Tripathi, P., Chandra, A., Shukla, S. K., & Solomon, S. (2018). Response of silixol sugarcane to growth and physio-biochemical characteristics of sugarcane. *Sugar Tech, 20*, 39–44. https://doi.org/10.1007/s12355-017-0557-z.

Jain, R., Singh, S. P., Singh, A., Singh, S., Tripathi, P., Kishor, R., & Solomon, S. (2016). Changes in growth, yield, juice quality and biochemical attributes of sugarcane in response to orthosilicic acid granules. *Sugar Tech, 19*, 300–304. https://doi.org/10.1007/s12355-016-0469-3.

Kim, Y. -H., Khan, A. L., Waqas, M., & Lee, I.-J. (2017). Silicon Regulates Antioxidant Activities of Crop Plants under Abiotic-Induced Oxidative Stress: A Review. *Frontiers in Plant Science, 8*, 256717. https://doi.org/10.3389/fpls.2017.00510.

Klotzbücher, T., Klotzbücher, A., Kaiser, K., Vetterlein, D., Jahn, R., & Mikutta, R. (2018). Variable silicon accumulation in plants affects terrestrial carbon cycling by controlling lignin synthesis. *Global Change Biology, 24*, e183–e189. https://doi.org/10.1111/gcb.13845.

Kostic, L., Nikolic, N., Bosnic, D., Samardzic, J., & Nikolic, M. (2017). Silicon increases phosphorus (P) uptake by wheat under low P acid soil conditions. *Plant and Soil, 419*, 447–455. https://doi.org/10.1007/S11104-017-3364-0/FIGURES/4.

Kumar, S., Kumar, S., & Mohapatra, T. (2021). Interaction between macro- and micro-nutrients in plants. *Frontiers in Plant Science, 12*, 665583. https://doi.org/10.3389/fpls.2021.665583.

Li, Z., Song, Z., Yang, X., Song, A., Yu, C., Wang, T., Xia, S., & Liang, Y. (2018). Impacts of silicon on biogeochemical cycles of carbon and nutrients in croplands. *Journal of Integrative Agriculture, 17*, 2182–2195. https://doi.org/10.1016/S2095-3119(18)62018-0.

Liang, Y., Zhang, W., Chen, Q., Liu, Y., & Ding, R. (2006). Effect of exogenous silicon (Si) on H+-ATPase activity, phospholipids and fluidity of plasma membrane in leaves of salt-stressed barley (*Hordeum vulgare* L.). *Environmental and Experimental Botany, 57*, 212–219. https://doi.org/10.1016/j.envexpbot.2005.05.012.

Liu, P., Yin, L., Wang, S., Zhang, M., Deng, X., Zhang, S., & Tanaka, K. (2015). Enhanced root hydraulic conductance by aquaporin regulation accounts for silicon alleviated salt-induced osmotic stress in *Sorghum bicolor* L. *Environmental and Experimental Botany, 111*, 42–51. https://doi.org/10.1016/j.envexpbot.2014.10.006.

Majumdar, S., & Prakash, N. B. (2020). An overview on the potential of silicon in promoting defence against biotic and abiotic stresses in sugarcane. *Journal of Soil Science and Plant Nutrition, 20*, 1969–1998. https://doi.org/10.1007/s42729-020-00269-z.

Manivannan, A., & Ahn, Y. -K. (2017). Silicon regulates potential genes involved in major physiological processes in plants to combat stress. *Frontiers in Plant Science, 8*, 275701. https://doi.org/10.3389/fpls.2017.01346.

Maoka, T. (2020). Carotenoids as natural functional pigments. *Journal of Natural Medicines, 74*, 1–16. https://doi.org/10.1007/s11418-019-01364-x.

Martineau, E., Domec, J. C., Bosc, A., Dannoura, M., Gibon, Y., Bénard, C., & Jordan-Meille, L. (2017). The role of potassium on maize leaf carbon exportation under drought condition. *Acta Physiologiae Plantarum, 39*, 1–13. https://doi.org/10.1007/s11738-017-2515-5.

Mengel, K., Kirkby, E. A., Kosegarten, H., & Appel, T. (2001). Plant Water Relationships. In: Mengel K, Kirkby EA, Kosegarten H, Appel T (eds) Principles of Plant Nutrition. Springer Netherlands, Dordrecht, pp. 181–242.

Misra, P., Nath, K., & Tandon, P. K. (2010). Effect of heavy metals (Ni and Pb) stress on sugarcane (*Saccharum officinarum* L.). *Research in Environment and Life Sciences, 3*, 193–188.

Mitani, N., Jian, F. M., & Iwashita, T. (2005). Identification of the silicon form in xylem sap of rice (*Oryza sativa* L.). *Plant, Cell & Environment, 46*, 279–283. https://doi.org/10.1093/pcp/pci018.

Moda, L. R., Prado, R. M., Caione, G., Campos, C. N. S., Silva, E. C., & Flores, R. A. (2015). Effect of sources and rates of phosphorus associated with filter cake on sugarcane nutrition and yield. *Australian Journal of Crop Science, 9*, 477–485.

Nawi, N. M., Chen, G., & Jensen, T. (2014). In-field measurement and sampling technologies for monitoring quality in the sugarcane industry: A review. *Precision Agriculture, 15*, 684–703. https://doi.org/10.1007/s11119-014-9362-9.

Neethirajan, S., Gordon, R., & Wang, L. (2009). Potential of silica bodies (*phytoliths*) for nanotechnology. *Trends in Biotechnology, 27*, 461–467. https://doi.org/10.1016/j.tibtech.2009.05.002.

Neu, S., Schaller, J., & Dudel, E. G. (2017). Silicon availability modifies nutrient use efficiency and content, C:N:P stoichiometry, and productivity of winter wheat (*Triticum aestivum* L.). *Scientific Reports, 7*, 1–8. https://doi.org/10.1038/srep40829.

Nowaki, R. H. D., Parent, S. -É., Cecílio Filho, A. B., Rozane, D. E., Meneses, N. B., Silva, J. A. dos S. da, Natale, W., & Parent, L. E. (2017). Phosphorus over-fertilization and nutrient misbalance of irrigated tomato crops in Brazil. *Frontiers in Plant Science, 8*, 825. https://doi.org/10.3389/fpls.2017.00825.

Oliveira Filho, A. S. B., de Mello Prado, R., Teixeira, G. C. M., de Cássia Piccolo, M., & Rocha, A. M. S. (2021a). Water deficit modifies C:N:P stoichiometry affecting sugarcane and energy cane yield and its relationships with silicon supply. *Scientific Reports, 11*, 20916. https://doi.org/10.1038/s41598-021-00441-0.

Oliveira Filho, A. S. B., de Mello Prado, R., Teixeira, G. C. M., Rocha, A. M. S., de Souza Junior, J. P., de Cássia Piccolo, M., & Rocha, J. R. (2021b). Silicon attenuates the effects of water deficit in sugarcane by modifying physiological aspects and C:N:P stoichiometry and its use efficiency. *Agricultural Water Management, 255*, 107006. https://doi.org/10.1016/J.AGWAT.2021.107006.

Oliveira Filho, A. S. B., Prado, R. de M., Teixeira, G. C. M., Cássia Piccolo, M., & Rocha, A. M. S. (2021c). Water deficit modifies C:N:P stoichiometry affecting sugarcane and energy cane yield and its relationships with silicon supply. *Scientific Reports, 11*, 1–10. https://doi.org/10.1038/s41598-021-00441-0.

Oliveira Filho, A. S. B., Prado, R. de M., Teixeira, G. C. M., Rocha, A. M. S., Souza Junior, J. P., Cássia Piccolo, M., & Rocha, J. R. (2021d). Silicon attenuates the effects of water deficit in sugarcane by modifying physiological aspects and C:N:P stoichiometry and its use efficiency. *Agricultural Water Management, 255*, 107006. https://doi.org/10.1016/J.AGWAT.2021.107006.

Oliveira, K. S., de Mello Prado, R., Checchio, M. V., & Gratão, P. L. (2021). Silicon via nutrient solution modulates deficient and sufficient manganese sugar and energy cane antioxidant systems. *Scientific Reports, 11*, 16900. https://doi.org/10.1038/s41598-021-96427-z.

Oliveira, K. S., de Mello Prado, R., Checchio, M. V., & Gratão, P. L. (2022). Interaction of silicon and manganese in nutritional and physiological aspects of energy cane with high fiber content. *BMC Plant Biology, 22*, 374. https://doi.org/10.1186/s12870-022-03766-8.

Parent, L. E., & Dafir, M. (1992). A theoretical concept of compositional nutrient diagnosis. *Journal of the American Society for Horticultural Science, 117*, 239–242. https://doi.org/10.21273/JASHS.117.2.239.

Parveen, A., Liu, W., Hussain, S., Asghar, J., Perveen, S., & Xiong, Y. (2019). Silicon priming regulates morpho-physiological growth and oxidative metabolism in maize under drought stress. *Plants, 8*, 431. https://doi.org/10.3390/plants8100431.

Pei, Z. F., Ming, D. F., Liu, D., Wan, G. L., Geng, X. X., Gong, H. J., & Zhou, W. J. (2010). Silicon improves the tolerance to water-deficit stress induced by polyethylene glycol in wheat (*Triticum aestivum* L.) seedlings. *Journal of Plant Growth Regulation, 29*, 106–115. https://doi.org/10.1007/s00344-009-9120-9.

Pissolato, M. D., Silveira, N. M., Machado, E. C., Zambrosi, F. C. B., Sodek, L., & Ribeiro, R. V. (2019). Photosynthesis and biomass accumulation in young sugarcane plants grown under increasing ammonium supply in nutrient solution. *Theoretical and Experimental Plant Physiology, 31*, 401–411. https://doi.org/10.1007/s40626-019-00154-w.

Rampazzo, M. V., Cunha, M. L. O., de Oliveira, L. C. A., Silva, V. M., Lanza, M. G. D. B., de Melo, A. A. R., & dos Reis, A. R. (2022). Physiological roles of nickel on antioxidant and nitrogen metabolism increasing the yield of sugarcane plants. *Journal of Soil Science and Plant Nutrition, 22*, 4438–4448. https://doi.org/10.1007/s42729-022-01045-x.

Rizwan, M., Ali, S., Ibrahim, M., Farid, M., Adrees, M., Bharwana, S. A., Zia-ur-Rehman, M., & Qayyum, M. F. (2015). Mechanisms of silicon-mediated alleviation of drought and salt stress in plants: A review. *Environmental Science and Pollution Research, 22*, 15416–15431. https://doi.org/10.1007/s11356-015-5305-x.

Rocha, J. R., de Mello Prado, R., Teixeira, G. C. M., & de Oliveira Filho, A. S. B. (2021). Si fertigation attenuates water stress in forages by modifying carbon stoichiometry, favouring physiological aspects. *Journal of Agronomy and Crop Science, 207*, 631–643. https://doi.org/10.1111/jac.12479.

Samuels, G., & Alexander, A. G. (1969). Influence of variable manganese and silicon on the nutrition, sugar production, and enzyme activity of immature sugarcane. *The Journal of Agriculture of the University of Puerto Rico, 53*, 14–27. https://doi.org/10.46429/jaupr.v53i1.11173.

Santos, S. M. M., Prado, R. de M., Souza Júnior, J. P., Teixeira, G. C. M., Duarte, J. C. dos S., & Medeiros, R. L. S. (2021). Silicon supplied via foliar application and root to attenuate potassium deficiency in common bean plants. *Scientific Reports, 11*, 19690. https://doi.org/10.1038/s41598-021-99194-z.

Santos, S. M. M., Prado, R. de M., Teixeira, G. C. M., Souza Júnior, J. P., Medeiros, R. L. S., & Barreto, R. F. (2022). Silicon supplied via roots or leaves relieves potassium deficiency in maize plants. *Silicon, 14*, 773–782. https://doi.org/10.1007/s12633-020-00908-1.

Savant, N. K., Korndörfer, G. H., Datnoff, L. E., & Snyder, G. H. (1999). Silicon nutrition and sugarcane production: A review. *Journal of Plant Nutrition, 22*, 1853–1903. https://doi.org/10.1080/01904169909365761.

Schaller, J., Brackhage, C., Gessner, M. O., Bäuker, E., & Dudel, E. G. (2012). Silicon supply modifies C:N:P stoichiometry and growth of *Phragmites australis*. *Plant Biology, 14*, 392–396. https://doi.org/10.1111/j.1438-8677.2011.00537.x.

Schaller, J., Cramer, A., Carminati, A., & Zarebanadkouki, M. (2020). Biogenic amorphous silica as main driver for plant available water in soils. *Scientific Reports, 10*, 1–7. https://doi.org/10.1038/s41598-020-59437-x.

Schaller, J., Schoelynck, J., Struyf, E., & Meire, P. (2016). Silicon affects nutrient content and ratios of wetland plants. *Silicon, 8*, 479–485. https://doi.org/10.1007/s12633-015-9302-y.

Segato, S. V. A. S., Pinto, E., & Jendiroba, J. C. M. (2006). *Atualização em produção de cana-de-açúcar.* Gráfica Prol. Piracicaba.

Shen, X. F., Zhao, Z. H., & Chen, Y. (2019). Effects of intercropping with peanut and silicon application on sugarcane growth, yield and quality. *Sugar Tech, 21*, 437–443. https://doi.org/10.1007/s12355-018-0667-2.

Silva, D. L. da, Prado, R. de M., Tenesaca, L. F. L., Silva, J. L. F. da, & Mattiuz, B.-H. (2021). Silicon attenuates calcium deficiency by increasing ascorbic acid content, growth and quality of cabbage leaves. *Scientific Reports, 11*, 1770. https://doi.org/10.1038/s41598-020-80934-6.

Silva, G. P. da, & Chiaia, H. L. J. (2021). Limitation due to nutritional deficiency and excess in sugarcane using the integral diagnosis and recommendation system (DRIS) and nutritional composition diagnosis (CND). *Communications in Soil Science and Plant Analysis, 52*, 1458–1467. https://doi.org/10.1080/00103624.2021.1885690.

Silva, G. P. da, Prado, R. de M., Wadt, P. G. S., Moda, L. R., & Caione, G. (2020). Accuracy of nutritional diagnostics for phosphorus considering five standards by the method of diagnosing nutritional composition in sugarcane. *Journal of Plant Nutrition, 43*, 1485–1497. https://doi.org/10.1080/01904167.2020.1730902.

Silva, V. S. G., Oliveira, M. W. de, Ferreira, V. M., Oliveira, T. B. A., Galvão, E. R., Silva, A. F. da, & Machado, P. A. S. (2018). Nutritional requirement of sugarcane cultivars. *Journal of Agricultural Science, 10*, 361. https://doi.org/10.5539/jas.v10n4p361.

Smirnoff, N. (1993). The role of active oxygen in the response of plants to water deficit and desiccation. *New Phytologist, 125*, 27–58. https://doi.org/10.1111/j.1469-8137.1993.tb03863.x.

Song, X. P., Verma, K. K., Tian, D. D., Zhang, X. Q., Liang, Y. J., Huang, X., Li, C. N., & Li, Y. R. (2021). Exploration of silicon functions to integrate with biotic stress tolerance and crop improvement. *Biological Research, 54*, 19. https://doi.org/10.1186/s40659-021-00344-4.

Song, Z., Liu, H., Zhao, F., & Xu, C. (2014). Ecological stoichiometry of N:P:Si in China's grasslands. *Plant and Soil, 380*, 165–179. https://doi.org/10.1007/s11104-014-2084-y.

Souri, Z., Khanna, K., Karimi, N., & Ahmad, P. (2020). Silicon and plants: current knowledge and future prospects. *Journal of Plant Growth Regulation, 40*, 906–925. https://doi.org/10.1007/S00344-020-10172-7.

Sousa Junior, G. da S., Hurtado, A. C., Souza Junior, J. P. de, Prado, R. de M., Piccolo, M. de C., & Santos, D. M. M. dos. (2022). Beneficial role of silicon on regulating C, N, and P stoichiometric homeostasis and the growth of sugarcane seedlings under aluminum toxicity. *Journal of Soil Science and Plant Nutrition. 22, 4138*–4152. https://doi.org/10.1007/s42729-022-01013-5.

Sousa Junior, G. da S., Hurtado, A. C., Souza Junior, J. P., Prado, R. de M., & Santos, D. M. M. dos. (2021). Nutritional and structural role of silicon in attenuating aluminum toxicity in sugarcane plants. *Silicon. 14, 5041*–5055. https://doi.org/10.1007/s12633-021-01294-y.

Souza Júnior, J. P., Oliveira, T. L., Mello Prado, R., Oliveira, K. R., & Soares, M. B. (2022). Analyzing the role of silicon in leaf C:N:P stoichiometry and its effects on nutritional efficiency and dry weight production in two sugarcane cultivars. *Journal of Soil Science and Plant Nutrition. 22, 2687*–2694. https://doi.org/10.1007/s42729-022-00836-6.

Teixeira, G. C. M., Mello Prado, R., Rocha, A. M. S., Santos, L. C. N., Santos, S. M. M., Gratão, P. L., & Fernandes, C. (2020a). Silicon in pre-sprouted sugarcane seedlings mitigates the effects of water deficit after transplanting. *Journal of Soil Science and Plant Nutrition, 20*, 849–859. https://doi.org/10.1007/S42729-019-00170-4.

Teixeira, G. C. M., Prado, R. de M., de Oliveira, L. T., de Castro Souza, J. V., & Rocha, A. M. S. (2022a). Silicon fertigation with appropriate source reduces water requirement of maize under water deficit. *Plant and Soil, 477*, 83–97. https://doi.org/10.1007/s11104-022-05446-w.

Teixeira, G. C. M., Prado, R. de M., Rocha, A. M. S., & Piccolo, M. de C. (2022b). Silicon as a sustainable option to increase biomass with less water by inducing carbon:nitrogen:phosphorus stoichiometric homeostasis in sugarcane and energy cane. *Frontiers in Plant Science, 13*, 826512. https://doi.org/10.3389/fpls.2022.826512.

Teixeira, G. C. M., Prado, R. M., & Rocha, A. M. S. (2021b). Low absorption of silicon via foliar in comparison to root application has an immediate antioxidant effect in mitigating water deficit damage in sugarcane. *Journal of Agronomy and Crop Science*, *Journal of Agronomy and Crop Science*, *208*(6), 805–814. https://doi.org/10.1111/JAC.12511.

Teixeira, G. C. M., Prado, R. M., Rocha, A. M. S., & Piccolo, M. de C. (2020b). Root- and foliar-applied silicon modifies C:N:P ratio and increases the nutritional efficiency of pre-sprouted sugarcane seedlings under water deficit. *PLOS ONE, 15*(12), e0240847. https://doi.org/10.1371/JOURNAL.PONE.0240847.

Teixeira, G. C. M., Prado, R. M., Rocha, A. M. S., Sousa Junior, G. da S., & Gratão, P. L. (2021a). Beneficial effect of silicon applied through fertigation attenuates damage caused by water deficit in sugarcane. *Journal of Plant Growth Regulation, 41*, 1–16. https://doi.org/10.1007/S00344-021-10510-3/FIGURES/10.

Tombeur, F. de, vander Linden, C., Cornélis, J. -T., Godin, B., Compère, P., & Delvaux, B. (2020). Soil and climate affect foliar silicification patterns and silica-cellulose balance in sugarcane (*Saccharum officinarum*). *Plant and Soil, 452*, 529–546. https://doi.org/10.1007/s11104-020-04588-z.

Tubaña, B. S., & Heckman, J. R. (2015). Silicon in soils and plants. In: B. S. Tubaña & J. R. Heckman (Eds.), *Silicon and Plant Diseases* (pp. 7–51). Springer International Publishing.

Ulloa, M., Nunes-Nesi, A., da Fonseca-Pereira, P., Poblete-Grant, P., Reyes-Díaz, M., & Cartes, P. (2021). The effect of silicon supply on photosynthesis and carbohydrate metabolism in two wheat (*Triticum aestivum* L.) cultivars contrasting in response to phosphorus nutrition. *Plant Physiology and Biochemistry, 169*, 236–248. https://doi.org/10.1016/j.plaphy.2021.11.022.

Verma, K. K., Anas, M., Chen, Z., Rajput, V. D., Malviya, M. K., Verma, C. L., Singh, R. K., Singh, P., Song, X. P., & Li, Y. R. (2020). Silicon supply improves leaf gas exchange, antioxidant defense system and growth in sugarcane responsive to water limitation. *Plants, 9*, 1032. https://doi.org/10.3390/plants9081032.

Verma, K. K., Kováč, J., Fialová, I., Fiala, R., Jašková, K., & Luxová, M. (2021). Multiple effects of silicon on alleviation of nickel toxicity in young maize roots. *Journal of Hazardous Materials, 415*, 125570. https://doi.org/10.1016/j.jhazmat.2021.125570.

Verma, K. K., Li, D. M., Singh, M., Rajput, V. D., Malviya, M. K., Minkina, T., Singh, R. K., Singh, P., Song, X. P., & Li, Y. R. (2020). Interactive role of silicon and plant-rhizobacteria mitigating abiotic stresses: A new approach for sustainable agriculture and climate change. *Plants, 9*, 1055. https://doi.org/10.3390/plants9091055.

Verma, K. K., Liu, X. H., Wu, K. C., Singh, R. K., Song, Q. Q., Malviya, M. K., Song, X. P., Singh, P., Verma, C. L., & Li, Y. R. (2020). The impact of silicon on photosynthetic and biochemical responses of sugarcane under different soil moisture levels. *Silicon, 12*, 1355–1367. https://doi.org/10.1007/s12633-019-00228-z.

Verma, K. K., Singh, P., Song, X. P., Malviya, M. K., Singh, R. K., Chen, G. L., Solomon, S., & Li, Y. R. (2020). Mitigating climate change for sugarcane improvement: Role of silicon in alleviating abiotic stresses. *Sugar Tech, 22*, 741–749. https://doi.org/10.1007/S12355-020-00831-0.

Verma, K. K., Singh, R. K., Song, Q. Q., Singh, P., Zhang, B. Q., Song, X. P., Chen, G. L., & Li, Y. R. (2019). Silicon alleviates drought stress of sugarcane plants by improving antioxidant responses. *Biomedical Journal of Scientific & Technical Research, 17*, 1–7. https://doi.org/10.26717/BJSTR.2019.17.002957.

Verma, K. K., Song, X. P., Budeguer, F., Nikpay, A., Enrique, R., Singh, M., Zhang, B., Wu, J. M., & Li, Y. R. (2022). Genetic engineering: an efficient approach to mitigating biotic and abiotic stresses in sugarcane cultivation. *Plant Signaling & Behavior, 17*(1), e2108253.

Verma, K. K., Song, X. P., Li, D. M., Singh, M., Wu, J. M., Singh, R. K., Sharma, A., Zhang, B. Q., & Li, Y. R. (2022). Silicon and soil microorganisms improve rhizospheric soil health with bacterial community, plant growth, performance and yield. *Plant Signaling & Behavior, 17*(1), e2104004.

Verma, K. K., Song, X. P., Lin, B., Guo, D. J., Singh, M., Rajput, V. D., Singh, R. K., Singh, P., Sharma, A., Malviya, M. K., Chen, G. L., & Li, Y. R. (2021). Silicon induced drought tolerance in crop plants: physiological adaptation strategies. *Silicon*, *14*, 2473–2487. DOI: 10.1007/s12633-021-01071-x.

Verma, K. K., Song, X. P., Rajput, V. D., Solomon, S., Li, Y. R., & Rao, G. P. (Eds.). (2023). Agro-industrial Perspectives on Sugarcane Production under Environmental Stress. Springer, Singapore. https://doi.org/10.1007/978-981-19-3955-6_1.

Verma, K. K., Song, X. P., Tian, D. D., Guo, D. J., Chen, Z. L., Zhong, C. S., Nikpay, A., Singh, M., Rajput, V. D., Singh, R. K., Minkina, T., & Li, Y. R. (2021). Influence of silicon on biocontrol strategies to manage biotic stress for crop protection, performance and improvement. *Plants, 10*, 2163. https://doi.org/10.3390/plants10102163.

Verma, K. K., Song, X. P., Tian, D. D., Singh, M., Verma, C. L., Rajput, V. D., Singh, R. K., Sharma, A., Singh, P., Malviya, M. K., & Li, Y. R. (2021). Investigation of defensive role of silicon during drought stress induced by irrigation capacity in sugarcane: physiological and biochemical characteristics. *ACS Omega, 6*, 19811–19821.

Verma, K. K., Song, X. P., Verma, C. L., Chen, Z. L., Rajput, V. D., Wu, K. C., Liao, F., Chen, G. L., & Li, Y. R. (2021). Functional relationship between photosynthetic leaf gas exchange in response to silicon application and water stress mitigation in sugarcane. *Biological Research, 54*, 15. https://doi.org/10.1186/s40659-021-00338-2.

Verma, K. K., Song, X. P., Verma, C. L., Malviya, M. K., Guo, D. J., Rajput, V. D., Sharma, A., Wei, K. J., Chen, G. L., Solomon, S., & Li, Y. R. (2021). Prediction of photosynthetic leaf gas exchange of sugarcane (Saccharum spp.) leaves in response to leaf positions to foliar spray of potassium salt of active phosphorus under limited water irrigation. *ACS Omega, 6*, 2396–2409.

Verma, K. K., Song, X. P., Yadav, G., Degu, H. D., Parvaiz, A., Singh, M., Huang, H. R., Mustafa, G., Xu, L., & Li, Y. R. (2022). Impact of agroclimatic variables on proteogenomics in sugarcane (Saccharum spp.) plant productivity. *ACS Omega, 7*, 22997–23008.

Verma, K. K., Song, X. P., Zeng, Y., Guo, D. J., Singh, M., Rajput, V. D., Malviya, M. K., Wei, K. J., Sharma, A., Li, D. P., Chen, G. L., & Li, Y. R. (2021). Foliar application of silicon boosts growth, photosynthetic leaf gas exchange, antioxidative response and resistance to limited water irrigation in sugarcane (Saccharum officinarum L.). *Plant Physiology and Biochemistry, 166*, 582–592. https://doi.org/10.1016/j.plaphy.2021.06.032.

Verma, K. K., Song, X. P., Zeng, Y., Li, D. M., Guo, D. J., Rajput, V. D., Chen, G. L., Barakhov, A., Minkina, T. M., & Li, Y. R. (2020). Characteristics and correlation of leaf stomata and its relationship with photosynthesis on Saccharum spp. under different irrigation and silicon application. *ACS Omega, 5*, 24145–24153.

Verma, K. K., Wu, K. C., Singh, P., Malviya, M. K., Singh, R. K., Song, X. P., & Li, Y. R. (2019). The protective role of silicon in sugarcane under water stress: Photosynthesis and antioxidant enzymes. *Biomedical Journal of Scientific & Technical Research, 15*, 1–7. https://doi.org/10.26717/BJSTR.2019.15.002685.

Verma, K. K., Wu, K. C., Verma, C. L., Li, D. M., Malviya, M. K., Singh, R. K., Singh, P., Chen, G. L., Song, X. P., & Li, Y. R. (2020). Developing mathematical model for diurnal dynamics on photosynthetic responses in sugarcane responsive to different irrigation and silicon application. *PeerJ, 8*, e10154. DOI: 10.7717/peerj.10154.

Viciedo, D. O., de Mello Prado, R., Lizcano Toledo, R., dos Santos, L. C. N., Calero Hurtado, A., Nedd, L. L. T., & Castellanos Gonzalez, L. (2019). Silicon supplementation alleviates ammonium toxicity in sugar beet (*Beta vulgaris* L.). *Journal of Soil Science and Plant Nutrition, 19*, 413–419. https://doi.org/10.1007/s42729-019-00043-w.

Woesz, A., Weaver, J. C., Kazanci, M., Dauphin, Y., Aizenberg, J., Morse, D. E., & Fratzl, P. (2006). Micromechanical properties of biological silica in skeletons of deep-sea sponges. *Journal of Materials Research, 21*, 2068–2078. https://doi.org/10.1557/jmr.2006.0251.

Wu, J. W., Shi, Y., Zhu, Y. -X., Wang, Y. -C., & Gong, H. -J. (2013). Mechanisms of enhanced heavy metal tolerance in plants by silicon: A review. *Pedosphere, 23*, 815–825. https://doi.org/10.1016/S1002-0160(13)60073-9.

Yoshida, S. (1965). Chemical aspects of the role of silicon in physiology of the rice plant. *Bulletin of the National Institute of Agricultural Sciences, Series B15*, 1–58.

Zahedi, S. M., Moharrami, F., Sarikhani, S., & Padervand, M. (2020). Selenium and silica nanostructure-based recovery of strawberry plants subjected to drought stress. *Scientific Reports, 10*, 17672. https://doi.org/10.1038/s41598-020-74273-9.

Zhu, Y. -X., Gong, H. -J., & Yin, J. -L. (2019). Role of silicon in mediating salt tolerance in plants: A review. *Plants, 8*, 147. https://doi.org/10.3390/plants8060147.

Zhu, Y., & Gong, H. (2014). Beneficial effects of silicon on salt and drought tolerance in plants. *Agronomy for Sustainable Development, 34*, 455–472. https://doi.org/10.1007/s13593-013-0194-1.

Zhu, Y., Guo, J., Feng, R., Jia, J., Han, W., & Gong, H. (2016). The regulatory role of silicon on carbohydrate metabolism in *Cucumis sativus* L. under salt stress. *Plant and Soil, 406*, 231–249. https://doi.org/10.1007/s11104-016-2877-2.

CHAPTER 8

Improvement of Sugarcane Crop Quality by Using Some PGPRs Under Biotic and Abiotic Stresses

TAMILSELVAN ARUMUGAM,[1] M. JAGADESH,[2], VINAY HEGDE,[1], JAGADISH RANE,[1] CHERUKUMALLI SRINIVASARAO,[3] A. KRISHNAVENI,[2] SANTOSH KUMAR SINGH,[4] ARADHANA KUMARI,[5] MUNMUN DASH,[2,] and VINOD KUMAR GARG[5]

[1]*ICAR–National Institute of Abiotic Stress Management, Baramati, Pune, Maharashtra, India*

[2]*Tamil Nadu Agricultural University, Coimbatore, Tamil Nadu, India*

[3]*ICAR–National Academy of Agricultural Research Management (NAARM), Hyderabad, Telangana, India*

[4]*Dr. Rajendra Prasad Central Agricultural University, Pusa, Samastipur, Bihar, India*

[5]*College of Agriculture, Jawaharlal Nehru Krishi Vishwavidyalaya, Ganj Basoda, Vidisha, Madhya Pradesh, India*

#Authors having equal contribution.

ABSTRACT

Agricultural sustainability has emerged as one of the major issues affecting agricultural systems in the modern period. By the end of this century, there will be a substantial increase in the frequency and magnitude of biotic and abiotic stresses like extreme soil moisture deficit, high temperature, and soil mineral status, particularly salinity and alkalinity, which will have a significant impact on income-assuring but resource-intensive, long-duration crops like sugarcane. These adverse conditions limit plant growth, which substantially

Sugarcane Cultivation and Management: Challenges and Opportunities. Krishan K. Verma, Xiu-Peng Song, Munna Singh, Renato de Mello Prado, Jian-Ming Wu, and Yang-Rui Li (Eds.)

negatively affects the quantity and quality of sugarcane yield. A group of microorganisms known as plant growth-promoting bacteria (PGPR) has been utilized to improve crop quality, biological control, and plant growth promotion. PGPRs promise to encourage the growth and development of sugarcane (*Saccharum* spp.) by supplying nutrients like nitrogen through biological nitrogen fixation, phosphate, and potassium through solubilization of their insoluble forms, inducing phytohormone production and plant resistance to microbial pathogens, and siderophore production aiding plant nutrition by chelation and improving biotic and abiotic stress tolerance. Their use in sugarcane agriculture can encourage the use of fewer agricultural inputs and boost the growth of environmentally friendly crops. In nutrient-poor environments, PGPR can promote plant establishment and crop growth. Through PGPR crops, rhizospheric interactions, soil, and plant health can be improved. Thus, the PGPRs have tremendous potential to sustainably enhance the sugarcane yield while preserving the health of the soil from an agricultural perspective. To accomplish this, a better insight into native PGPR for sugarcane rhizospheric environments is required. A sustainable agricultural ecosystem is developed when PGPRs are utilized in combination with inorganic and organic nutrition sources in long-duration crops like sugarcane. However, the benefit depends on the appropriate choice of PGPR to facilitate biotic and abiotic stress tolerance and the successful establishment of microbes with the required effects under field conditions. The future applications of native multi-stress-tolerant PGPR for enhancing crop quality in combination with current strategies for heat, drought, salinity, heavy metal, and biotic stress tolerance may help to ensure global food security.

8.1 INTRODUCTION

The introduction of the Green Revolution in the latter half of the 20th century spurred a global agricultural boom. Introducing new high-yielding seed varieties and boosting the use of inorganic fertilizers, pesticides, and other agrochemicals dramatically enhanced plant productivity and agricultural production (Kesavan et al., 2018). Since then, the world's agricultural environment has seen a significant transformation. Over the last few decades, there has been a global decline in technological gains in agricultural production due to the excessive usage of synthetic agrochemicals, which has negatively impacted the biological, structural, and compositional health of arable soil (Pingali et al., 2012; Bishnoi et al., 2018). In the current situation, land resources are becoming depleted, and biological diversity is diminishing. The yield and productivity of crops must be raised simultaneously as the production of goods is connected to agriculture to meet the growing need for sustainable agriculture. The complex mentioned above, ecological, socioeconomic, and technical issues with promoting sustainable agriculture cannot be solved in a single, simple way (Kesavan et al., 2018). A potent strategy to break off the hasty environmental deterioration while increasing high agricultural productivity and better soil health is to promote sustainable agriculture with an incremental reduction in synthetic agrochemicals and a greater emphasis on biowaste-derived substances (Fascella et al., 2016; Liu et al., 2016; Verma et al., 2022a, 2023). In addition to genetic alteration of crop physiology and plant metabolism for crop improvement, certain groups of the microbiota, particularly those obtained from plant rhizosphere, may help plants resist or partially overcome environmental challenges (Ilangumaran et al., 2017). Sugarcane is a key tropical and subtropical vegetatively propagated crop farmed on over 27 million hectares in over 120 countries worldwide, accounting for more than 75% of the total global sugar supply (Aslam et al., 2018; Verma et al., 2020a, 2022b).

Due to its high biomass output and widespread ethanol production using molasses, sugarcane is considered a vital energy and biofuel crop in addition to the conventional sugar industry. Furthermore, it is the most efficient feedstock for bio-butanol and bio-diesel, accounting for 40% of global biofuel production. Other valuable by-products from sugarcane production are paper, acetic acid, plywood, and industrial enzymes (Rahman et al., 2019; Verma et al., 2020b). Modern sugarcane cultivars were produced by crossing *Saccharum officinarum* (2n = 80), a noble sugar-producing species, and *S. spontaneum* (2n = 40 128), a wild species with excellent fiber content and stress resistance (Piperidis et al., 2010; Vieira et al., 2018). Breeding strategies improved sugarcane yield until the 1970s when it remained static across countries (Viswanathan et al., 2022). Climate change is expected to have significant implications on sugarcane production globally, especially in developing countries, because of the relatively low ability for adaptation, high susceptibility to extreme weather events, inadequate forecasting systems, and inadequate mitigating measures. By the end of this century, biotic and abiotic stresses like severe soil moisture deficit, high temperatures, and soil mineral status, particularly salinity and alkalinity, will occur more frequently and with greater intensity. These stresses will significantly influence crops with long developmental stages that require many resources, like sugarcane, which provides income but is resource-intensive. These unfavorable conditions restrict plant growth, negatively affecting sugarcane yield in quantity and quality.

8.2 BIOTIC AND ABIOTIC STRESS IMPACTS ON SUGARCANE

In sugarcane, more than 100 pathogens (including bacteria, fungi, viruses, phytoplasmas, and nematodes) have been recognized as causal agents of several diseases (Rott, 2000; Govindaraju et al., 2019). Many biotic constraints seriously affect the productivity of the crop, which is specific to cane-growing countries. Diseases such as smut, ratoon stunting, yellow leaf, and mosaic are significant constraints in most countries. Diseases such as red rot and wilt significantly impact the cane output of South and Southeast Asian countries, with considerable historic red rot epiphytotic causing significant crop losses in India. Similarly, in the Asian region, phytoplasma diseases, grassy shoots, and white leaves are severe constraints. Some of the impacts of diseases on the cane and its quality, such as red rot and smut, can cause considerable changes in the juice quality. Because of a compromised metabolism of sucrose, diseased cane recovers sugar insufficiently. Red rot-affected canes had a 25–75% lower sucrose content than healthy canes (Viswanathan & Samiyappan, 1999a). Due to smut, the yield will be reduced by up to 50% (Viswanathan & Rao, 2011). Besides yield loss, a reduction in juice extraction will be up to 15–30%, up to 20% in sugar recovery, and a decline in germination caused by wilt. Combined infection of red rot and wilt causes more loss to the crop than their infection alone. Leaf scalding causes the leaf lamina to dry up, causing the leaves to wither. The yield can be reduced by 10–90% due to viral infections such as the sugarcane mosaic virus (SCMV), sugarcane streak mosaic virus (SCSMV), and potyvirus (Viswanathan & Rao, 2011). Interveinal chlorotic flecks, streaks, or stripes, especially on immature sugarcane leaves, are the typical symptoms of viral diseases.

The most common insect pests are stalk borers, which cause significant economic losses. Each country or region has its particular subset of borer bugs. An infestation of sugarcane borer can reduce sugar output and sugar quality by interfering with the quality of the raw material (such as sucrose concentration, soluble solids, purity, juice acidity, and fiber) (Clarke & Legendre, 1999). In addition to the borer pests, various sucking pests, and root grubs pose a severe threat to sugarcane cultivation (Viswanathan et al., 2022).

8.3 IMPORTANCE OF PGPRS

PGPRs (plant growth-promoting rhizobacteria) are typically free-living microorganisms that colonize the rhizosphere (Verma et al., 2024). The microbial population around the rhizosphere varies from its surroundings due to the abundance of root exudates, which primarily supply the nutrients for their growth and development (Subba Rao & Dommergues, 1998), causing the selection of specific rhizospheric communities (Huang et al., 2014). A total of 350 distinct bacterial strains with various activities were identified from sugarcane plant rhizospheric soil samples (Singh et al., 2020). Among the several genera of PGPBs found or associated with sugarcane crops are *Azotobacter, Azospirillum, Bacillus, Burkholderia, Gluconacetobacter, Herbaspirillum, Pseudomonas,* and *Rhizobium* (Tayade et al., 2019). PGPRs promote plant development and health either directly or indirectly under stressed and non-stressed conditions (Glick et al., 2007; Nadeem et al., 2010; Verma et al., 2023, 2024) through nitrogen fixation, phosphorus solubilization, the formation of siderophores, plant growth regulators, and organic acids, as well as defense by enzymes like ACC-deaminase, chitinase, and glucanase. These are some of the mechanisms that encourage plant growth (Berg, 2009; Glick et al., 2007; Hayat et al., 2010). Moreover, it has been proven effective under abiotic conditions by causing systemic resistance (Dutta & Khurana, 2015; Etesami & Beattie, 2018; Verma et al., 2021a, b, c, d) (Figure 8.1).

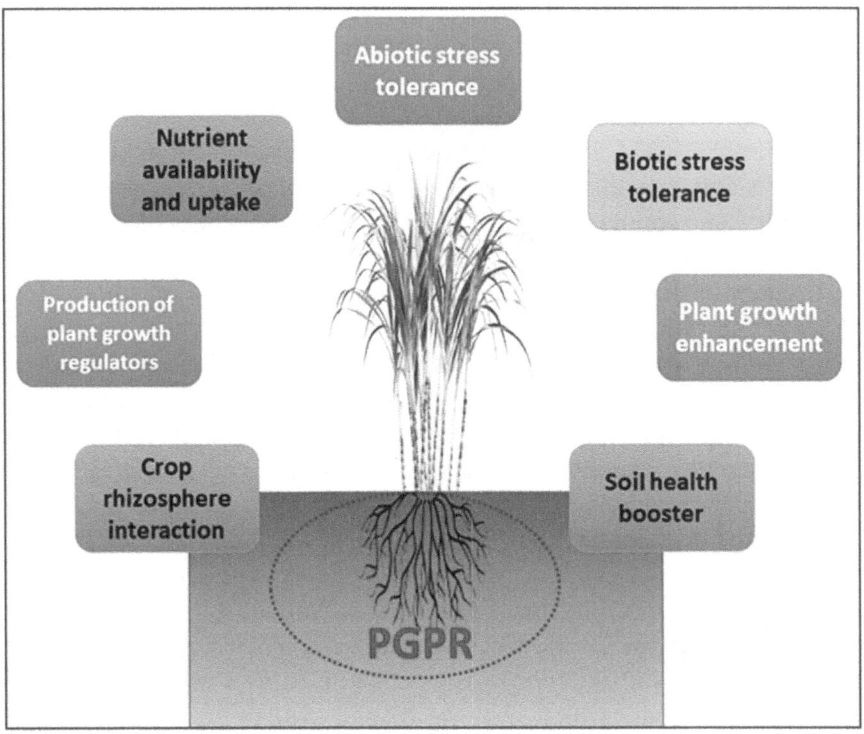

FIGURE 8.1 Importance of PGPR in sugarcane crop ecosystem.

8.4 ROLE OF PGPR IN SUGARCANE

8.4.1 SOIL HEALTH

Soil is a microbial biodiversity hotspot and the ultimate source of plant nutrients and food. The ability of soil to support all life forms and promote their growth and development in the ecosystem is referred to as "soil health" broadly (Doran & Safely, 1997). Some characteristics of healthy soil include nutrient recycling and storage, preservation of water and soil quality, suppression of pathogens, decomposition of soil organic matter, and detoxification of harmful substances (Dubey et al., 2016). The biological activity of soil microorganisms and their presence are the most promising and reliable indicators of soil health (Jain & Pandey, 2016a; Takoutsing et al., 2016). Due to long-term usage of chemical fertilizers, may negatively impact the environment, soil quality, and plant growth potential (Worlanyo & Jiangfeng, 2021), and consecutive monoculture and lack of crop rotation lead to the accumulation of soil-borne pathogens or autotoxic substances (Tayyab et al., 2021), which cause significant constraints to sugarcane production due to poor and insufficient fertile soil resulting from the deterioration of physiochemical properties of soil. This fails to meet nutritional and growth needs, restricting high production (Caione et al., 2015). PGPR has been considered a potential component to address soil sustainability issues due to its ability to enhance nutrient availability and soil health (Ju et al., 2020). The advantages of PGPR over other approaches, which can optimize sugarcane production by lowering the quantity of fertilizer required by exhibiting significant metabolic flexibility for carrying out functions including nitrogen fixation, phosphate, potassium, zinc, silica, and other substrate solubilization or mineralization, the release of phytohormones (PGPSs), the production of antibiotics, and the biodegradation of soil organic matter. These economical and environmentally friendly processes help maintain soil health (Kalia et al., 2020). For example, phosphorus-solubilizing rhizobacteria (PGPR) produce organic acids such as carboxylic acid, which lowers the pH in the plant rhizosphere (Deubel & Merbach, 2005). By solubilizing insoluble phosphates, these PGPRs can improve the availability of phosphate to plants, increasing the plants' access to nutrients (Gyaneshwar et al., 2002). Thus, beneficial soil microorganisms such as PGPRs are often referred to as 'biofertilizers,' which increase plants' availability and absorption (Vessey, 2003).

8.4.2 CROP-MICROBE INTERACTION

The rhizosphere has emerged as a diverse and dynamic ecological habitat with intense plant-microbe interactions capable of extracting critical micro- and macronutrients that impact crop growth (Mayak et al., 2004). The crop root colonization process in the rhizosphere is influenced by several variables, including bacterial traits, root exudates, and numerous other biotic and abiotic factors (Benizri et al., 2001). According to Andrews & Harris (2000), PGPRs are known to colonize plant roots in many rhizospheric associations and promote crop growth. The ability of a PGPR inoculant to survive and colonize plant roots is also higher in its immobilized form compared to free forms. Consequently, rhizospheric colonization is seen as a crucial process in realizing the advantages of PGPR for purposes such as biofertilization, phytostimulation, biocontrol, and phytoremediation.

The most frequently found PGPRs in sugarcane plants are *Azospirillum brasilense, Azospirillum lipoferum, Azospirillum amazonense, Acetobacter diazotrophicus, Bacillus tropicalis, Bacillus borstelensis, Herbaspirillum rubrisubalbicans,* and *Herbaspirillum seropedicae* (Hayat et al., 2012). Sugarcane crop response to inoculation with several soil microorganisms has been associated with the potential of inoculum to synthesize and excrete substances in the rhizosphere, such as various plant growth regulators, including auxins, gibberellins, cytokinins, ethylene, and abscisic acid (Tilak et al., 2006). The generation of total indole in sugarcane was substantially high in the species of *Ochrobactrum anthropi* (strains IMP311 and N208) and *Pseudomonas luteola* IMPCA244 (Morgado et al., 2015), showing the importance of crop-rhizosphere interaction for the productive use of PGPRs in plant growth.

PGPRs can also take part in the immobilization, organic phosphate mineralization, and inorganic phosphate solubilization processes through beneficial crop-rhizosphere interaction (Bhattacharyya et al., 2012). The phosphate-solubilizing bacteria (PSB) are a diverse group of bacteria that produce a wide range of organic acids, including citric, butyric, malonic, lactic, succinic, malic, and gluconic acids, among others, to solubilize phosphate and make it accessible for plant absorption (Pérez et al., 2007). The species of *Stenotrophomonas maltophilia* 79 found in sugarcane has the most significant levels of phosphate solubilization (Morgado et al., 2015). In addition to this, *B. subtilis* and *B. pumilus* enhance soil fertility, improve the quality of sugarcane by-products, and promote plant growth, providing strong evidence that microbes and sugarcane by-products can be combined to produce high yield and productivity in sugarcane production through effective crop-rhizosphere interaction (Santos et al., 2018).

8.4.3 ROLE OF PGPR IN BIOTIC STRESS TOLERANCE OF SUGARCANE

Limited studies have shown that crop plants can develop plant resistance to insect pests. *Pseudomonas, Bacillus, Burkholderia, Xenorhabdus, Photorhabdus, Agrobacterium, Streptomyces*, and other PGPR genera are widely used in insect pest control. PGPR inhibits insect pest activity by eliciting systemic resistance that produces phytochemical compounds (terpenes, siderophores, hydrogen cyanide, and so on), and some exhibit direct pest pathogenic potential (Maheshwari et al., 2019). The indirect effect of PGPR against insects begins with the host plant's recognition of the microbes and the activation of specific hormonal signal pathways, which may lead to the biosynthesis of defense-related biologically active compounds, enzymes, proteins, secondary metabolites, and volatile organic compounds (VOCs) against insect pests (Pineda et al., 2010, 2013; Van Oosten et al., 2008; Zamioudis & Pieterse, 2012; Zebelo et al., 2016). By enhancing the structural and mechanical strength of the cell wall and altering the physiological and biochemical response of the host, PGPR develops induced systemic resistance and triggers the production of defense-related chemicals against the pathogen (Viswanathan & Samiyappan, 1999a, b). The PGPR-induced sugarcane resistance will last 90 days of crop growth (Viswanathan, 1999). Generally, the durability of resistance imparted by PGPR differs from crop to crop due to different bacterial strains. Among the biocontrol agents, the strain of PGPR, *P. fluorescens*, is the promising one (Commarea et al., 2002). They activate systemic resistance by inducing plants' latent defense mechanisms and controlling insect pests (Commarea et al., 2002), in addition to exerting a beneficial effect on plant growth promotion (Gandhi et al., 2006). Several examples of plants treated with PGPR have shown decreased insect herbivory (Herman et al., 2008), especially the spotted cucumber beetle, *Diabrotica undecimpunctata Howardi* Barber, shown reduced feeding (Zehnder et al., 1997). In addition, the population growth of the green peach aphid, Myzuspersicae, was significantly reduced by the plants treated with defense elicitors compared to control plants.

Similarly, Herman et al. (2008) acknowledged that *B. subtilis* and *B. amyloliquefaciens* might be useful in managing Myzuspersicae for pepper plants grown in areas with high aphid pressure. Additionally, white clover and Medicago plants were more resilient to the effects of blue-green aphids when cultivated in the presence of Pseudomonas-like PGPR (Kempster et al., 2002). The role of PGPR in pest management was studied in many crops. Rhizobacteria promote plant growth (PGPR) and protect against disease by directly inhibiting the pathogen or eliciting resistance in the host plant, enabling the plant to withstand pathogen attacks on its aerial portions. As far as our knowledge goes, in sugarcane crops, there are no comprehensive management strategies for pest management. However, by using the transfer of cryIA(c) gene cloned from native *Bacillus thuringiensis* strain to *Pseudomonas fluorescens* (*P. fluorescens* 14 Omegon-Km-cry) showing activity against the sugarcane borer, *Eldana saccharina* (Herrera et al., 1994). Thus, PGPR is a potential eco-friendly mode of pest management in several crops, which has to be tested for sugarcane crop pest control in the near future.

Against major diseases of sugarcane, the role of biocontrol agents has been widely documented in managing plant diseases by generating a range of secondary metabolites and using several different mechanisms of action. By lowering soil-borne *C. falcatum* inoculum, some bioagents, such as *Trichoderma harzianum, T. viride,* and *Pseudomonas* species, were effective in managing sugarcane red rot (Singh et al., 2008; Hassan et al., 2011). Biological control using PGPR is an eco-friendly and sustainable practice to control red rot disease (Sharma et al., 2018). Under *in vitro* and *in vivo* settings, it was discovered that *Bacillus amyloliquefaciens* (RB 19), which was isolated from the sugarcane clone CoA92081's rhizoplane, had a solid antagonistic impact against *C. falcatum*. This isolate (RB 19) was shown to be superior to all other isolates in encouraging plant growth and disease suppression, as evidenced by its capacity to create IAA, HCN, and protease, as well as to solubilize phosphate and cellulose. HCN, siderophores, extracellular metabolites, diffusible, volatile antibiotics, and the broad-spectrum active antifungal metabolite pyoluteorin were produced by the bacterial strain *P. putida* NH-50. In both types of trials, where the fungus was either injected at above-ground nodes or inoculated in soil, effective control of red rot disease was achieved, i.e., secondary infection of red rot under field conditions on both varieties Co-1148 and SPF-234 (Hassan et al., 2011). The effectiveness of endophytic *P. fluorescens* strain EP1, isolated from sugarcane stalk tissues, was demonstrated by Viswanathan (1999) and Viswanathan & Samiyappan (1999a) against *C. falcatum*. According to Arencibia et al. (2006), *G. diazotrophicus* has a specific framework that induces systemic resistance against the sugarcane leaf scald disease caused by *Xanthomonas albilineans*. In order to control the spread of the disease to developing agamic shoots, *G. diazotrophicus* transmits elicitor molecules that activate the sugarcane defensive response and confer resistance to *X. albilineans* on the plant. In the presence of *G. diazotrophicus*, the infection was not seen. Their findings support that the plant is protected against *X. albilineans* attack by induced systemic resistance.

8.5 ROLE OF PGPR IN ABIOTIC STRESS TOLERANCE OF SUGARCANE

8.5.1 DROUGHT STRESS

Due to their semi-perennial development cycle, sugarcane plants are susceptible to seasonal dryness (FlackPrain et al., 2021). Droughts, which have become more frequent and severe due to changing climate, negatively affect sugarcane's physiological, morphological, and biochemical characteristics. It involves the accessibility and movement of soil nutrients because water is critical for the transport of nutrients from the soil to the roots. Soil moisture deficit reduces nutrient dispersion and movement of water-soluble nutrients such as nitrate, sulfate, Ca, Mg, and Si (Selvakumar et al., 2012). Additionally, it results in declining photosynthesis, which is strongly influenced by stomatal conductance and results in lower leaf water potential and increasing declines in CO_2 assimilation rates (Ribeiro et al., 2013; Sales et al., 2015; Verma et al., 2021a, d). Reactive oxygen species (ROS) from drought stress, including superoxide, hydrogen peroxide, and hydroxyl radicals, can cause oxidative stress and impair antioxidant defenses. Elevated ROS concentrations can therefore be detrimental to an organization on various levels, causing membrane disintegration and damaging plant proteins, lipids, and nucleic acids (Hendry, 2005; Nair et al., 2008; Fonseca et al., 2022; Verma et al., 2022a, b).

PGPR plays a crucial function in conferring drought tolerance and acclimatization in plants and has the potential to solve future food security challenges. Under drought conditions, the interplay between plants and PGPR affects the plant and soil attributes. The mechanisms triggered by PGPR, such as activating new genes and initiating an osmotic response, are crucial for sustaining plant survival in drought conditions (Vurukonda et al., 2016) and adjust the physiological reactions of plants to water deprivation, ensuring plant survival in such harsh situations (Marasco et al., 2012). The hormone that controls plant development in response to water, salinity, or drought is ethylene. The 1-aminocyclopropane-1-carboxylate (ACC) deaminase activity found in PGPR from extreme circumstances lowers the levels of ACC and endogenous ethylene, minimizing the negative consequences of stress on overall plant growth. The PGPR-inoculated plants with ACC deaminase are comparatively more resilient to environmental stress (Naveed et al., 2008).

Furthermore, it has been demonstrated that plants that have been inoculated with PGPR show significant increases in growth and yield as well as drought resistance in arid and semiarid locations (Marulanda et al., 2007). Arabidopsis thaliana inoculated with *Phyllobacterium brassicacearum* strain STM196 enhanced resistance to a moderate water deficit by delaying reproduction and altering the transpiration rate (Bresson et al., 2013). Inoculating cucumber plants with BBS (*Bacillus cereus* AR156, *Bacillus subtilis* SM21, and *Serratia* sp. XY21), a trio of plant-growth-promoting rhizobacterium strains, undergoing drought stress resulted in a decrease in the amount of monodehydroascorbate (MDA) in the leaf tissue and an increase in superoxide dismutase (SOD) activity, leaf proline content (Wang et al., 2012).

The advantageous microbiome is linked with roots and plant tissues through diverse mechanisms that counteract plant stress (Rolli et al., 2014). PGP bacteria containing the ACC deaminase enzyme (ACCd) can break down the plant ethylene precursor ACC, significantly reducing the level of ethylene in stressed plants (Glick, 2004). Actinobacter and Pseudomonas also boost shoot and leaf biomass, as well as the photosynthetic activity of plants exposed to stress, with an amplified effect in drought-sensitive rootstock through water stress-induced promotion, showing PGP activity as stress-dependent (Rolli et al., 2014). When compared to untreated settings, two promising bacterial strains (BSSC11 and BMSE7) from the *B. subtilis* and *B. megaterium* genera, respectively, have shown enhanced drought resistance in addition to growth-promoting activity through deeper root growth, expedited secondary bud formation, less decline in the relative water content, and increased accumulation of compatible solutes (Chandra et al., 2018).

8.5.2 WATER LOGGING

Water logging or excess water is an abiotic stress that limits potential crop productivity by impairing various physiological processes, such as reduced leaf photosynthesis, stomatal closure, and growth inhibition of the leaves, stems, and roots, leading to decreased yield. However, the severity of the problem depends on the duration of waterlogging, soil types, and prevailing drainage facilities. Water logging during the cane's formative stage has resulted in reductions in plant height by 13%, tiller production by 21.63%, and leaf area by 26.52% (Gomathi et al., 2015). Depending on the crop stage and the duration of flooding, the average yield loss due to water logging is estimated to be 15–25%, although it can exceed 40% (Gomathi & Chandran,

2009). Juice quality degrades due to reduced sucrose levels, excessive invert sugars, gums, non-protein nitrogen content, and higher total colloids in juice, even if waterlogged conditions hasten maturity (Gomathi et al., 2014).

In order to boost plant metabolism, biochar, and PGPR are both ecologically safe technologies that complement one another. The generation of secondary metabolites, ion regulation, enhanced ion uptake via roots, and redistribution to shoots are some of the functional aspects of PGPR (Kazerooni et al., 2021; Danish et al., 2019). On the other hand, biochar rejuvenates the soil, improves carbon capture, and promotes microbial growth (Abideen et al., 2020; Ren et al., 2019). Numerous research studies have documented the effectiveness of PGPR treatment for plant development and growth (Masciarelli et al., 2014). In addition, Abideen et al. (2020) showed that biochar application improved the growth of *Phragmites karka* under waterlogging conditions, and Kammann et al. (2011) reported that appropriate doses of biochar improved the growth of *Chenopodium quinoa*, also subjected to water logging. PGPR and biochar application control diverse physiological attributes of waterlogged soybean plants, including the normal function of intrinsic growth regulators, amino acids, and the antioxidant system.

8.5.3 HEAT STRESS

High temperature is hazardous to the ecosystem and causes abiotic stress, which limits agricultural productivity. Heat stress has the most destructive impact on plant metabolism and development among all abiotic stresses. Plants inhibit growth, development, and physiological cellular metabolism as the temperature exceeds (Santos et al., 2022; Haider et al., 2022; Perrella et al., 2022). In sugarcane crop production, heat stress affects the aggregation and denaturation of proteins, the fluidity of membrane lipids, the inactivation of enzymes in mitochondria and chloroplasts, the loss of membrane integrity, and the suppression of protein synthesis (Howarth et al., 2005; Wahid et al., 2007). According to recent studies, PGPR application provides agricultural plants with tolerance to abiotic stresses, including high temperatures (Ali et al., 2009; Dai et al., 2019). They promote plant development and reduce the negative consequences of stress (Bednarek et al., 2009; Thomas et al., 2022).

In sugarcane crop, the potential of numerous genera, including *Pseudomonas* and *Bacillus*, to produce resistance against various stressful conditions, such as high temperatures, has been examined (Morgado et al., 2015; Xia et al., 2020). Only a few studies have reported the role of PGPR in sugarcane under heat stress. The role of *B. xiamenensis* PM14 to withstand multiple stresses, including heat stress, has been reported. This strain can be used in extreme environments to demonstrate rhizosphere proficiency and saprophytic competitive ability, benefiting the sugarcane plants with improved health and yield (Xia et al., 2020). These beneficial PGPRs reduce the harmful effects of high-temperature stress by producing phytohormones and metabolites. They improve plant defenses against heat stress through plant-microbial interactions that activate the antioxidant defense system. They have also been found to enhance ion homeostasis by preserving osmoprotectant levels (Shaffique et al., 2022). Therefore, isolating and identifying useful PGPRs in sugarcane protection against heat stress would be extremely valuable.

8.5.4 SALINITY STRESS

Salinity significantly impacts sugarcane yield, influencing various physiological and morphological functions and inhibiting growth (de Andrade et al., 2017). Due to its high susceptibility to salinity stress (EC > 2 dSm^{-1}), sugarcane exhibits phytotoxicity that may have a direct or indirect impact on its physiological and metabolic functions, nutritional value, and growth (15–86%) (Gomathi et al., 2014). Salts that are soluble in water degrade soil productivity and have a negative impact on plant growth and development (Munns & Tester, 2008). The result of toxic ions (Na^+ and Cl^-) building up in the cells causes ion toxicity, which develops over several hours to days and weeks. Osmotic stress is the immediate impact of salinity (occurs within minutes) due to hypertonic conditions. Ion homeostasis, transpiration, photosynthesis, hormone state, and other metabolic activities are all affected by disturbed water balance and ion homeostasis (Munns, 2002). Crop plant inoculation with beneficial microorganisms is gaining agronomic importance because it facilitates cultivation in saline-prone conditions by enhancing salt tolerance and yield restoration (Lugtenberg et al., 2013). Bacterial strains isolated from harsh environments like deserts and oceans have been found to improve salt tolerance in crop plants. The introduction of antioxidant enzymes in lettuce increases a PGPR's efficiency more than arbuscular mycorrhizal (AM) fungi for enhanced tolerance to extreme salinity stress (Kohler et al., 2009). As a result, microbial inoculation to mitigate salinity stress could provide a novel and economical solution to soil salinity stress (Bano & Fatima, 2009). Many bacterial genera, such as *Alcaligenes, Azospirillum, Bacillus, Clostridium, Klebsiella, Pseudomonas, Rhizobium, Thiobacillus, Serratia*, and *Streptomyces*, which are plant growth promoters under saline conditions are being used and tested (Whipps, 2001).

Some PGPR isolates like ASN-1, a gram-positive, rod-shaped bacterium, have shown enhanced salinity tolerance and improved the sugarcane crop's growth through IAA production at about 196.48 ± 1.1 μg/mL. The ROS-scavenging antioxidative enzyme defense systems, i.e., CAT, SOD, PPO, POD, and APX, were significantly increased, with low MDA content and electrolyte leakage, and upregulation of stress-related genes such as ScDREB2A, Sucat, and SuSOD, which may confer the salt stress tolerance of the sugarcane crop (Sharma et al., 2021).

8.6 CONCLUSION

Plant growth-promoting rhizobacteria (PGPR) facilitate stress alleviation in sugarcane plants by producing metabolites that can improve the physiology of crop plants and thus offer an environmentally friendly alternative to existing synthetic agrochemicals and conventional agricultural practices in achieving sustainable agriculture. However, most of the strains of PGPR are not well characterized in a long-duration crop like sugarcane; hence, there is a lot of scope for studies on its impact on crop growth promotion and tolerance to biotic and abiotic stresses. Because of climate change, combined stresses are becoming more common, necessitating future testing of PGPR for the role of multiple combined stress tolerance.

KEYWORDS

- **abiotic stress**
- **abiotic stress tolerance**
- **crop plants**
- **crop quality**
- **PGPRs**
- **sugarcane**
- **sustainability**
- **synthetic agrochemicals**
- **waterlogging**

REFERENCES

Abideen, Z., Koyro, H. W., Huchzermeyer, B., Ansari, R., Zulfiqar, F., & Gul, B. J. P. B. (2020). Ameliorating effects of biochar on photosynthetic efficiency and antioxidant defence of *Phragmites karka* under drought stress. *Plant Biology, 22*, 259–266.

Ali, S. Z., Sandhya, V., Grover, M., Kishore, N., Rao, L. V., & Venkateswarlu, B. (2009). *Pseudomonas* sp. strain AKM-P6 enhances tolerance of sorghum seedlings to elevated temperatures. *Biology and Fertility of Soils, 46*(1), 45–55.

Andrews, J. H., & Harris, R. F. (2000). The ecology and biogeography of microorganisms on plant surfaces. *Annual Review of Phytopathology, 38*, 145–180.

Arencibia, A. D., Vinagre, F., Estevez, Y., Bernal, A., Perez, J., Cavalcanti, J., Santana, I., & Hemerly, A. S. (2006). *Gluconacetobacter diazotrophicus* elicits a sugarcane defense response against a pathogenic bacteria *Xanthomonas albilineans. Plant Signaling & Behavior, 1*(5), 265–273.

Backer, R., Rokem, J. S., Ilangumaran, G., Lamont, J., Praslickova, D., Ricci, E., & Smith, D. L. (2018). Plant growth-promoting rhizobacteria: context, mechanisms of action, and roadmap to commercialization of biostimulants for sustainable agriculture. *Frontiers in Plant Science, 9*, 1473.

Balloi, A., Rolli, E., Marasco, R., Mapelli, F., Tamagnini, I., Cappitelli, F., Borin, S., & Daffonchio, D. (2010). The role of microorganisms in bioremediation and phytoremediation of polluted and stressed soils. *Agrochimica, 54*, 353–369.

Bano, A., & Fatima, M. (2009). Salt tolerance in *Zea mays* (L.) following inoculation with *Rhizobium* and *Pseudomonas*. *Biology and Fertility of Soils, 45*, 405–413.

Bednarek, P., & Osbourn, A. (2009). Plant-microbe interactions: Chemical diversity in plant defense. *Science, 324*, 746–748.

Benizri, E., Baudoin, E., & Guckert, A. (2001). Root colonization by inoculated plant growth-promoting rhizobacteria. *Biocontrol Science and Technology, 11*(5), 557–574.

Berg, G. (2009). Plant microbe interactions promoting plant growth and health: perspectives for controlled use of microorganisms in agriculture. *Applied Microbiology and Biotechnology, 84*, 11–18.

Bhattacharyya, P. N., & Jha, D. K. (2012). Plant growth-promoting rhizobacteria (PGPR): emergence in agriculture. *World Journal of Microbiology and Biotechnology, 28*(4), 1327–1350.

Bishnoi, U. (2018). Agriculture and the dark side of chemical fertilizers. *Environmental Analysis and Ecology Studies, 3*, 198-201. DOI: 10.31031/EAES.2018.03.000552.

Boughton, A. J., Hoover, K., & Felton, G. W. (2006). Impact of chemical elicitor applications on greenhouse tomato plants and population growth of the green peach aphid, *Myzus persicae*. *Entomologia Experimentalis et Applicata, 120*, 175–188.

Bresson, J., Varoquaux, F., Bontpart, T., Touraine, B., & Vile, D. (2013). The PGPR strain *Phyllobacterium brassicacearum* STM196 induces a reproductive delay and physiological changes that result in improved drought tolerance in *Arabidopsis*. *New Phytologist, 200*, 558–567.

Carson, K. C., Meyer, J. M., & Dilworth, M. J. (2000). Hydroxamate siderophores of root nodule bacteria. *Soil Biology and Biochemistry, 32*, 11–21.

Chandra, P., Tripathi, P., & Chandra, A. (2018). Isolation and molecular characterization of plant growth-promoting *Bacillus spp.* and their impact on sugarcane (*Saccharum spp.*

hybrids) growth and tolerance towards drought stress. *Acta Physiologiae Plantarum, 40*, 1–15.

Clarke, M. A., & Legendre, B. R. (1999). Sugarcane quality: Impacts on sugar yield and quality factors. *STAB–Sugar, Alcohol and Byproducts*, Piracicaba, São Paulo, Brazil.

Commare, R. R., Nandakumar, R., Kandan, A., Suresh, S., Bharathi, M., Raguchander, T., & Samiyappan, R. (2002). *Pseudomonas fluorescens* based bio-formulation for the management of sheath blight disease and leaffolder insect in rice. *Crop Protection, 21*, 671–677.

Cook, R. J. (2002). Advances in plant health management in the twentieth century. *Annual Review of Phytopathology, 38*, 95–116.

Dai, L., Zhang, G., Yu, Z., Ding, H., Xu, Y., & Zhang, Z. (2019). Effect of drought stress and developmental stages on microbial community structure and diversity in peanut rhizosphere. *International Journal of Molecular Sciences, 20*, 2265.

Danish, S., & Zafar-ul-Hye, M. (2019). Co-application of ACC-deaminase producing PGPR and timber-waste biochar improves pigments formation, growth and yield of wheat under drought stress. *Scientific Reports, 9*, 1–13.

de Andrade, L. M., dos Santos Brito, M., Fávero Peixoto Junior, R., Marchiori, P. E. R., Nóbile, P. M., Martins, A. P. B., & Creste, S. (2017). Reference genes for normalization of qPCR assays in sugarcane plants under water deficit. *Plant Methods, 13*(1), 1–9.

Deubel, A., & Merbach, W. (2005). Influence of microorganisms on phosphorus bioavailability in soils. In: Buscot F., Varma A. (Eds), Microorganisms in soils: roles in genesis and functions. Springer-Verlag, Berlin, Germany, pp. 177–191.

Dodd, I. C., & Perez-Alfocea, F. (2012). Microbial amelioration of crop salinity stress. *Journal of Experimental Botany, 63*, 3415–3428.

Doran, J. W., & Safley, M. (1997). Defining and assessing soil health and sustainable productivity. Biological Indicators of Soil Health. CAB International, New York.

Dos Santos, T. B., Ribas, A. F., de Souza, S. G. H., Budzinski, I. G. F., & Domingues, D. S. (2022). Physiological responses to drought, salinity, and heat stress in plants: A review. *Stresses, 2*, 113–135.

Dubey, R. K., Tripathi, V., Dubey, P. K., Singh, H. B., & Abhilash, P. C. (2016). Exploring rhizospheric interactions for agricultural sustainability: the need of integrative research on multi-trophic interactions. *Journal of Cleaner Production, 115*, 362–365.

Dutta, S., & Khurana, S. M. P. (2015). Plant growth-promoting rhizobacteria for alleviating abiotic stresses in medicinal plants. In: Plant-Growth-Promoting Rhizobacteria (PGPR) and Medicinal Plants, Egamberdieva, D., Shrivastava, S., & Varma, A. (eds.). (Berlin: Springer), pp. 167–200. DOI: 10.1007/978-3-319- 13401-7_8.

Elliot, L. F., & Lynch, J. M. (1984). Pseudomonads as a factor in the growth of winter wheat (*Triticum aestivum* L.). *Soil Biology and Biochemistry, 16*, 69–71.

El-Tarabily, K. A., & Sivasithamparam, K. (2006). Non-streptomycete actinomycetes as biocontrol agents of soil-borne fungal plant pathogens and as plant growth promoters. *Soil Biology and Biochemistry, 38*, 1505–1520.

Etesami, H., & Beattie, G. A. (2017). Plant-microbe interactions in adaptation of agricultural crops to abiotic stress conditions. In: Probiotics and Plant Health, eds V. Kumar et al. (Berlin: Springer), pp. 163–200. DOI: 10.1007/978-981-10-3473- 2_7.

Fascella, G., Montoneri, E., Ginepro, M., & Francavilla, M. (2015). Effect of urban biowaste-derived soluble substances on growth, photosynthesis, and ornamental value of *Euphorbia × lomi*. *Scientia Horticulturae, 197*, 90–98.

Flack-Prain, S., Shi, L., Zhu, P., da Rocha, H. R., Cabral, O., Hu, S., & Williams, M. (2021). The impact of climate change and climate extremes on sugarcane production. *GCB Bioenergy, 13*(3), 408–424.

Fonseca, M. D. C. D., Bossolani, J. W., de Oliveira, S. L., Moretti, L. G., Portugal, J. R., Scudeletti, D., & Crusciol, C. A. C. (2022). *Bacillus subtilis* inoculation improves nutrient uptake and physiological activity in sugarcane under drought stress. *Microorganisms, 10*(4), 809.

Galal, Y. G. M. (2003). Assessment of nitrogen availability to wheat (*Triticum aestivum* L.) from inorganic and organic N sources as affected by *Azospirillum brasilense* and *Rhizobium leguminosarum* inoculation. *Egyptian Journal of Microbiology, 38*, 57–73.

Gandhi, P. I., Gunasekaran, K., & Tongmin, S. (2006). Neem oil as a potential seed dresser for managing homopterous sucking pests of okra (*Abelmoschus esculentus* (L.) Moench). *Journal of Pest Science, 79*, 103–111.

Glick, B. R. (2004). Bacterial ACC deaminase and the alleviation of plant stress. *Advances in Applied Microbiology, 56*, 291–312.

Glick, B. R. (2012). Plant growth-promoting bacteria: mechanisms and applications. *Scientifica*, 2012, 963401. DOI: 10.6064/2012/963401.

Glick, B. R., Cheng, Z., Czarny, J., & Cheng, Z. (2007). Promotion of plant growth by ACC deaminase-producing soil bacteria. *European Journal of Plant Pathology, 119*, 329–339.

Goswami, D., Thakker, J. N., & Dhandhukia, P. C. (2016). Portraying mechanics of plant growth-promoting rhizobacteria (PGPR): a review. *Cogent Food & Agriculture, 2*(1), 1127500.

Grattan, S. R., & Grieve, C. M. (1999). Salinity–mineral nutrient relations in horticultural crops. *Scientia Horticulturae, 78*, 127–157.

Gyaneshwar, P., Naresh, K. G., & Parekh, L. J. (2002). Effect of buffering on the phosphate solubilizing ability of microorganisms. *World Journal of Microbial Biotechnology, 14*, 669–673.

Haider, S., Raza, A., Iqbal, J., Shaukat, M., & Mahmood, T. (2022). Analyzing the regulatory role of heat shock transcription factors in plant heat stress tolerance: A brief appraisal. *Molecular Biology Reports, 49*(6), 5771–5785.

Hassan, M. N., Afghan, S., & Hafeez, F. Y. (2011). Biological control of red rot in sugarcane by native pyoluteorin-producing *Pseudomonas putida* strain NH-50 under field conditions and its potential modes of action. *Pest Management Science, 67*(9), 1147–1154.

Hayat, R., Ahmed, I., & Sheirdil, R. A. (2012). An overview of plant growth-promoting rhizobacteria (PGPR) for sustainable agriculture. In Crop Production for Agricultural Improvement (pp. 557–579).

Hayat, R., Ali, S., Amara, U., Khalid, R., & Ahmed, I. (2010). Soil beneficial bacteria and their role in plant growth promotion: a review. *Annals of Microbiology, 60*, 579–598.

Hellriegel, H., & Wilfarth, H. (1888). Untersuchungen uber die Stickstoffnahrung der Gramineen und Leguminosen. *Beilageheftzu der Zeitschrift des Vereins fur Rubenzucker-Industrie Deutschen Reichs*, pp. 234.

Hendry, G. A. (2005). Oxygen free radical process and seed longevity. *Seed Science Journal, 3*, 141–147.

Herman, M. A. B., Nault, B. A., & Smart, C. D. (2008). Effects of plant growth-promoting rhizobacteria on bell pepper production and green peach aphid infestations in New York. *Crop Protection, 27*, 996–1002.

Herrera, G., Snyman, S. J., & Thomson, J. A. (1994). Construction of a bioinsecticidal strain of *Pseudomonas fluorescens* active against the sugarcane borer, *Eldana saccharina*. *Applied and Environmental Microbiology, 60*(2), 682–690.

Howarth, C. J. (2005). Genetic improvements of tolerance to high temperature. In M. Ashraf, & P. J. C. Harris (Eds.), *Abiotic Stresses: Plant Resistance through Breeding and Molecular Approaches* (pp. 277–300). Haworth Press Inc., New York.

Huang, X. F., Chaparro, J. M., Reardon, K. F., Zhang, R., Shen, Q., & Vivanco, J. M. (2014). Rhizosphere interactions: root exudates, microbes, and microbial communities. *Botany, 92*(4), 267–275.

Ilangumaran, G., & Smith, D. L. (2017). Plant growth-promoting rhizobacteria in amelioration of salinity stress: A systems biology perspective. *Frontiers in Plant Science, 8*, 1768.

Ju, W., Jin, X., Liu, L., Shen, G., Zhao, W., Duan, C., & Fang, L. (2020). Rhizobacteria inoculation benefits nutrient availability for phytostabilization in copper contaminated soil: drivers from bacterial community structures in rhizosphere. *Applied Soil Ecology, 150*, 103450.

Kalia, A., Sharma, S. P., Kaur, S., & Kaur, H. (2020). Bacterial inoculants: how can these microbes sustain soil health and crop productivity? In *Soil Health* (pp. 337–372). Springer, Cham.

Kammann, C. I., Linsel, S., Gößling, J. W., & Koyro, H. (2011). Influence of biochar on drought tolerance of *Chenopodium quinoa* Willd and on soil–plant relations. *Plant and Soil, 345*, 195–210.

Kaymak, H. C. (2011). Potential of PGPR in Agricultural Innovations. In D. K. Maheshwari (Ed.), *Plant Growth and Health Promoting Bacteria* (pp. 45–79). Microbiology Monographs, Springer, Berlin.

Kazerooni, E. A., Maharachchikumbura, S. S. N., Adhikari, A., Al-Sadi, A. M., Kang, S. M., Kim, L. R., & Lee, I. -J. (2021). Rhizospheric *Bacillus amyloliquefaciens* protects *Capsicum annuum* cv. Geumsugangsan from multiple abiotic stresses via multifarious plant growth-promoting attributes. *Frontiers in Plant Science, 12*, 669693.

Kempster, V. N., Scott, E. S., & Davies, K. A. (2002). Evidence for systemic, cross-resistance in white clover (*Trifolium repens*) and annual medic (*Medicago truncatula var truncatula*) induced by biological and chemical agents. *Biocontrol Science and Technology, 12*, 615–623.

Kesavan, P. C., & Swaminathan, M. S. (2018). Modern technologies for sustainable food and nutrition security. *Current Science, 115*, 1876–1883.

Khalid, A., Arshad, M., & Zahir, Z. A. (2004). Screening plant growth promoting rhizobacteria for improving growth and yield of wheat. *Journal of Applied Microbiology, 96*(3), 473–480.

Kloepper, J. W., Lifshitz, R., & Zablotowicz, R. M. (1989). Free-living bacterial inocula for enhancing crop productivity. *Trends in Biotechnology, 7*(2), 39–44.

Kohler, J., Caravaca, F., Carrasco, L., & Roldan, A. (2006). Contribution of *Pseudomonas mendocina* and *Glomus intraradices* to aggregate stabilization and promotion of biological properties in rhizosphere soil of lettuce plants under field conditions. *Soil Use and Management, 22*, 298–304.

Liu, J., Ma, K., Ciais, P., & Polasky, S. (2016). Reducing human nitrogen use for food production. *Scientific Reports, 6*, 30104.

Lugtenberg, B. J., Dekkers, L., & Bloemberg, G. V. (2001). Molecular determinants of rhizosphere colonization by *Pseudomonas*. *Annual Review of Phytopathology, 39*, 461–490.

Lugtenberg, B. J., Malfanova, N., Kamilova, F., & Berg, G. (2013). Plant growth promotion by microbes. *Molecular Microbial Ecology of the Rhizosphere, 1–2*, 559–573.

Maheshwari, D. K., & Dheeman, S. (2019). *Field Crops: Sustainable Management by PGPR.* Switzerland: Springer. pp. 458.

Marasco, R., Rolli, E., Ettoumi, B., Vigani, G., Mapelli, F., Borin, S., AbouHadid, A. F., El-Behairy, U. A., Sorlini, C., Cherif, A., Zocchi, G., & Daffonchio, D. (2012). A drought resistance promoting microbiome is selected by root system under desert farming. *PLoS One, 7*, e48479.

Marulanda, A., Porcel, R., Barea, J. M., & Azcon, R. (2007). Drought tolerance and antioxidant activities in lavender plants colonized by native drought tolerant or drought sensitive *Glomus* species. *Microbial Ecology, 54*(3), 543–552.

Masciarelli, O., Llanes, A., & Luna, V. (2014). A new PGPR co-inoculated with *Bradyrhizobium japonicum* enhances soybean nodulation. *Microbiological Research, 169*, 609–615.

Mayak, S., Tirosh, T., & Glick, B. R. (2004). Plant growth-promoting bacteria confer resistance in tomato plants to salt stress. *Plant Physiology and Biochemistry, 42*(6), 565–572.

Montesinos, E. (2003). Plant-associated microorganisms: a view from the scope of microbiology. *International Microbiology, 6*(4), 221–223.

Morgado González, A., Espinosa Victoria, D., & Gómez Merino, F. C. (2015). Efficiency of plant growth promoting rhizobacteria (PGPR) in sugarcane. *Terra Latinoamericana, 33*(4), 321–330.

Munns, R. (2002). Comparative physiology of salt and water stress. *Plant, Cell & Environment, 25*, 239–250.

Munns, R., & Tester, M. (2008). Mechanisms of salinity tolerance. *Annual Review of Plant Biology, 59*, 651–681.

Nadeem, S. M., Zahir, Z. A., Naveed, M., & Ashraf, M. (2010). Microbial ACC-deaminase: prospects and applications for inducing salt tolerance in plants. *Critical Reviews in Plant Sciences, 29*, 360–393.

Nair, A., Abraham, T. K., & Jaya, D. S. (2008). Studies on the changes in lipid peroxidation and antioxidants in drought stress induced Cowpea (*Vigna unguiculata* L.) varieties. *Journal of Environmental Biology, 29*, 689–691.

Olanrewaju, O. S., Glick, B. R., & Babalola, O. O. (2017). Mechanisms of action of plant growth promoting bacteria. *World Journal of Microbiology and Biotechnology, 33*(11), 1–16.

Paul, D., & Nair, S. (2008). Stress adaptations in a plant growth promoting rhizobacterium (PGPR) with increasing salinity in the coastal agricultural soils. *Journal of Basic Microbiology, 48*(5), 378–384.

Pérez, E., Sulbarán, M., Ball, M. M., & Yarzábal, L. A. (2007). Isolation and characterization of mineral phosphate-solubilizing bacteria naturally colonizing a limonitic crust in the south eastern Venezuelan region. *Soil Biology and Biochemistry, 39*, 2905–2914.

Perrella, G., Bäurle, I., & van Zanten, M. (2022). Epigenetic regulation of thermos morphogenesis and heat stress tolerance. *New Phytologist, 234*, 1144–1160.

Pingali, P. L. (2012). Green revolution: Impacts, limits, and the path ahead. *Proceedings of the National Academy of Sciences of the United States of America, 109*, 12302–12308.

Ramamoorthy, V., Viswanathan, R., Raguchander, T., Prakasam, V., & Samiyappan, R. (2001). Induction of systemic resistance by plant growth promoting rhizobacteria in crop plants against pests and diseases. *Crop Protection, 20*(1), 1–11.

Rangarajan, S., Saleena, L. M., Vasudevan, P., & Nair, S. (2003). Biological suppression of rice diseases by *Pseudomonas* spp. under saline soil conditions. *Plant and Soil, 251*, 73–82.
Ren, H., Lv, C., Fernández-García, V., Huang, B., Yao, J., & Ding, W. (2019). Biochar and PGPR amendments influence soil enzyme activities and nutrient concentrations in a eucalyptus seedling plantation. *Biomass Conversion and Biorefinery, 11*, 1865–1874.
Ribeiro, R. V., Machado, R. S., Machado, E. C., Machado, D. F. S. P., Magalhães Filho, J. R., & Landell, M. G. A. (2013). Revealing drought-resistance and productive patterns in sugarcane genotypes by evaluating both physiological responses and stalk yield. *Experimental Agriculture, 49*(2), 212–224.
Rolli, E., Marasco, R., Vigani, G., Ettoumi, B., Mapelli, F., Deangelis, M. L., Gandolfi, C., Casati, E., Previtali, F., Gerbino, R., Cei, F. P., Borin, S., Sorlini, C., Zocchi, G., & Daffonchio, D. (2014). Improved plant resistance to drought is promoted by the root-associated microbiome as a water stress-dependent trait. *Environmental Microbiology, 17*, 316–331.
Ryan, P. R., Dessaux, Y., Thomashow, L. S., & Weller, D. M. (2009). Rhizosphere engineering and management for sustainable agriculture. *Plant and Soil, 321*, 363–383.
Sales, C. R., Marchiori, P. E. R., Machado, R. S., Fontenele, A. V., Machado, E. C., Silveira, J. A. G., & Ribeiro, R. V. (2015). Photosynthetic and antioxidant responses to drought during sugarcane ripening. *Photosynthetica, 53*, 547–554.
Santos, R. M., Kandasamy, S., & Rigobelo, E. C. (2018). Sugarcane growth and nutrition levels are differentially affected by the application of PGPR and cane waste. *Microbiology Open, 7*(6), e00617.
Sekar, S., & Kandavel, D. (2010). Interaction of plant growth promoting rhizobacteria (PGPR) and endophytes with medicinal plants – new avenues for phytochemicals. *Journal of Phytology, 2*, 91–100.
Selvakumar, G., Panneerselvam, P., & Ganeshamurthy, A. N. (2012). Bacterial mediated alleviation of abiotic stress in crops. In D. K. Maheshwari (Ed.), *Bacteria in Agrobiology: Stress Management* (pp. 205–224). Springer-Verlag.
Shaffique, S., Khan, M. A., Wani, S. H., Pande, A., Imran, M., Kang, S. M., & Lee, I. J. (2022). A review on the role of endophytes and plant growth promoting rhizobacteria in mitigating heat stress in plants. *Microorganisms, 10*(7), 1286.
Shaharoona, B., Arshad, M., Zahir, Z. A., & Khalid, A. (2006). Performance of *Pseudomonas* spp. containing ACC-deaminase for improving growth and yield of maize (*Zea mays* L.) in the presence of nitrogenous fertilizer. *Soil Biology and Biochemistry, 38*(9), 2971–2975.
Sharma, A., Singh, R. K., Singh, P., Vaishnav, A., Guo, D. J., Verma, K. K., & Li, Y. R. (2021). Insights into the Bacterial and Nitric Oxide-Induced Salt Tolerance in Sugarcane and Their Growth-Promoting Abilities. *Microorganisms, 9*(11), 2203.
Sharma, R., Sindhu, S., & Sindhu, S. S. (2018). Suppression of Alternaria blight disease and plant growth promotion of mustard (*Brassica juncea* L.) by antagonistic rhizosphere bacteria. *Applied Soil Ecology, 129*, 145–150.
Singh, R. K., Singh, P., Li, H. B., Song, Q. -Q., Guo, D. -J., Solanki, M. K., Verma, K. K., Malviya, M. K., Song, X. -P., Lakshmanan, P., Yang, L. -T., & Li, Y. -R. (2020). Diversity of nitrogen-fixing rhizobacteria associated with sugarcane: a comprehensive study of plant-microbe interactions for growth enhancement in *Saccharum spp. BMC Plant Biology, 20*, 220. doi:10.1186/s12870-020-02400-9.
Stainforth, D. A., Aina, T., Christensen, C., Collins, M., Faull, N., Frame, D. J., & Allen, M. R. (2005). Uncertainty in predictions of the climate response to rising levels of greenhouse gases. *Nature, 433*(7024), 403–406.

Steenhoudt, O., & Vanderleyden, J. (2000). Azospirillum, a free-living nitrogen-fixing bacterium closely associated with grasses: genetic, biochemical and ecological aspects. *FEMS Microbiology Reviews, 24*, 487–506.

Subba Rao, N. S., & Dommergues, Y. R. (1998). *Microbial Interactions in Agriculture and Forestry*. Science Publishers, Inc.

Tayyab, M., Yang, Z., Zhang, C., Islam, W., Lin, W., & Zhang, H. (2021). Sugarcane monoculture drives microbial community composition, activity and abundance of agricultural-related microorganisms. *Environmental Science and Pollution Research, 28*, 48080–48096.

Thomas, A. S. S., Pongprayoon, W., Cheenkachorn, K., & Sriariyanun, M. (2022). Plant-Microbe Interactions-Insights and Views for Applications in Sustainable Agriculture. *Applied Science & Engineering Progress*.

Tilak, K., Anganayakin, R., & Anoharacharic, M. (2006). Synergistic effects of plant-growth promoting rhizobacteria and Rhizobium on nodulation and nitrogen fixation by pigeonpea (*Cajanus cajan*). *European Journal of Soil Science, 57*, 67–71.

Trivedi, P., Pandey, A., & Palni, L. M. S. (2005). Carrier-based preparations of plant growth-promoting bacterial inoculants suitable for use in cooler regions. *World Journal of Microbiology & Biotechnology, 21*, 941–945.

Verma, K. K., Li, D. M., Singh, M., Rajput, V. D., Malviya, M. K., Minkina, T., Singh, R. K., Singh, P., Song, X. P., & Li, Y. R. (2020a). Interactive role of silicon and plant-rhizobacteria mitigating abiotic stresses: A new approach for sustainable agriculture and climate change. *Plants, 9*, 1055. DOI: 10.3390/plants9091055.

Verma, K. K., Song, X. P., Budeguer, F., Nikpay, A., Enrique, R., Singh, M., Zhang, B., Wu, J. M., & Li, Y. R. (2022a). Genetic engineering: an efficient approach to mitigating biotic and abiotic stresses in sugarcane cultivation. *Plant Signaling & Behavior, 17*(1), e2108253.

Verma, K. K., Song, X. P., Li, D. M., Singh, M., Wu, J. M., Singh, R. K., Sharma, A., Zhang, B. Q., & Li, Y. R. (2022b). Silicon and soil microorganisms improve rhizospheric soil health with bacterial community, plant growth, performance, and yield. *Plant Signaling & Behavior, 17*(1), e2104004.

Verma, K. K., Song, X. P., Lin, B., Guo, D. J., Singh, M., Rajput, V. D., Singh, R. K., Singh, P., Sharma, A., Malviya, M. K., Chen, G. L., & Li, Y. R. (2022). Silicon-induced drought tolerance in crop plants: physiological adaptation strategies. *Silicon, 14,* 2473–2487. DOI: 10.1007/s12633–021–01071-x.

Verma, K. K., Song, X. P., Rajput, V. D., Solomon, S., Li, Y. R., & Rao, G. P. (Eds.) (2023). *Agro-industrial Perspectives on Sugarcane Production under Environmental Stress*. Springer, Singapore. https://doi.org/10.1007/978-981-19-3955-6_1.

Verma, K. K., Song, X. P., Tian, D. D., Singh, M., Verma, C. L., Rajput, V. D., Singh, R. K., Sharma, A., Singh, P., Malviya, M. K., & Li, Y. R. (2021a). Investigation of the defensive role of silicon during drought stress induced by irrigation capacity in sugarcane: physiological and biochemical characteristics. *ACS Omega, 6*, 19811–19821.

Verma, K. K., Song, X. P., Verma, C. L., Chen, Z. L., Rajput, V. D., Wu, K. C., Liao, F., Chen, G. L., & Li, Y. R. (2021b). Functional relationship between photosynthetic leaf gas exchange in response to silicon application and water stress mitigation in sugarcane. *Biological Research, 54*, 15. https://doi.org/10.1186/s40659-021-00338-2.

Verma, K. K., Song, X. P., Zeng, Y., Guo, D. J., Sing, M., Rajput, V. D., Malviya, M. K., Wei, K. J., Sharma, A., Li, D. P., Chen, G. L., & Li, Y. R. (2021d). Foliar application of silicon boosts growth, photosynthetic leaf gas exchange, antioxidative response, and resistance to limited water irrigation in sugarcane (*Saccharum officinarum* L.). *Plant Physiology and Biochemistry, 166*, 582–592.

Verma, K. K., Song, X. P., Zeng, Y., Li, D. M., Guo, D. J., Rajput, V. D., Chen, G. L., Barakhov, A., Minkina, T. M., & Li, Y. R. (2020b). Characteristics and correlation of leaf stomata and its relationship with photosynthesis on *Saccharum spp.* under different irrigation and silicon application. *ACS Omega, 5*, 24145–24153.

Verma, R. R., Srivastava, T. K., & Singh, P. (2019). Climate change impacts on rainfall and temperature in sugarcane growing Upper Gangetic Plains of India. *Theoretical and Applied Climatology, 135*(1), 279–292.

Verma, K. K., Joshi. A., Song. X-P, Singh. S., Kumari, A., Arora, J., Singh, S. K., Solanki, M. K., Seth, C. S., & Li, Y-R. (2024) Synergistic interactions of nanoparticles and plant growth promoting rhizobacteria enhancing soil-plant systems: a multigenerational perspective. *Frontiers in Plant Science 15*, 1376214. doi: 10.3389/fpls.2024.1376214

Vessey, J. K. (2003). Plant growth-promoting rhizobacteria as biofertilizers. *Plant and Soil, 255*, 571–586.

Viswanathan, R., & Rao, G. P. (2011). Disease scenario and management of major sugarcane diseases in India. *Sugar Tech, 13*(4), 336–353.

Viswanathan, R., & Samiyappan, R. (1999a). Induction of systemic resistance by plant growth promoting rhizobacteria against red rot disease caused by *Collectotrichum falcatum* went in sugarcane. *Proceedings of the Sugar Tech. Assoc. India, 61*, 24–39.

Viswanathan, R., & Samiyappan, R. (1999b). Red rot disease of sugarcane: major constraint for Indian sugar industry. *Sugar Cane, 5*, 9–15.

Viswanathan, R., et al. (2022). Genomic Designing for Biotic Stress Resistance in Sugarcane. In: Kole, C. (Eds.), *Genomic Designing for Biotic Stress Resistant Technical Crops*. Springer, Cham. https://doi.org/10.1007/978-3-031-09293-0_9.

Vurukonda, S. S. K. P., Vardharajula, S., Shrivastava, M., & SkZ, A. (2016). Enhancement of drought stress tolerance in crops by plant growth promoting rhizobacteria. *Microbiological Research, 184*, 13–24.

Wahid, A., & Close, T. J. (2007). Expression of dehydrins under heat stress and their relationship with water relations of sugarcane leaves. *Biologia Plantarum, 51*, 104–109.

Wang, C. J., Yang, W., Wang, C., Gu, C., Niu, D. -D., Liu, H. -X., Wang, Y. -P., & Guo, J. -H. (2012). Induction of drought tolerance in cucumber plants by a consortium of three plant growth-promoting rhizobacterium strains. *PLoS ONE, 7*(12), e52565.

Whipps, J. M. (2001). Microbial interactions and biocontrol in the rhizosphere. *Journal of Experimental Botany, 52*, 487–511.

Worlanyo, A. S., & Jiangfeng, L. (2021). Evaluating the environmental and economic impact of mining for post-mined land restoration and land-use: A review. *Journal of Environmental Management, 279*, 111623.

Xia, Y., Farooq, M. A., Javed, M. T., Kamran, M. A., Mukhtar, T., Ali, J., & Chaudhary, H. J. (2020). Multi-stress tolerant PGPR *Bacillus xiamenensis* PM14 activating sugarcane (*Saccharum officinarum* L.) red rot disease resistance. *Plant Physiology and Biochemistry, 151*, 640–649.

Zehnder, G., Kloepper, J., Tuzun, S., Yao, C., Wei, G., Chambliss, O., & Shelby, R. (1997). Insect feeding on cucumber mediated by rhizobacteria-induced plant resistance. *Entomologia Experimentalis et Applicata, 83*, 81–85.

CHAPTER 9

Role of Plant Secondary Metabolism in Sugarcane During Unfavorable Environmental Conditions

R. G. VYSHNAVI,[1] KARISHMA BEHERA,[2] ARADHANA KUMARI,[3] R. K. SAMAIYA,[1] and P. K. MISHRA[4]

[1]*Department of Plant Physiology, Jawaharlal Nehru Krishi Vishwavidyalaya, Jabalpur, Madhya Pradesh, India*

[2]*Department of Plant Breeding and Genetics, Jawaharlal Nehru Krishi Vishwavidyalaya, Jabalpur, Madhya Pradesh, India*

[3]*Department of Plant Physiology, College of Agriculture, Jawaharlal Nehru Krishi Vishwavidyalaya, Ganj Basoda, Madhya Pradesh, India*

[4]*Department of Agronomy, College of Agriculture, Jawaharlal Nehru Krishi Vishwavidyalaya, Jabalpur, Madhya Pradesh, India*

ABSTRACT

In nature, the plant endures several challenging and stressful conditions. To attain the adaptive capacity in response to adverse environmental conditions, the plant produces diverse organic molecules that are biosynthesized from carbohydrates, lipids, and amino acids, called secondary metabolites (SMs). The versatile role of plant secondary metabolites ranges from pollination to protection and tolerance against biotic and abiotic stress factors. Besides their role in plants' ecological functions, SMs play an apparent shielding role over cellular components during multiple stress conditions. Accumulation of several secondary metabolites depends on the conditions the plant endures and its physiological and developmental stages. The secondary plant

Sugarcane Cultivation and Management: Challenges and Opportunities. Krishan K. Verma, Xiu-Peng Song, Munna Singh, Renato de Mello Prado, Jian-Ming Wu, and Yang-Rui Li (Eds.)

metabolites in recent decades have witnessed commensurable importance in their repertoire contribution towards unfavorable conditions. Sugarcane (*Saccharum officinarum* L.) experiences several abiotic and biotic stress conditions. In response, it develops secondary metabolites, a thick waxy layer, trichomes on leaves, and glandular hairs. The SMs produced in sugarcane include terpenes, phenolic compounds, and nitrogen-containing compounds. Terpenes include Arundoin, Crusgallin, Cylindrin, etc. The Phenolic compounds include Apigeninidin, Luteolinidin, Piceatannol, *p*-coumaric acid, Caffeic acid, Ferulic acid, 3-indole acrylic acid, 6-deoxyfagomine, Trigonelline, Quercetin, and Tricin, etc. This chapter summarizes the major types of secondary metabolites involved in sugarcane ramified under unfavorable conditions and their biosynthesis mechanisms.

9.1 INTRODUCTION

In nature, plants grow and reproduce in harsh, unfavorable environmental conditions subjected to stress. The multitude of stress is caused by abiotic (non-living) and biotic (living) factors. Stress resulting from abiotic factors includes high temperature (heat stress), water deficit (drought stress), cold injury (chilling and freezing), salt stress, and heavy metal stress, while biotic stress is caused by a plethora of microorganisms which include fungi, bacteria, and viruses. Being sessile, plants cannot avoid unfavorable conditions, and while enduring these stressful conditions, plants evolved several strategies to cope with and mitigate adverse conditions at the metabolic level, modulating the physiological mechanism (Song et al., 2019, 2021, 2022; Verma et al., 2022, 2023). In this line and length, plants produce a group of secondary compounds that play a defensive and protective role against unfavorable conditions. These metabolites have no direct role in the growth and development of the plant. They are known as secondary products, natural products, or secondary metabolites (SMs). Biochemist Albrecht Kossel coined the term secondary metabolite. SMs are synthesized from primary metabolites (proteins, carbohydrates, and lipids) or their biosynthetic intermediates, either constitutively or inductively, in response to several environmental stimuli (Piasecka et al., 2015). Primary metabolites are directly involved in particular plant functions such as cell division, assimilation, photosynthesis, and respiration. On the other hand, secondary metabolites serve a wide range of ecological functions, such as defending and protecting the plants against herbivores and microbial pathogens, attracting the pollinators for seed dispersal mechanism, fruit set, and abscission. Apart from their protective role in attenuating unfavorable conditions, they defend many primary metabolites under stress conditions.

The phylogenetic occurrence of secondary metabolites *in planta* is confined to particular lineages of the family or genus, indicating the relevant biosynthetic pathways (Weng et al., 2012). For many years, scientists have assumed secondary metabolites as functionless metabolic wastes and neglected them until their importance was recognized as medicinal drugs, flavors, poisons, nutritive and cosmetic purposes, and industrial materials. Over the last 10 years, SMs attained an enormous interest in scientific know-how of their valuable role in plant stress physiology. Sugarcane (*Saccharum officinarum* L.), an economically productive cash crop cultivated worldwide, contributes to 75% of sugar (Kumar et al., 2019; Deng et al., 2022; Xu et al., 2022). It is a perennial, monocotyledonous plant belonging to the genus (*Saccharum* spp.) and family Poaceae, which consists of several hybrid species. Some of the cultivated species include *S. officinarum* (noble canes, 2n = 80), *S. sinense* (Chinese cane, 2n = 111–120), *S. barberi* (Indian species, 2n = 81–124), and wild species *S. spontaneum* (2n = 40–128) and *S. robustum* (2n = 60–80). It is an efficient C_4 plant mainly grown in tropical and subtropical regions. Globally, besides the essence of sucrose, it is recognized as a source of sustainable renewable bioenergy (Huang et al., 2021; Anas et al., 2019a, b). Global warming and climate change lead to consistent temperature fluctuations with different precipitation patterns, dry spells, wet spells, etc. The average temperature for the period of 1800–2012 has increased by 0.85°C (IPCC, 2013). The rise in 1°C temperature exerts drastic effects on crop growth. In this scenario of climate risk in seasonality, profound investigation and research must be done to obtain better-performing genotypes and cultivars. Sugarcane is highly vulnerable to extreme weather conditions such as temperature, precipitation, and elevated CO_2. The cane yield and production have fluctuated with the changing environment in sugarcane (Chandiposa et al., 2013; Hussain et al., 2018; Ruan et al., 2018; Verma et al., 2022; Oliveira et al., 2022). The tolerance capacity of sugarcane to stress through morphological and physiological adaptations and biochemical adjustments directly influences plant growth and development, dry matter accumulation, and photosynthesis (Verma et al., 2023).

The functions of SMs in plants involve serving as attractants for pollination and herbivore deterrents. Besides that, their indispensable role lies in the phytoprotection of primary metabolites, in regulating the plant growth hormones, and in the interaction of plants and the environment (microbes, insects, herbivores, and unfavorable conditions; Figure 9.1). The defense activity of SMs can directly instigate the hormonal signaling pathways and induce resistance against several phytopathogens, insects, and herbivores. They are also involved in regulating the target of rapamycin, auxin-dependent ROS accumulation, and auxin-mediated transcriptional regulation. The phytoprotection role of SMs in alleviating photoinhibition and scavenging the ROS species involves either as an antioxidant or in quenching the antioxidants. SMs can reinstate or rejuvenate the primary metabolites when required. The study of secondary metabolites and their putative functions has robust, practical applications in improving plant tolerance and defense mechanisms under different biotic and abiotic stresses. Further, in this chapter, we will discuss the significant plant secondary metabolites produced in sugarcane, their biosynthesis, and their functions explicitly under unfavorable conditions.

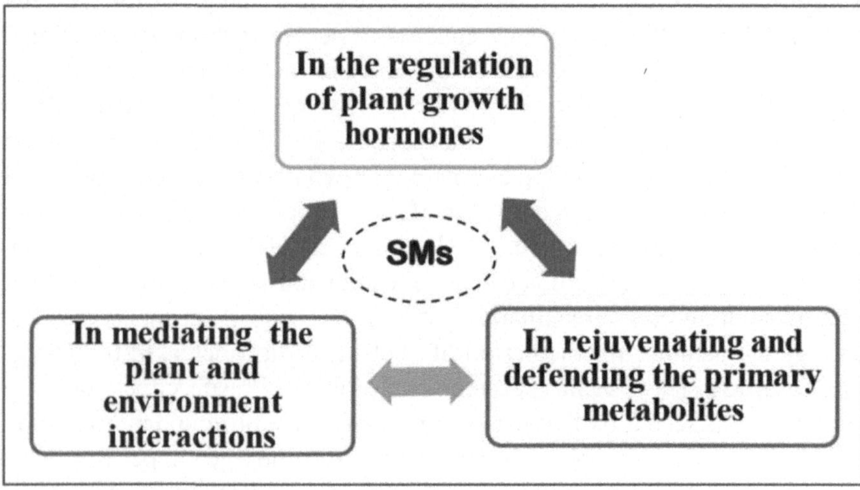

FIGURE 9.1 The functional trichotomy of secondary metabolites in plants.

9.2 CLASSIFICATION OF SECONDARY METABOLITES

Based on their existence and diverse structure, several criteria exist for the classification of secondary metabolites.

9.2.1 BASED ON THE OCCURRENCE OF SECONDARY METABOLITES BASED UPON THE RESPONSE TO STIMULI OR ELICITATION

Depending upon the development of secondary metabolite occurrence in response to stimuli, they are categorized into two types, i.e., phytoanticipins and phytoalexins.

9.2.1.1 PHYTOANTICIPINS

Phytoanticipins are antimicrobial compounds that are constitutively developed and accumulated in response to microbial invasion (VanEtten et al., 1994). Framing the apt role of phytoanticipins as the guardians that never sleep, they serve as defensive plant metabolites. The structural diversity of phytoanticipins involves a subset of compounds, including saponins, cyanogenic glucosides, and glucosinolates. The development of saponins in plants is an adaptive strategy of resistance in response to diseases, insects, and herbivores (Table 9.1).

1. **Saponins:** These are steroidal triterpene glycosides. Most of the saponins in plants are produced from roots. The basic structure of saponins includes the presence of both hydrophobic terpenoid parts and hydrophilic sugars, aiding the detergent-like features of saponins. When saponins enter the bloodstream of herbivores/insects, they act as toxins that disrupt the cell membranes and interfere with the digestive system. The diverse functions of saponins include antiherbivore, antimicrobial, anti-insect, and allelopathic properties. They are synthesized from mevalonic acid through the isoprenoid pathway. Based on their nature, aglycone skeleton saponins are categorized into two groups, such as steroidal saponins and triterpenoid saponins. Steroidal saponins are abundant in monocotyledons, whereas triterpenoid saponins are predominantly depicted in the dicotyledonous family (Bruneton, 1995). Steroidal saponins are structurally classified into spirostanol and furostanol saponins (Sidana et al., 2018). The presence of saponins in the sugarcane stem barks is reported by (Abago et al., 2014; Pathal et al., 2017). The role of saponins as antimicrobial compounds corresponding to plant immunity is well-evidenced. The functions of saponins as crucial players in the salt tolerance mechanism are reported by Sari et al. (2021).

2. **Glucosinolates:** These (ß-D-thioglucoside-N-hydroxysulfates) are one of the constitutive groups of secondary metabolites derived from amino acids like alanine, methionine, tryptophan, tyrosine, leucine, and isoleucine. They are detected mainly in the Brassicaceae family. In *Arabidopsis thaliana*, the glucosinolates are of two types depending on the structural diversity and the amino acid they are derived from, i.e., Methionine-derived aliphatic glucosinolates and tryptophan-derived indolic glucosinolates. The loss of cellular integrity stimulates the rapid hydrolysis of glucosinolates in the presence

of enzymes thioglucoside glucohydrolases, also known as myrosinases, resulting in the products glucose and an unstable aglycone, which can rearrange into isothiocyanates and other products. Some of the specific glucosinolate metabolites present in sugarcane include (2-(2′-Methylthio) ethylmalic acid), Kaempferol-3-O-glucoside, cyanidin 3-glucoside, S-(4-Methylthiobutylthiohydroximoyl)-L-cysteine, Indolylmethyl-desulfoglucosinolate, 3,4-Dihydroxyphenylglycol, Cis-beta-D-Glucosyl-2-hydroxycinnamate, 2-(3′-Methylthio) propylmalic acid, 3-(3,4-Dihydroxy-5-methoxy)-2-propenoic acid, Pantothenic acid, 5-methyl tetrahydrofolic acid, 3,4-Dihydroxy mandelaldehyde, 2-(3′-methyl thio) propylmalic acid, and spermidine (Wijma et al., 2021). Two other metabolites are 2,3-Dihydroxy-isovalerate, an intermediate from valine, leucine, and isoleucine biosynthesis, and 12-oxo-cis-dodec-9-enoic acid, a metabolite of both leucine and isoleucine, have been discovered. The functional role of glucosinolates involves abiotic and biotic stresses (Chowdhury, 2022). Plants resist stress conditions by altering the levels of glucosinolates either by increasing or decreasing their quantity. Their accumulation depends on the intensity and duration of the unfavorable conditions. The increased levels of glucosinolates are observed in the case of salt, drought, heat, heavy metal, and UV-radiation stresses. The exact mechanism by which glucosinolates are involved in the tolerance of stresses is unclear, as their biosynthetic pathways are complex and linked to various other pathways.

3. **Cyanogenic Glucosides:** These (α-hydroxynitrile glycosides) are widely distributed constitutive secondary metabolites that have been reported in ferns, gymnosperms, and angiosperms. Cyanogenic glycosides have been detected in more than 3,000 plant species. The first and the newest discovered cyanogenic glucosides from plants are amygdalin (attained from bitter almonds) and prunasin methacrylate. The biosynthesis pathway of cyanogenic glucosides and glucosinolates is interlinked. Cyanogenic glycosides are not chemically functional until hydrolyzed by respective cyanogenic ß-glycosidases resulting in the formation of hydrogen cyanide (HCN). Cyanide in plants is also released as a co-product during ethylene biosynthesis. The α-hydroxynitrile glycosides chemical structure comprises an aglycon part and sugar moiety. The aglycone part consists of a nitrile group linked with the aliphatic, aromatic, or

cyclic moieties. They are derived from amino acids, such as valine, glycine, leucine, isoleucine, tyrosine, and phenylalanine. The sugar moiety generally consists of glucose in addition to that allose, apiose, arabinose, rhamnose, and xylose have been reported. A cyanogenic glucoside named dhurrin was initially discovered from sorghum (*Sorghum bicolor*). They contribute to the defense against plant-insect and plant-microbe interactions. Hence, Cyanogenic glycosides' role in sugarcane includes biotic and abiotic stresses.

TABLE 9.1 Secondary Metabolites which are Characterized in Sugarcane under Adverse Conditions

Terpenes	Mode of Activity	References
1. Terpenes		
i. Triterpenes		
Arundoin, Crusgallin, Cylindrin, Swamilletin	Biotic stress	Smith et al. (1978); Nuissier et al. (2002); Bryce et al. (1967)
Friedelin	Anti-herbivore, salinity stress	Sarri et al. (2021)
α-amyrin	Secondary metabolite	Teixeira et al. (2021)
ii. Cyclo triterpene	Secondary metabolite	Franca et al. (2001)
• Cycloartenol		
iii. Sesquiterpenes	Repellent to *Sitophillus granaries*	Riffel et al. (2021)
• (E)-Caryophyllene		
2. Phenolics Compounds		
i. Hydroxybenzoic acid		
• Benzoic	Defense against soil-borne pathogens	Windisch et al. (2021)
• Gallic	antimicrobe	Kong et al. (2015)
• Catechin	Highlight/UV irradiation	Carvalho et al. (2010)
• Ellagic acid	Antioxidant/Salinity stress	Khan et al. (2017)
• Vanillic	Heavy metal stress (Cu, Pb, Cd), salt stress, and drought stress	Li et al. (2018)
ii. Hydroxycinnamic acid		
• Ferulic	Drought stress, antimicrobial, reduces salt stress	Linic et al. (2021); Kong et al. (2015)
• Coumarin	Reduces electrolyte leakage in salt stress	Kaur et al. (2017)
• Caffeic	Drought stress and heavy metal stress reduce the effect of chilling stress and salt stress	Wan et al. (2015); Kong et al. (2015)
• Spermidine	Drought tolerance	Wijma et al. (2021)

TABLE 9.1 *(Continued)*

Terpenes	Mode of Activity	References
• Chlorogenic	Resistant to diseases induced by exposure to *Diatraea saccharalis*	Sabino et al. (2019)
A. Flavones/flavone glycosides		
• Apigeninidin	Phytoalexin in red rot/salinity tolerance and antioxidant activity	Malathi et al. (2008); Mekawy (2018)
• Luteolinidin	Phytoalexin/salinity tolerance and antioxidant activity	Mekawy et al. (2018); Vafadar et al. (2020)
• Naringenin	Phytoalexins and drought stress	Hu et al. (2019)
• Piceatannol	Phytoalexin red rot disease	Malathi et al. (2008)
B. Anthocyanins • Cyanidin • Petunidin	Resistance against cold stress	Zhao et al. (2018)
C. Flavanols • Quercetin and afzelin rutin	Enhanced under drought stress and is a powerful antioxidant that detoxifies the produced H_2O_2, and inhibits virus replication.	Wijma et al. (2021)
D. Stilbene • Resveratrol	Phytoalexin under fungal infection and antioxidant in excess UV light	Grayer & Harborne (1994)
3. Nitrogen-containing compounds i. Alkaloids		
• Trigonelline	Salinity and oxidative stress/ anti-herbivore, anti-insecticidal (oviposition deterrent)	Sabino et al. (2019)
ii. **Cyanogenic glucosides**		
• Dhurrin	Anti-herbivore and drought resistance	Wijma et al. (2021)

9.2.1.2 PHYTOALEXINS

The term Phytoalexin was coined by Muller & Borger (1940). They are antimicrobial metabolites produced instantly in response to phytopathogens and accumulate highly in leaves. Phytoalexins are the troops that arrive and take part in intricate defense systems enabling plants to resist the invading microbes. Several hundreds of phytoalexins have been isolated and characterized in different species (Schmelz et al., 2014). Phytoalexins belong to various chemical compounds, such as terpenoids, phenolics, glycoalkaloids, sulfur compounds, indoles, and furanoacetylenes (Jeandet, 2015). Sugarcane, in response to fungal attacks, results in the hypersensitive response mechanism. It involves the accumulation of red-colored pigments through the hypersensitive response mechanism. Such accumulated compounds include phytoalexins. Phytoalexins in sugarcane comprise apigeninidin, luteolinidin, piceatannol, sakuranetin, swamilletin, resveratrol, arundoin, crusgallin, and cylindrin. In this context, the selected groups of phytoalexins in sugarcane whose functions are unequivocally proven have been discussed in Table 9.1.

1. **Apigeninidin and Luteolinidin:** The red rot disease caused by *Colletotrichum falcatum* in sugarcane results in the accumulation of reddish pigment in stalks indicating the phytoalexins. These accumulated compounds mainly contain apigeninidin, luteolinidin, apigeninidine-5-glucoside, luteolinidin-5-glucoside, and caffeic acid ester of arabinosyl-5-o-apigenidin. They belong to the 3-deoxy anthocyanidin group. The multifold increase of these compounds in stalk tissues of resistant plants is observed. Apigeninidin, in the presence of flavonoid-3-hydroxylase, transforms into luteolinidin. Their biosynthetic pathway originates from the flavanone naringenin (Figure 9.2). The function of these two involves the growth reduction of fungal pathogens, including affecting the hyphal formation and the retardation of pathogen reproduction of *C. falcatum* (Kumar et al., 2015). The induction of Apigeninidin is confined only to the resistant genotype of sugarcane, whereas the accumulation of luteolinidin is observed in both susceptible and resistant genotypes. However, the quantity of luteolinidin accumulation is higher in the resistant genotype compared to the susceptible one (Malathi et al., 2008).
2. **Piceatannol:** The antifungal stilbene, was isolated from sugarcane infected with *C. falcatum*. The presence of piceatannol was not observed in healthy plants or wounded ones but was only seen in sugarcane plants infected with the pathogenic fungus, representing its distinct role in the antifungal mechanism. The role of piceatannol involves inhibiting the spore germination and germ tube growth of *C. falcatum* (Brinker et al., 1991).
3. **Resveratrol:** It is a polyphenolic phytoalexin known for its occurrence in response to fungal elicitation. In the plant kingdom, its existence is limited to certain specific species consisting of stilbene synthase enzymes. Resveratrol is synthesized through the phenylpropanoid pathway from the *p-coumaric*-CoA precursor in the presence of stilbene synthase. In plants, resveratrol accumulation increases in response to UV light and fungal infection, revealing its function in biotic and abiotic stress conditions.

9.2.1.3 TERPENES

Terpenes and terpenoids are the largest class of secondary metabolites, and isoprenoids are the basic structural units of terpenes. Terpenes are suitably classified based on the number of carbon atom repeats. Monoterpenoids (C10), sesquiterpenoids (C15), diterpenoids (C20), triterpenoids (C30), sesterpenoids (C25), tetraterpenoids (C40), polyterpenoids (C>40). A five-membered lactone ring characterizes the structure of sesquiterpenes. Triterpenes in sugarcane include arundoin, crusgallin, cylindrin, swamilletin, simiarenol, friedelin, and α-amyrin, cycloterpenes such as cycloartenol, and sesquiterpenes like (E)-caryophyllene. They are acknowledged for their functions against herbivores as feed repellants (Kseller et al., 2001). Besides that, they can act as transcriptional activators and are involved in the initiation of bud dormancy and preservation of seed (Table 9.1).

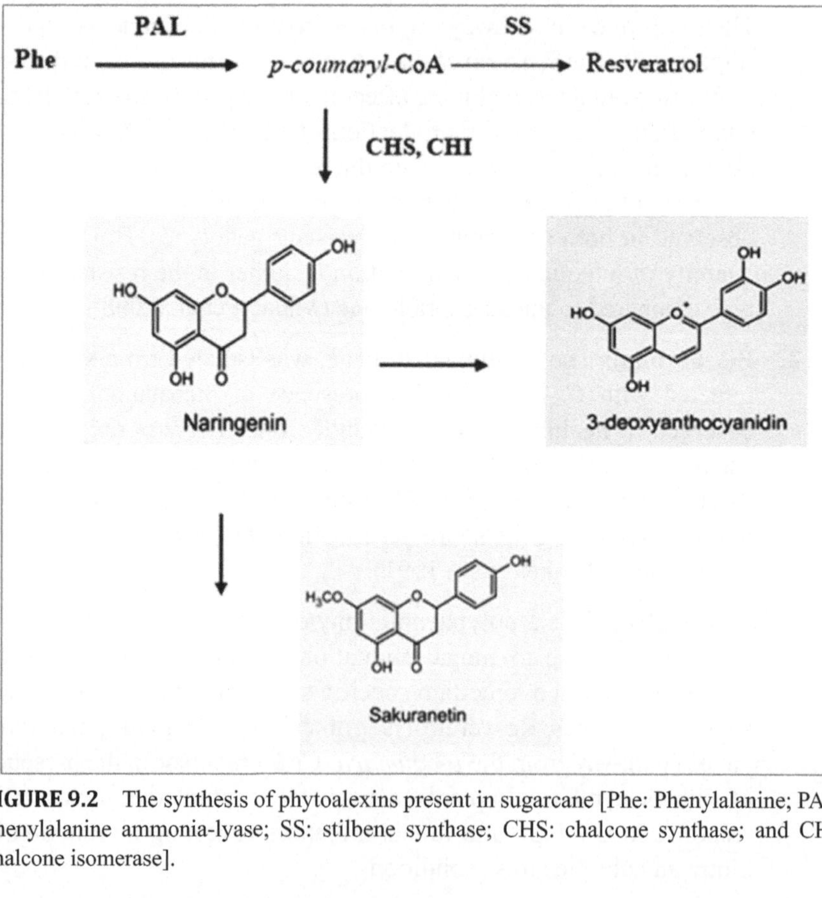

FIGURE 9.2 The synthesis of phytoalexins present in sugarcane [Phe: Phenylalanine; PAL: phenylalanine ammonia-lyase; SS: stilbene synthase; CHS: chalcone synthase; and CHI: chalcone isomerase].

9.2.1.3.1 Biosynthesis of Terpenes: Isoprenoid Pathway

Terpenes are biosynthesized through the mevalonic acid pathway and methyl erythritol phosphate pathway. The mevalonic acid pathway occurs in the cytosol. The three molecules of acetyl CoA result in the formation of mevalonic acid. The resulting mevalonic acid undergoes decarboxylation, dehydration, and phosphorylation to form the 5C compound Isopentenyl diphosphate (IPP). This pathway occurs in chloroplasts. On the other side, glyceraldehyde-3-phosphate and pyruvate yield methylerythritol phosphate. By undergoing phosphorylation and dehydration, it results in the formation of the 5C Dimethylallyl diphosphate (DMAPP). The IPP and DMAPP combine to give the 10C compound Geranyl diphosphate (GPP), a precursor of almost all monoterpenes. GPP combines with other molecules of IPP, leading to Farnesyl diphosphate (15C), from which sesquiterpenes arise. FPP combining with another IPP leads to the formation of Geranyl Geranyl diphosphate (GGPP, 20C), which can give rise to Diterpenes and Tetraterpenes. Some of the well-known triterpenes occurring in sugarcane and their simplified biosynthesis pathway are illustrated in Figure 9.3.

9.2.1.3.2 Annotation of Terpenes Associated with Unfavorable Environmental Conditions

The function of terpenes as feeding deterrents and antimicrobial compounds is well described. Terpenes play a major role in mitigating stresses, curtailing herbivores, and regulating the plant's physiological responses through phytohormones, such as Gibberellic acid (GA_3) and Abscisic acid (ABA). Plant and insect interactions represent a perpetual arms race, a continuous process of outsmarting one another, which involves a small fraction (1%) of low molecular weight compounds (300 Da) known as plant volatiles or Volatile organic compounds (VOCs) or volatile terpenes. Volatile organic compounds work in a manner that benefits plants and is detrimental to insects/animals. Overall VOCs contribute to stress alleviation, plant microbes, and herbivores trine. Some of the VTs are modified by glycosylation for storage and are considered "Hidden terpenoids," which undergo hydrolysis on exposure to unfavorable conditions (Yazaki et al., 2017). VTs can mitigate oxidative stress by modulating the two signature responses, such as oxidative status and volatile emission profile of the plant under unfavorable conditions. Some terpene synthases induced during UV exposure include squalene synthases and sesquiterpene cyclases (Franca et al., 2001).

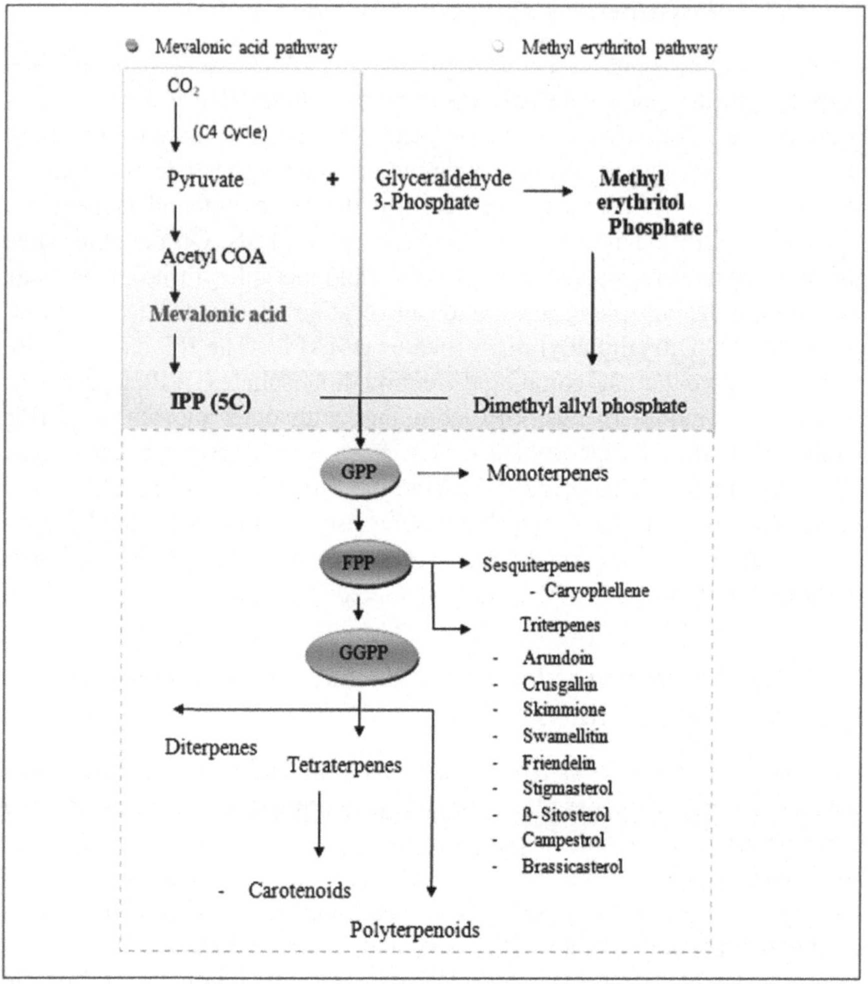

FIGURE 9.3 A simplified view of terpene biosynthesis pathway in sugarcane. [IPP: Isopentenyl diphosphate (5C); FPP: farnesyl diphosphate (15C); and GGPP: geranylgeranyl diphosphate (20C)].

The plant-insect interactions are mediated by Volatile Organic Compounds (VOCs) involving toxic, repellent, and attracting activities. In VOCs, a substantial fraction of terpenes act as elicitors in triggering the defense mechanisms against pests. Besides their role as secondary metabolites, terpenes assist in essential primary functions such as growth and development. For instance, the principal plant growth hormone gibberellic acid is a diterpene compound, in the formation of abscisic acid, a sesquiterpene, and in the case of indispensable cellular components such as sterols formed by triterpenes. The terpene-derived phytol side chain of chlorophyll, in the formation of carotenoids which function as accessory pigments, and in photoprotection against photooxidation. Carotenoids are potential antioxidants, scavengers, quenchers, and membrane stabilizers under stress conditions (Uarrota et al., 2018). Some of the triterpenoids in sugarcane include arundoin, crusgallin, cylindrin, Skimmione, Swamilletin, simiarenol, Friedelin, and α-amyrin.

9.2.1.4 PHENOLIC COMPOUNDS

Phenolics are widely diversified secondary plant products. Their basic structure consists of a phenol group, i.e., a hydroxyl group attached directly to one or more aromatic/benzene rings. The plant phenolics class consists of a substantial heterogeneous compound based on their structures; they are categorized into different groups, i.e., phenolics, flavonoids (flavones, flavonols, isoflavones, anthocyanins, and stilbenoids), lignins, and tannins. Based on their distinct structure, phenolics include hydroxybenzoic (HBAs) and hydroxycinnamic (HCAs) acids and their derivatives (Table 9.1).

9.2.1.4.1 Biosynthesis of Phenolic Compounds: Phenylpropanoid Pathway

The biosynthesis mechanism of phenols is synthesized from the primary metabolites such as erythrose-4-phosphate and phosphoenolpyruvate which are derived from the reductive pentose phosphate pathway and Hexose diphosphate pathway. The shikimic acid pathway converts these simple carbohydrate precursors into aromatic amino acids, resulting in phenylalanine. Phenylalanine is converted into cinnamic acid in the presence of the enzyme phenylalanine ammonia-lyase (PAL). Relevant phenylpropanoids are synthesized from cinnamic acid by hydration, dehydration, and methylation. Subsequent reactions catalyzed by PAL result in the formation of *p*-coumaric acid and its phenylpropanoid derivatives. The most studied enzyme in plant secondary metabolism is PAL. The increment of PAL activity is observed during unfavorable environmental factors such as increased light intensity, microbial infection, and reduced nutrient levels. Hydroxybenzoic acids are formed from early intermediates of the shikimic acid pathway. They include salicylic acid, gallic acid, vanillic acid, and various alternative forms of dihydroxybenzoic acids (DHBA) such as 2,3-DHBA, 3,5-DHBA, 2,5-DHBA, 3,4-DHBA. HCAs comprise ferulic acid, *p*-coumaric acid, sinapic acid, chlorogenic acid, and spermidine. The rate-limiting flavonoid genes include Chalcone synthase and Chalcone isomerase. The protein degradation emanated by abiotic stress conditions induces NADPH oxidation, and the increased NADP+/NADPH steers up the pentose phosphate pathway. The accelerated phenylpropanoid pathway provides precursors to the shikimic acid pathway, which in turn renders the biosynthesis of the phenylpropanoid pathway (Lattanzio et al., 2013) (Figure 9.4).

9.2.1.4.2 Annotation of Phenolic Compounds Associated with Unfavorable Environmental Conditions

Phenolic compounds are the pivotal secondary compounds that arise in response to plant stress conditions, which have the virtuous capacity to scavenge the reactive oxygen species that evolve during unfavorable conditions. They are powerful antioxidants triggered during stress conditions through the phenylpropanoid biosynthetic pathway. Some of the phenolic compounds that occur in sugarcane include Apigenin, Luteolin, Piceatannol, Ferulic acid, Coumaric acid, Caffeic acid, Ellagic acid, Gallic acid, Chlorogenic acid, Quercetin, isoquercetin, Naringenin, Rutin, Swertecin, Epicatechin, Schaftoside, Isoschaftoside, Orientin, Vitexin, Tricin, and their derivatives such as arabinose-5-O-apigenin, Luteolin-8-c-(rhamnosyl-glucoside), 4′,5′-dimethyl-luteolin-8-c-glycoside, Tricin-7-O-neohesperidoxide, Tricin-7-O-glucoside, Diosmetin-8-C-glycoside. Phenolic compounds reverse the negative effects caused by heavy metal stress. They act as antimicrobial and antioxidant compounds. The antioxidant capacity of flavonoids dominates the well-known major reactive nitrogen species, such as Peroxynitrite (Singh et al., 2021).

The accretion of plant phenolic compounds would result in the inflation of plant tolerance over high temperature, ultraviolet light, drought, cold, heavy metal, and nutrient stress (Naikoo et al., 2019). Apart from that, they play an essential role in pollination and seed dispersal by serving as attractants and protecting chloroplasts. In sugarcane, the enzymes such as sesquiterpene cyclases and chalcone synthase are induced due to ultraviolet radiation. Isoflavone reductase and Isoflavone synthase during plant defense and root nodulation, caffeic acid O-methyltransferase in the biosynthesis of lignin (Franca et al., 2001). Stress factor in case of metal stress and Reactive oxygen species like hydroxyl free radicals (OH^-), superoxide free radicals (O_2^-), and non-free radicals like hydrogen peroxide (H_2O_2) and singlet oxygen (O_2^*) along with methylglyoxal (cytotoxic compound) evolved during oxidative stress can serve as signaling factors in the abiotic elicitation of phenols production. The yielded phenolic compounds in sugarcane are present in leaves and mature culm tissue. The plants exposed to high irradiance of UV light are observed to increase flavonoid biosynthesis. Flavonoids act as ultraviolet sunscreens by preventing the entry of UV-B into the mesophyll cells. The hydroxybenzoic phenol compounds, such as vanillic acid, correspond to drought stress (Li et al., 2018) and heavy metal stress tolerance (Kisa et al., 2016). The pigment group of flavonoids, such as anthocyanins synthesis and accumulation, is initiated by UV-B light (Table 9.1).

Flavonoids retain the power to modify proteins under unfavorable conditions. The specific anthocyanins in sugarcane include cyanidin and petunidin (Zhao et al., 2018). The high anthocyanin contents in sugarcane leaves exposed to cold stress represent their functions as antioxidants under low-temperature stress (Yang et al., 2021). The elevated phenolics and flavonoids in response to drought stress increase the tolerance mechanism of plants. In particular, Quercetin is an efficient antioxidant that regulates plant physiological processes during drought. Plants under highly fluctuating environmental conditions adjust their physiological mechanism by regulating growth hormones. High temperature increases endogenous auxin levels. It is also reported that enhanced IAA levels in a plant contribute to more tolerance towards drought and oxidative stress. It also prevents male sterility (Cheng et al., 2006). The surge in IAA levels upshots the production of flavonols, reducing the IAA radicals developed during IAA oxidation under stress conditions. The increase in IAA results in the increased production of Quercetin, which modulates the transport and signaling of auxin by interrupting the auxin efflux proteins (PIN) (Adamowski & Friml, 2015). Quercetin regulates the ABA-induced stomatal closure mechanism under stress (Agati et al., 2009). Dihydroxy B ring flavonoids like luteolin 7-O glycosides and Quercetin 3-O-glycosides serve as UV-stimulated ROS scavenging in sugarcane (Agati et al., 2012).

9.2.1.5 NITROGEN-CONTAINING COMPOUNDS

They are plant secondary metabolites containing nitrogen in their structure derived from amino acids. These metabolites are categorized into four types such as alkaloids, cyanogenic glycosides, glucosinolates, and non-protein amino acids. Their function mainly involves acting as toxic antifeedants and as precursors of defense systems in the resistant mechanism against pests, pathogens, and herbivores. In the cell, they are spatially isolated from cellular compounds (Table 9.1).

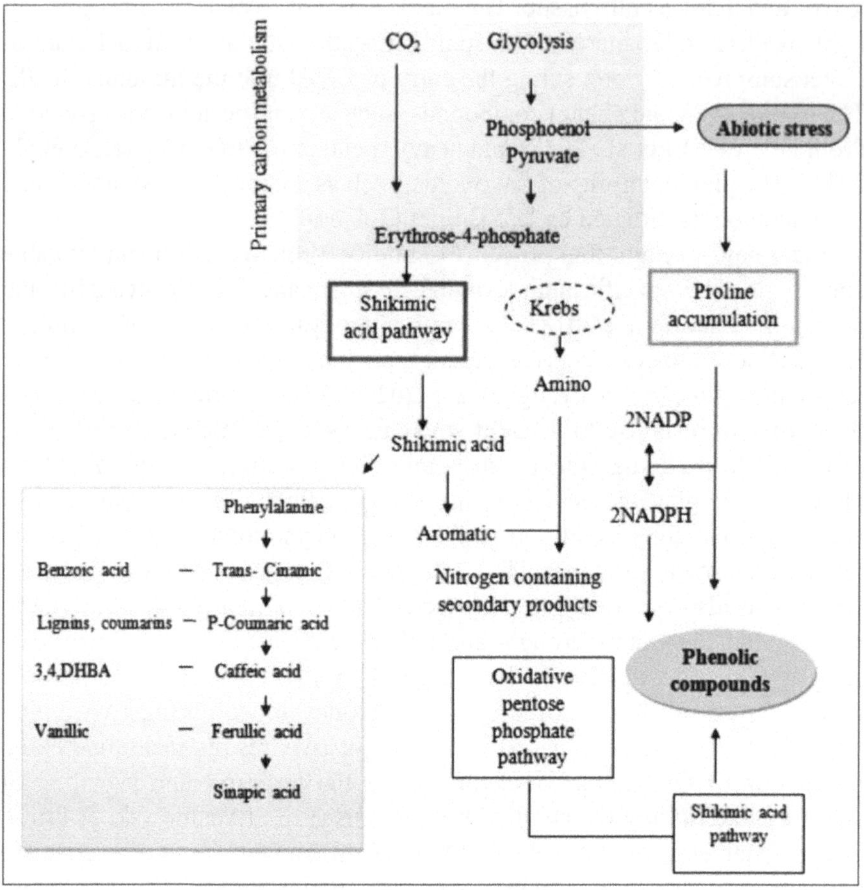

FIGURE 9.4 The schematic representation of the stress-induced phenylpropanoids pathway.

9.2.1.5.1 Annotation of Nitrogen-Containing Compounds Associated with Unfavorable Environmental Conditions

In sugarcane, four alkaloids are discovered, i.e., trigonelline, piperidine, 3-indole acrylic acid, and 6-deoxyfagomine (Sabino et al., 2019). The induced alkaloids at different seasons in sugarcane include Secologanin, Hypoxanthine, 7-methyluric acid, and 4-hydroxyphenyl acetaldehyde (Wijma et al., 2021). Trigonelline in plants has several regulatory functions, including cell cycle regulation, in the nutrition processes of nodulation and methylation. Besides, it has a protective role against salinity and oxidative stress (Minorsky et al., 2022). Trigonelline acts as an anti-insect compound by deterring the oviposition of female insects belonging to Lepidoptera and Papilionidae. One of the cyanogenic glycosides in sugarcane is Dhurrin (Wijma et al., 2021). Dhurrin plays an important role in drought resistance and as an antiherbivore. It is synthesized from tyrosine. This compound highly exists at the internodes contributing to the delay in plant maturation under unfavorable conditions (Figure 9.5).

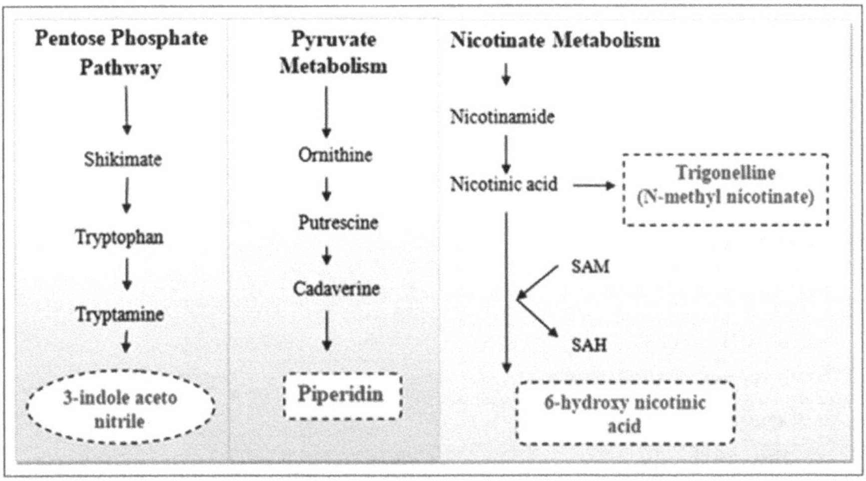

FIGURE 9.5 Different biosynthesis pathways of alkaloid compounds that are present in sugarcane. [SAM: S-adenosyl methionine; and SAH: S-adenosyl homocysteine].

9.3 CONCLUSION

Like many other crops, sugarcane (*Saccharum officinarum* L.) encounters unfavorable conditions. To tolerate these conditions, plants develop certain physiological, biochemical, and molecular mechanisms. Metabolomic mechanisms include the production of secondary metabolites. Based on the literature, some specific SMs employed in the biotic stress resistance of sugarcane include Arundoin, Piceatannol, Luteolinidin, (E)-Caryophyllene, Chlorogenic acid, and Apigeninidin that are developed in response to pathogens, insects, and herbivore trine. On the other hand, SMs involved in abiotic stress resistance include Dhurrin, Resveratrol, Spermidine, Trigonelline, Caffeic acid, Vanillin, Ellagic, and Quercetin. The potent key role of SMs in resisting abiotic stress involves defending primary metabolites, scavenging the reactive oxygen species, and regulating the plant growth hormones that alter the plant's physiological responses in accordance with stress. Certain SMs, such as quercetin, resveratrol, dhurrin, naringenin, apigeninidin, and luteolinidin resist both abiotic and biotic stresses. However, the role of SMs is not much exploited in sugarcane. Therefore, delving into the diverse role of SMs in sugarcane plants could contribute towards the production of tolerant genotypes as well as in the reduction of harmful pesticide load in the environment.

ACKNOWLEDGMENT

We would like to acknowledge and express our gratitude towards the Department of Plant Physiology, Jawaharlal Nehru Krishi Vishwavidyalaya, Jabalpur, Madhya Pradesh, India.

KEYWORDS

- **abiotic stress**
- **adverse conditions**
- **apigeninidin**
- **biosynthesis**
- **genotypes**
- **luteolinidin**
- **phenolic compounds**
- **piceatannol**
- **secondary metabolites**
- **sugarcane**
- **terpenes**

REFERENCES

Adamowski, M., & Friml, J. (2015). PIN-dependent auxin transport: action, regulation and evolution. *Plant Cell, 27*(1), 20–32.

Agati, G., Aarello, E., Pollastri, S., & Tattini, M. (2012). Flavonoids as antioxidants in plants: location and functional significance. *Plant Science, 196*, 67–76.

Agati, G., Stefano, G., Biricolti, S., & Tattini, M. (2009). Mesophyll distribution of antioxidant flavonoid glycosides in *Ligustrum vulgare* leaves under contrasting sunlight irradiance. *Annals of Botany, 104*(5), 853–861.

Aimé-Jhustelin, A. M., Roger, O. N., Serge, O. A. A., Marc, E. N., Constant, S. F., Aubin, O. J., & François, E. (2014). Origin of some chemical compounds of a sugarcane wine of *Saccharum officinarum* L. from centreville, Gabon. *Journal of Environmental Science, Toxicology and Food Technology, 8*(11), 70–72.

Anas, M., Verma, K. K., Haq, I. U., Naeem, M., Li, Q., Liao, F., Xiang, Y. X., & Li, Y. R. (2021). Characterization of exotic, native and wild-type genotypes of sugarcane (*Saccharum spp.* Hybrids) for internal nitrogen use efficiency under different nitrogen levels. *Sugar Tech. 23, 1258*–1267. https://doi.org/10.1007/s12355-021-01014-1.

Anas, M., Verma, K. K., Riaz, M., Qiang, L., Liao, F., Liu, Y., & Li, Y. R. (2021). Physio-morphological and biochemical mechanism of nitrogen use efficiency in sugarcane (*Saccharum spp.*) genotypes under different growth stages and nitrogen levels. *Journal of Plant Interactions, 16*, 332–343.

Brinker, A. M., & Seigler, D. S. (1991). Isolation and identification of piceatannol as a phytoalexin from sugarcane. *Phytochemistry, 30*(10), 3229–3232.

Bruneton, J. (1995). Pharmacognosy, phytochemistry and medicinal plants. Paris: Lavoisier Publishing.

Bryce, T. A., Martin-Smith, M., Osske, G., Schreiber, K., & Subamanian, G. (1967). Sterols and triterpenoids—XI: Isolation of arundoin and sawamilletin from cuban sugar cane wax. *Tetrahedron, 23*(3), 1283–1296.

Carvalho, I. S., Cavaco, T., Carvalho, L. M., & Duque, P. (2010). Effect of photoperiod on flavonoid pathway activity in sweet potato (*Ipomoea batatas (L.) Lam.*) leaves. *Food Chemistry, 118*(2), 384–390.

Cheng, Y. (2006). Auxin biosynthesis by the YUCCA flavin monooxygenases controls the formation of floral organs and vascular tissues in Arabidopsis. *Genes & Development, 20*(13), 1790–1799.

Deng, Z. N., Huang, H. R., Huang, C. M., Verma, K. K., Luo, H. B., Jiang, S. L., Li, Y. J., Cao, H. Q., Wu, K. C., & Xu, L. (2022). Short-term limited water irrigation influences the photosynthetic pigments, enzymatic and non-enzymatic activities in *Saccharum spontaneum* L. at vegetative stage. *Sugar Tech. 25, 110*–118. https://doi.org/10.1007/s12355-022-01175-7.

Franca, S. C., Roberto, P. G., Marins, M. A., Puga, R. D., Rodrigues, A., & Pereira, J. O. (2001). Biosynthesis of secondary metabolites in sugarcane. *Genetics and Molecular Biology, 24*(4), 243–250.

Grayer, R. J., & Harborne, J. B. (1994). A survey of antifungal compounds from higher plants 1982–1993. *Phytochemistry, 37*(1), 19–42.

Hu, B., Yao, H., Gao, Y., Wang, R., Li, F., Guo, J., Li, K., Zhao, M., & Jin, L. (2019). Overexpression of chalcone synthase gene improves flavonoid accumulation and drought tolerance in tobacco. *Research Square. 2*(5), 1–11.

Huang, H. R., Gao, Y. J., Malviya, M. K., Verma, K. K., Solanki, M. K., Huang, Y. X., Li, X., Deng, Y., Yan, J., Tang, S. Y., Wang, L. W., & Xu, L. (2021). Genetic diversity analysis of sugarcane (*Saccharum spp.* Hybrids) among high-sucrose clones of GT series and commonly used parents by using microsatellite markers in Guangxi, China. *Sugar Tech. 24, 397*–407. https://doi.org/10.1007/s12355-021-01033-y.

Hussain, S., Khaliq, A., Mehmood, U., Qadir, T., Saqib, M., Iqbal, M. A., & Hussain, S. (2019). Sugarcane production under changing climate, effects of environmental vulnerabilities on sugarcane diseases, insects and weeds. In *Sugarcane Production – Agronomic, Scientific and Industrial Perspectives*, Published: IntechOpen, pp. 1–16.

IPCC. (2013). *The physical science basis. Contribution of Working Group I to the Fifth Assessment Report of the Intergovernmental Panel on Climate Change.* USA: Cambridge University Press.

Jeandet, P. (2015). Phytoalexins: Current Progress and Future Prospects. *Molecules, 20*(2), 2770–2774.

Kessler, A., & Baldwin, I. T. (2001). Defensive function of herbivore-induced plant volatile emissions in nature. *Science, 291*, 2141–2144. DOI: 10.1126/science.291.5511.2141.

Khan, A., Nazar, S., Lang, I., Nawaz, H., & Hussain, M. H. (2017). Effect of ellagic acid on growth and physiology of canola (*Brassica napus* L.) under saline conditions. *Journal of Plant Interactions, 12*(1), 520–525.

Kısa, D., Elmasta, S. M., Öztürk, L., & Kayır, O. (2016). Responses of the phenolic compounds of *Zea mays* under heavy metal stress. *Applied Biological Chemistry, 59*(6), 813–820.

Kong, F., Yu, S., Zeng, F., & Wu, X. (2015). Preparation of antioxidant and evaluation of the antioxidant activities of antioxidants extracted from sugarcane products. *Journal of Food and Nutrition Research, 3*(7), 458–463.

Kumar, D., Malik, N., & Sengar, R. S. (2019). Physio-biochemical insights into sugarcane genotypes under water stress. *Biological Rhythm Research, 52*(1), 92–115.

Kumar, V. G., Vishwanathan, R., Malathi, P., Nandakumar, M., & Sundar, A. R. (2015). Differential Induction of 3-deoxyanthocyanidin Phytoalexins in Relation to *Colletotrichum falcatum* Resistance in Sugarcane. *Sugar Tech., 17*(3), 314–321.

Kumari, A., & Singh, S. K. (2022). Role of plant hormones in combating biotic stress in plants. In Plant Perspectives to Global Climate Changes. In Tariq Aftab, Aryadeep Roychoudhury, eds. Academic Press, pp. 373–391.

Kumari, A., Lakshmi, G. A., Krishna, G. K., Patni, B., Prakash, S., Bhattacharyya, M., Singh, S. K., & Verma, K. K. (2022). Climate Change and Its Impact on Crops: A Comprehensive Investigation for Sustainable Agriculture. *Agronomy, 12*(12), 1–16.

Lattanzio, V. (2013). Phenolic compounds: introduction. In *Natural Products*, Edited by Ramawat, K. G., Merillon, J. M. Springer-Verlag, Berlin, Heidelberg, pp. 1543–1580.

Le Gall, H., Fontaine, J. X., Molinie, R., Pelloux, J., Mesnard, F., Gillet, F., & Fliniaux, O. (2017). NMR-based metabolomics to study the cold acclimation strategy of two miscanthus genotypes. *Phytochemical Analysis, 28*(1), 58–67.

Li, M., Li, Y., Zhang, W., Li, S., Gao, Y., Ai, X., Zhang, D., Liu, B., & Li, Q. (2018). Metabolomics analysis reveals that elevated atmospheric CO2 alleviates drought stress in cucumber seedling leaves. *Analytical Biochemistry, 2*(559), 71–85.

Linić, I., Mlinarić, S., Brkljačić, L., Pavlović, I., Smolko, A., & Salopek-Sondi, B. (2021). Ferulic Acid and Salicylic Acid Foliar Treatments Reduce Short-Term Salt Stress in Chinese Cabbage by Increasing Phenolic Compounds Accumulation and Photosynthetic Performance. *Plants (Basel), 10*(11), 2346.

Malathi, P., Viswanathan, R., Padmanaban, P., Mohanraj, D., Kumar, V. G., & Salin, K. P. (2008). Differential accumulation of 3-deoxy anthocyanidin phytoalexins in sugarcane varieties varying in red rot resistance in response to *Colletotrichum falcatum* infection. *Sugar Tech., 10*(2), 154–157.

Mekawy, A. M. M., Abdelaziz, M. N., & Ueda, A. (2018). Apigenin pretreatment enhances growth and salinity tolerance of rice seedlings. *Plant Physiology and Biochemistry, 130*(1), 94–104.

Misheck, C. (2013). Potential impact of climate change in sugarcane and mitigation strategies in Zimbabwe. *African Journal of Agricultural Research, 8*(23), 2814–2818.

Muller, K. O., & Borger, H. (1940). Experimentelle Untersuchungen uber die Phytophthora-Resistenz der Kartoffel. *Arbeiten aus dem Biologischen Reichsanstalt für Land- und Forstwirtschaft in Berlin, 23*(1), 189–231.

Ni, Y., Chen, H., Liu, D., Zheng, L., Chen, P., & Liu, C. (2021). Discovery of genes involved in anthocyanin biosynthesis from the rind and pith of three sugarcane varieties using integrated metabolic profiling and RNA-seq analysis. *BMC Plant Biology, 21*(214), 1–15.

Nuissiera, G., Bourgeoisa, P., Micheline Grignon-Duboisb, M., PatrickPardonb, M., & He´le`ne Lescureb, M. H. (2001). Composition of sugarcane waxes in rum factory wastes. *Phytochemistry, 61*(6), 721–726.

Oliveira, M. A., Verma, K. K., Bhatt, R., & Oliveira, T. B. A. (2022). Impact of green and organic fertilizers on soil fertility and sugarcane productivity. In K. K. Verma et al. (Eds.), *Agro-industrial perspectives on sugarcane production under environmental stress*. https://doi.org/10.1007/978-981-19-3955-6_11.

Pathak, V., & Tiwari, V. K. (2017). Phytochemical Screening of *Saccharum Officinarum* (Linn.) Stem. *International Journal of Innovative Science and Research Technology, 2*(8), 291–305.

Piasecka, A., Jedrzejczak-Rey, N., & Bednarek, P. (2015). Secondary metabolites in plant innate immunity: conserved function of divergent chemicals. *New Phytologist, 206*(3), 948–964.

Pulvento, C., Riccardi, M., Lavini, A., Iafelice, G., Marconi, E., & d'Andria, R. (2012). Yield and quality characteristics of quinoa grown in open field under different saline and non-saline irrigation regimes. *Journal of Agronomy and Crop Science, 198*(4), 254–263.

Ramakrishna, A., & Ravishankar, G. A. (2011). Influence of abiotic stress signals on secondary metabolites in plants. *Plant Signaling & Behavior, 6*(11), 1720–1731.

Riffel, A., Silva, B. F., dos Santos, S. P. A., da Silva, W. L., Ribeiro, T. F. L., Oliveira, D. J. A., de Oliveira, J. S., da Costa, J. G., Caulfield, J. C., Goulart, H. F., Santana, A. E. G., & Birkett, M. A. (2021). Exposure to sugarcane borer-induced plant volatile (E)-caryophyllene enhances parasitoid recruitment. *Entomologia Experimentalis et Applicata, 169*(10), 1–10.

Ruan, H., Feng, P., Wang, B., Xing, H., O'Leary, G. J., Huang, Z., Guo, H., & Liu, D. L. (2018). Future climate change projects positive impacts on sugarcane productivity in southern China. *European Journal of Agronomy, 96*(1), 108–119.

Sabino, A. R., Tavares, S. S., Riffel, A., Li, J. V., Oliveira, D. J. A., Feres, C. I. M. A., Henrique, L., Oliveira, J. S., Correia, G. D. S., Sabino, A. R., Nascimento, T. G., Hawkes, G., Santana, A. E. G., Holmes, E., & Bento, E. S. (2019). 1H NMR metabolomic approach reveals chlorogenic acid as a response of sugarcane induced by exposure to *Diatraea saccharalis*. *Industrial Crops and Products, 140*(111651), 1–6.

Sarri, E., Termentzi, A., Abraham, E., Papadopoulos, G. K., Baira, E., Machera, K., Loukas, V., Komaitis, F., & Tani, E. (2021). Salinity Stress Alters the Secondary Metabolic Profile of *M. sativa*, *M. arborea* and Their Hybrid (*Alborea*). *International Journal of Molecular Sciences, 22*(9), 1–19.

Schmelz, E. A., Huffaker, A., Sims, J. W., Christensen, S. A., Lu, X., Okada, K., & Peters, R. J. (2014). Biosynthesis, elicitation and roles of monocot terpenoid phytoalexins. *Plant Journal, 79*(4), 659–678.

Sidana, J., Singh, B., & Sharma, O. P. (2018). 5α/5ß Stereochemistry of Spirostanol and Furostanol Saponins. *Current Science, 115*(4), 618–621.

Singh, P., Arif, Y., Bajguz, B., & Hayat, S. (2021). The role of quercetin in plants. *Plant Physiology and Biochemistry, 166*, 10–19.

Smith, R. M., & Martin-Smith, M. M. (1978). Triterpene methyl esters in leaf waxes of *Saccharum* and related genera. *Phytochemistry, 17*(1), 1307–1312.

Song, X. P., Mo, F., Verma, K. K., Wei, J., Zhang, X., Yang, L., & Li, Y. R. (2019). Effect of sugarcane smut (*Ustilago scitaminea* Syd.) on ultrastructure and biochemical indices of sugarcane. *Biomed Journal of Scientific and Technical Research, 17*(1), 12546–12551. BJSTR. MS.ID.002950.

Song, X. P., Mo, F., Yan, M., Zhang, X., Zhang, B., Huang, X., Huang, D., Pan, Y., Verma, K. K., & Li, Y. R. (2022). Effect of smut infection on the photosynthetic physiological characteristics and related defense enzymes of sugarcane. *Life, 12*, 1201. https://doi.org/10.3390/life12081201.

Song, X. P., Verma, K. K., Tian, D. D., Zhang, X. Q., Liang, Y. J., Huang, X., Li, C. N., & Li, Y. R. (2021). Exploration of silicon functions to integrate with biotic stress tolerance and crop improvement. *Biological Research, 54*, 19. https://doi.org/10.1186/s40659-021-00344-4.

Teixeira, S., Vidigal, S. S. M. P., Pimentel, L. L., Costa, T. P., Pintado, M. E., & Rodríguez-Alcalá, L. M. (2021). Bioactive Sugarcane Lipids in a Circular Economy Context. *Foods, 10*(5), 1125.

Uarrota, V. G., Stefen, D. L. V., Leolato, L. S., Gindri, D. M., & Nerling, D. (2018). Revisiting carotenoids and their role in plant stress responses: From biosynthesis to plant signaling mechanisms during stress. In *Antioxidants and Antioxidant Enzymes in Higher Plants* (pp. 207–232). Springer International Publishing. https://doi.org/10.1007/978-3-319-75088-0_10.

Vafadar, F., Amooaghaie, R., Ehsanzadeh, P., & Ghanadian, M. (2020). Salinity stress alters ion homeostasis, antioxidant activities and the production of rosmarinic acid, luteolin and apigenin in *Dracocephalum kotschyi* Boiss. *Biologia, 75*(12), 1–14.

VanEtten, H. D., Mansfield, J. W., Bailey, J. A., & Farmer, E. E. (1994). Two classes of plant antibiotics – phytoalexins versus "phytoanticipins." *Plant Cell, 6*(9), 1191–1192.

Verma, K. K., Song, X. P., Rajput, V. D., Boldyreva, V., Zhang, B. Q., Minkina, T., & Li, Y. R. (2022). Morpho-physiological, biochemical and ultrastructural modifications on sugarcane to prolonged water deficit. In Verma KK., et al. (eds.), *Agro-industrial perspectives on sugarcane production under environmental stress*. https://doi.org/10.1007/978-981-19-3955-6_11.

Verma, K. K., Song, X. P., Rajput, V. D., Solomon, S., Li, Y. R., & Rao, G. P. (Eds.). (2023). *Agro-industrial Perspectives on Sugarcane Production under Environmental Stress*. Springer, Singapore. https://doi.org/10.1007/978-981-19-3955-6_1.

Wan, Y. Y., Zhang, Y., Zhang, L., Zhou, Z., Li, X., Shi, Q., Wang, X., & Bai, J. (2015). Caffeic acid protects cucumber against chilling stress by regulating antioxidant enzyme activity and proline and soluble sugar contents. *Acta Physiologiae Plantarum, 37*(1706), 1–10.

Weng, J. K., Philippe, R. N., & Noel, J. P. (2012). The rise of chemodiversity in plants. *Science, 336*(6089), 1667–1670.

Wijma, M., Lembke, C. G., Diniz, A. L., Santini, L., Zambotti-Villela, L., Colepicolo, P., Carneiro, M. S., & Souza, G. M. (2012). Planting Season Impacts Sugarcane Stem Development, Secondary Metabolite Levels, and Natural Antisense Transcription. *Cells, 10*(3451), 1–33.

Windisch, S., Walter, A., Moradtalab, N., Walker, F., Höglinger, B., El-Hasan, A., & Grosch, R. (2021). Role of Benzoic Acid and Lettucenin A in the Defense Response of Lettuce against Soil-Borne Pathogens. *Plants, 10*(11), 2336.

Xu, L., Deng, Z. N., Wu, K. C., Malviya, M. K., Solanki, M. K., Verma, K. K., Ludewig, U., Neumann, G., & Huang, H. R. (2022). Transcriptome analysis reveals a gene expression pattern that contributes to sugarcane bud propagation induced by Indole-3-Butyric Acid. *Frontiers in Plant Science, 13*, 852886. doi: 10.3389/fpls.2022.852886.

Yazaki, K., Arimura, G., & Ohnishi, T. (2017). *'HiddenTerpenoids'* in Plants: Their Biosynthesis, Localization and Ecological Roles. *Plant Cell Physiology, 58*(10), 1615–1621.

Zhao, Z., Yan, H., Zheng, R., Khan, M. S., Fu, X., Tao, Z., & Zhang, Z. (2018). Anthocyanins characterization and antioxidant activities of sugarcane (*Saccharum officinarum* L.) rind extracts. *Industrial Crops and Products, 113*, 38–45.

CHAPTER 10

Major Sugarcane Pests and Their Management in Sri Lanka: Current Status and Future Outlook

GAYANI K. M. CHANCHALA,[1] VKASM WANASINGHE,[1] and AMIN NIKPAY[2]

[1]*Sugarcane Research Institute, Udawalawe, Sri Lanka*

[2]*Department of Plant Protection, Sugarcane and By-Products Development Company, Salman Farsi Agro-Industry, Ahwaz, Iran*

ABSTRACT

Sugarcane (*Saccharum officinarum* L.) is one of the most important plantation crops currently grown in Sri Lanka. The sugarcane industry of Sri Lanka has been affected by several pest and disease problems over the decades. Commercial sugarcane plantations were recorded to be associated with 103 insect species, including tissue borers, sucking pests, and cane grubs. Sugarcane Wooly Aphid (SWA), *Pyrilla perpusilla,* termite, *Sesamia inferens* (Shoot borer; SB), *Chilo sacchariphagus* (Internode borer; INB), and *Deltocephalus menoni*; vector of Sugarcane White Leaf Disease (WLD) have become major pests throughout the period while sugarcane spider mite and armyworm were recorded as occasional pests. Except for the locally available pest, Fall Armyworm (FAW) was recorded as a newly introduced pest to the country, and sugarcane met sudden outbreak conditions. They were successfully addressed in an integrated manner using biological control agents, cultural control, resistant varieties, and other cultural management strategies. Within the integrated approach, using biological control agents, augmentation in the natural ecosystem, and cultural pest control have been the most adapted and successful strategies. In contrast, synthetic chemicals

Sugarcane Cultivation and Management: Challenges and Opportunities. Krishan K. Verma, Xiu-Peng Song, Munna Singh, Renato de Mello Prado, Jian-Ming Wu, and Yang-Rui Li (Eds.)

have been adapted minimally to the system. There are some advances to be added to the IPM programs with new machine learning techniques, biotechnological approaches, and volatile attractants.

10.1 INTRODUCTION

Sugarcane (*Saccharum officinarum* L.) is one of the most important plantation crops currently grown in Sri Lanka and used to manufacture sugar at the industrial level. Sugar is one of the major food commodities in Sri Lanka (Ariyarathna et al., 2007). In Sri Lanka, the sugarcane crop is cultivated on nearly 18,000 ha. It is an important sub-sector of the economy as it makes a significant contribution to the national balance of payments in addition to its contributions to income, employment generation, and development of rural areas in the dry and intermediate zones of the country, specifically in the districts of Monaragala and Badulla in Uva Province and Ampara and Trincomalee in the Eastern Province (Keerthipala, 2016). The sugarcane industry of Sri Lanka has been affected by several pest and disease problems over the decades (Wanasinghe & Chanchala, 2020). Commercial sugarcane plantations were recorded to be associated with 103 insect species (Kumarasinghe, 2003; Song et al., 2019, 2021, 2022), including tissue borers, sucking pests, and cane grubs.

Sugarcane Woolly Aphid (SWA), *Pyrilla perpusilla*, termite, *Sesamia inferens* (Shoot borer; SB), *Chilo schariphagus* (Internode borer; INB), and *Deltocephalus menoni*; vector of Sugarcane White Leaf Disease (WLD) have become major pests throughout the period, while Sugarcane spider mite and Armyworm are recorded as occasional pests (Wanasinghe & Chanchala, 2020). Except for the locally available pest, Fall Army Worm (FAW) was recorded as a newly introduced pest to the country, and sugarcane met sudden outbreak conditions (Wanasinghe et al., 2019). Integrated Pest Management (IPM) packages have been developed and introduced to the industry and farmers to manage sugarcane pests. Using agronomic, biological, cultural, and physical methods and resistant plants to suppress pest populations in sugarcane contributed to the maximization of production. Applying small amounts of selective pesticides according to the pest forecasting schedules also contributed to lowering the pest damages in case of new pests or pest outbreaks to manage the initial populations until other strategies are developed and natural enemies are introduced and adapted to the ecosystem (Verma et al., 2023).

10.2 MAJOR SUGARCANE PESTS IN SRI LANKA

10.2.1 SUGARCANE MOTH BORERS, INTERNODE BORERS (INB), AND SHOOT BORERS (SB)

Internode borers (INB), *Chilosacchariphagus* (Lepidoptera: Pyralidae) and Shoot borers (SB), Sesamiainferens (Lepidoptera: Noctuidae) are the major pest groups found in the area, and high to moderate infestations have been observed, mainly in the dry seasons of each year. An increasing trend of borer damage incidences has been found in sugarcane plantations in Sri Lanka since 2016, and it reached an injurious economic level in 2019. The larvae of both species bore young cane shoots close to the ground level resulting in the death of the host plants, and therefore both species can be considered as shoot borers. The symptoms of damage are the characteristic "dead heart," where the central whorl dries up. Larvae of *Chilosacchariphagus* are also responsible for the attack on the millable cane stalks causing poor quality seed-cane production in seed-cane nurseries and reduced sugar production due to low juice quality by reducing normal extraction (Brix, Pol, Purity) and increasing impurities (nitrogen, gums, ash) (Figure 10.1).

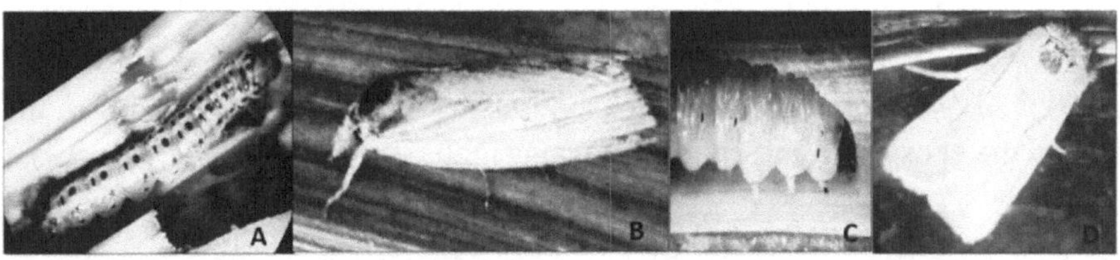

FIGURE 10.1 (A) and (B) Larva and adult of *Chilo sacchariphagus* (internode borer (INB)); (C) and (D) larva and adult of *Sesamia inferens* (pink borer).

Source: Photo gallery of the Division of Crop Protection, SRI.

10.2.2 WLD VECTOR; DELTOCEPHALUS MENONI (HEMIPTERA: CICADELLIDAE)

Deltocephalus menoni (Hemiptera: Cicadellidae) was identified as the vector of WLD (Senevirathne, 2008; Chanchala et al., 2014) which transmits the sugarcane White Leaf Disease (WLD), one of the most devastating phytoplasma diseases of sugarcane in Sri Lanka (Figure 10.2). *D. menoni* is a semi-persistent vector (Chanchala et al., 2022), where 58% of the natural population is viruliferous and can transmit for up to 11 days after acquisition. Both male and female adults and nymphs could also transmit the WLD phytoplasma at rates of 45%, 56%, and ≤33%, respectively (Chanchala et al., 2019).

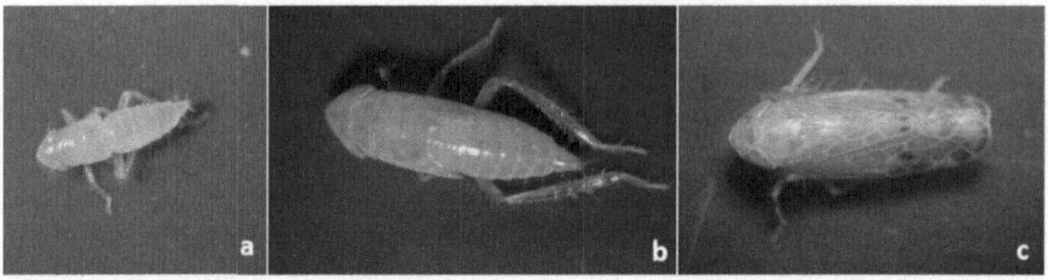

FIGURE 10.2 Stages of *D. menonisuch* as (a) 2nd instar nymphs; (b) 4th instar nymphs; and (c) adult.

Source: Photo Gallery of the Division of Crop Protection, SRI.

10.3 SUGARCANE TERMITES

Termites are the most common soil-borne pathogens, especially when sugarcane is planted in a rain-fed environment, which damages the sugarcane crop throughout its growing stages. Total crop losses have been observed due to termites in dry areas of Sri Lanka (Unpublished data). Furthermore, Reddish Brown Earths (RBE), the dominant soil type found in the present and potential commercial sugarcane growing regions on the island, are prone to termite infestation due to their sandy clay-loam texture. *Coptotermes ceylonicus, Heterotermes ceylonicus, Odontotermes ceylonicus, Odontotermes redimani,* and *Odontotermes horni* are the most common termite species in the sugarcane ecosystem in Sri Lanka (Wanasinghe et al., 2014) (Figure 10.3).

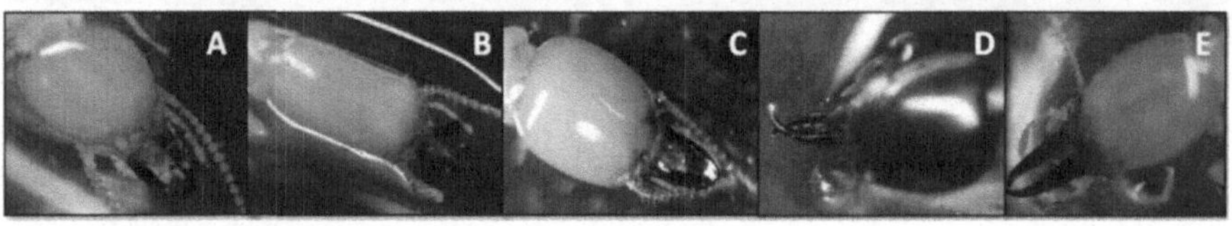

FIGURE 10.3 *Coptotermes ceylonicus*, *Heterotermes ceylonicus*, *Odontotermes ceylonicus*, *Odontotermes redimani*, and *Odontotermes horni*.

Source: Photo Gallery of the Division of Crop Protection, SRI.

10.3.1 SUGARCANE PLANTHOPPER: PYRILLA PERPUSILLA WALKER (HOMOPTERA: LOPHOPIDAE)

In Sri Lanka, the sugarcane leaf hopper *Pyrilla perpusilla* Walker (Homoptera: Lophopidae) was first reported in 1968 (Rajendra, 1979). It had been a serious sap-sucking pest during the late 1980s and the early 1990s in the sugarcane plantations at Sevanagala in the Uva Province of Sri Lanka, where sugarcane cultivation is mainly carried out (Senevirathne & Kumarasinghe, 2002) (Figure 10.4).

FIGURE 10.4 Stages of *Pyrilla perpusilla*, i.e., (a) egg cluster; (b) nymph; and (c) adult.

Source: Photo Gallery of the Division of Crop Protection, SRI.

10.3.2 SUGARCANE WOOLLY APHID (SWA): CERATOVACUNA LANIGERA (HOMOPTERA: APHIDIDAE)

Sugarcane Woolly Aphid (SWA): Ceratovacuna lanigera (Homoptera: Aphididae) is a serious sap-sucking pest of sugarcane all over the world, and it was first reported in Sri Lanka during the year 2006 from mid-country hill areas of Badulla (Kumarasinghe, 2007). Since then, it has turned out to be one of the major reasons for the decline in sugar production throughout sugarcane cultivating areas (Figure 10.5).

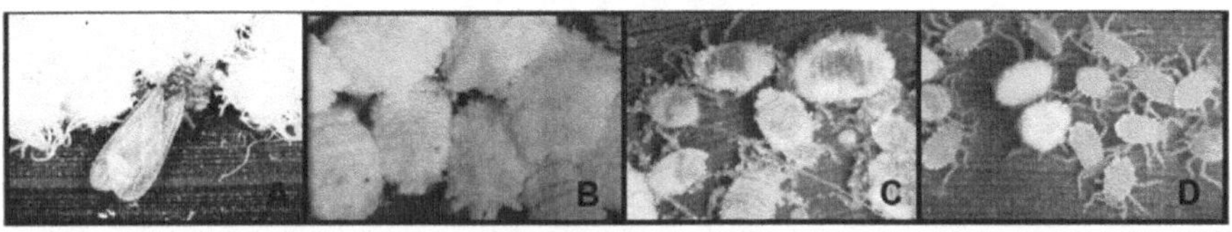

FIGURE 10.5 Stages of woolly sugarcane aphids, such as (A) adult-winged form; (B) adult-wingless form; and (C) and (D) nymphal stages.

Source: Photo Gallery of the Division of Crop Protection, SRI.

10.3.3 SUGARCANE SPIDER MITE: OLIGONYCHUS SACCHARI (TETRANYCHIDAE)

Sugarcane spider mite (SSM) is a seasonal pest of sugarcane in Sri Lanka that appears in the year's dry season (March-August). Usually, mites do not cause major economic damage to sugarcane. Recently, mite invasions occurred more frequently with high severity of infection in sugarcane growing areas belonging to the country's intermediate zone.

10.4 IPM STRATEGIES FOR SUGARCANE PESTS IN SRI LANKA

10.4.1 BIOLOGICAL CONTROL

Most economically important pests have been successfully managed using locally identified and imported biological control agents. One of the most significant achievements in the management of Pyrilla perpusilla Walker is *Epiricania melanoleuca* (Lepidoptera: Epipyropidae) imported from Pakistan. *Pyrilla perpusilla* was first reported in Sri Lanka in 1968 in commercial sugarcane plantations by Rajendra (1979). Several outbreaks of *Pyrilla perpusilla* have occurred in sugarcane plantations, especially during dry periods. Egg parasitoid *Parachrysocharis javensis* Girault, the entomogenous fungus *Metarhizium anisopliae* (Metsch) (Seneviratne & Kumarasinghe, 2002), *Brumoides suturalis* Fabricus, *Micraspis discolor* Fabricus, *Micraspis allardi* Mulsant (Coleoptera: Coccinellidae), *Cheiloneurus pyrillae* Mani (Hymenoptera: Encyrtidae), and *Richardsi dryinus pyrillae* Kiefer (Hymenoptera: Ryinidae) were naturally available in the ecosystem and contributed to managing the Pyrilla to some extent (Kumarasinghe, 2003) (Figure 10.6).

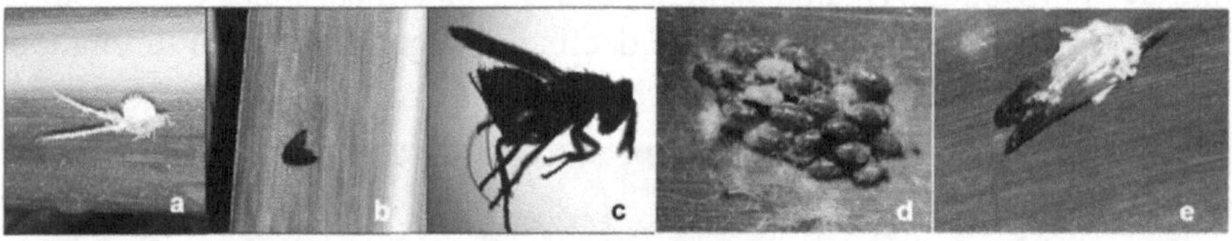

FIGURE 10.6 External larval and adult parasitoid: *Epiricania melanoleuca* Fletcher (Lepidoptera: Epipyropidae) ((a) and (b)); native egg parasitoids: *Parachrysocharis javensis* Girault (Hymenoptera: Eulophidae) ((c) and (d)); pathogenic fungi: *Metarhizium anisopliae*.

Source: Photo Gallery of the Division of Crop Protection, SRI.

Sugarcane Research Institute (SRI), Sri Lanka, imported the ectoparasitoid *Epiricania melanoleuca* (Fletcher) (Lepidoptera: Epipyropidae) from Pakistan in 1991. The insect was multiplied and released to the *Pyrilla*-infested sugarcane plantations (Seneviratne & Kumarasinghe, 2002) due to its short life cycle, higher reproductive rate, and tremendous searching ability (Chaudhary & Sharma, 1988). Subsequently, *Epiricania* was successfully established in sugarcane-growing areas and has maintained the population levels of *Pyrilla* below the Economic Threshold Level since 1993. High populations of *Pyrilla* observed in 2013 and 2015 at the seed cane nurseries at Kantale were also managed successfully with *Epiricania* as a long-lasting solution for managing *Pyrilla*. *Epiricania* cocoons (16,100) and eggs (1,207,500) were released to the infested plantation, and field cages were used for the augmentation of this parasitoid (Wanasinghe et al., 2019) (Figure 10.7).

FIGURE 10.7 Mass multiplication of *E. melanoleuca* in the field (A); field release of cocoons of *E. melanoleuca* (B); and the establishment of the parasitoid in the field (C).

Source: Photo Gallery of the Division of Crop Protection, SRI.

Sugarcane Woolly Aphid (SWA) is a successive story in using biological control agents to manage sugarcane pests in Sri Lanka. The spectrum of natural enemies of SWA that comprised several arthropod predators, i.e., *Diphaaphidivora* (Lepidoptera: Pyralidae), *Micraspis discolor* (Coleoptera: Coccinelidae), *Micraspis allardi* (Coleoptera: Coccinelidae) *Eupeodes* spp. (Diptera: Syrphidae), *Micraspis discolor*, *Synonycha* spp. (Coleoptera: Coccinelidae) (Kumarasinghe, 2007; Wanasinghe et al., 2012). Laboratory rearing protocols developed for *Diphaaphidivora* and *Micromus spp.*, released to the SWA-infested plantations at the rate of 1,000/ha of Dipha cocoons and 2,500/ha of micromus 2–3 times depending on the incidence of the pest. Spraying of the recommended insecticide was omitted during the releases of cocoons for better establishment and survival of *Diphaaphidivora* and *Micromus spp*. Currently, the spectrum of natural enemies of SWA is well established in the sugarcane ecosystem, where no spraying is required further (Figure 10.8).

FIGURE 10.8 Natural enemy spectrum of SWA – (A) and (B) *Eupeodes* spp.; (C) *Micromus* spp. (adult); (D) *Micromus* spp. (larva); (E): *Synonycha* spp.; (F) *Micraspis discolor*; (G) *Micraspis allardi*; and (H) larva of *Diphaaphidivora.*

Source: Photo Gallery of the Division of Crop Protection, SRI.

Four indigenous natural enemies were recorded on Sugarcane moth borers, INB and PB, i.e., larval parasitoid; *Cotesia flavipes* Cameron (Hymenoptera: Braconidae) (Kumarasinghe, 2003; Wanasinghe et al., 2014), two egg parasitoids; *Telenomus dignus* Gahan (Hymenoptera: Scelionidae), *Trichogramma chilonis* Ishii (Hymenoptera: Trichogrammatidae) (Kumarasinghe, 2003) and Larval, pupal parasitoid: *Xanthompla spp*. Field observations revealed the presence of earwigs in the borer holes on sugarcane stems, and laboratory studies confirmed that earwigs consume larvae of INB and PB. However, the contribution of generalist predators like earwigs has not yet been quantified (Wanasinghe et al., 2023) (Figure 10.9).

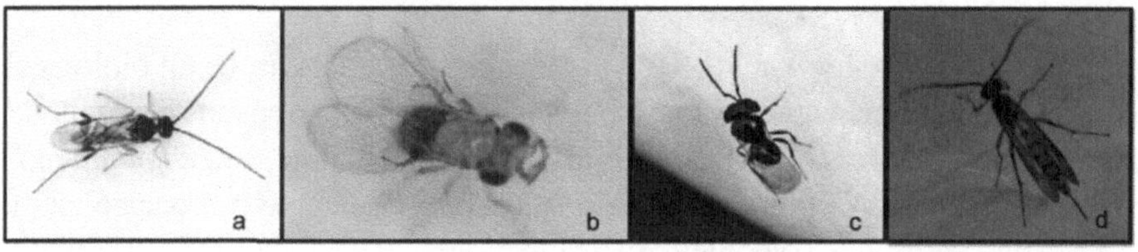

FIGURE 10.9 Larval parasitoid: *Cotesiaflavipes* Cameron (a); egg parasitoids: *Trichogramma chilonis Ishii* (b); *Telenomus dignus* (c); and larval pupal parasitoid: *Xanthompla* spp. (d).

Source: Photo Gallery of the Division of Crop Protection, SRI.

Among all natural enemies, *Cotesia flavipes* is reported to be a very efficient parasitoid in managing moth borer populations all over the sugarcane growing areas except Hingurana (Wanasinghe et al., 2023) due to excessive use of chemicals for other crops cultivated in the area. Many attempts at mass multiplication of the larval parasitoid in laboratory conditions were taken, and also releasing field-collected borer larvae infested with *Cotesia flavipes* to the sugarcane plantations in Eastern Province is being continued. Wanasinghe et al. (2019) reported establishing the *Cotesia*, particularly in sugarcane growing plantations, in 2018 after continuous field release and augmenting strategies, augmenting Cotesia using flowering plants act as companion plants to Cotesia, i.e., *Ocimum sanctum, Oxalis corimbosa, Cleome viscose,* and *Phylanthus viridis*.

Biological control agents contribute to managing WLD vectors, including a spectrum of predators (Figure 10.10) available naturally in the sugarcane ecosystem (Unpublished data). It is recommended to monitor the natural enemy population of the infested plantations and conserve those by trash mulching. Minimum trash burning with minimal use of pesticides may ensure a higher population of natural enemies (Wanasinghe et al., 2022).

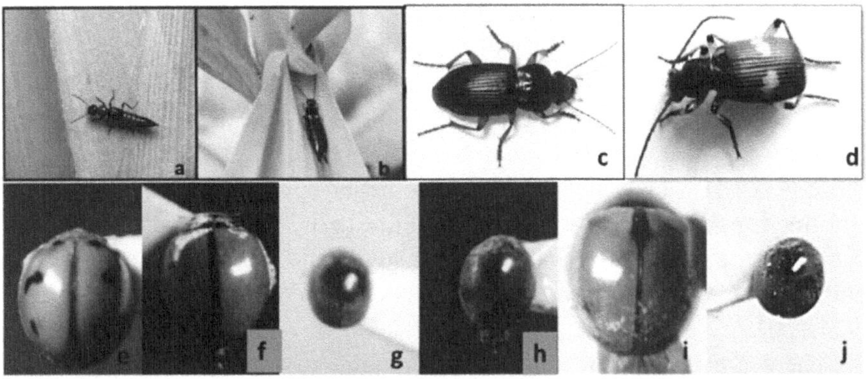

FIGURE 10.10 Rover beetle species (a) and *Euborellia* spp. (b); ground beetle spp., *Harpalus pensylvanicus* (Coleoptera: Carabidae) (c); and *Bembidion quadrimaculatum* (Coleoptera: Carabidae) (d) are efficient egg predators of WLD vector. Six predatory *Coccinellid beetle* spp.: *M. discolour* (e); *M. alardi* (f); *Scymnus nubilus* (g); *Jairavio dorsialis* (h); *Propylea dissector* (i); and *Pseudaspidimerus trinotatus* (j).

10.4.1.1 SPIDER MITE

Four predators of sugarcane spider mites were identified in Sri Lanka, and among them, *Stethorus* spp. indicate high potential in managing this pest population. Miticides are omitted in most outbreak conditions to ensure the conservation and augmentation of the natural enemies. This has caused the rapid build-up of natural enemies synchronized with the initial stage of spider mite infestation (Figure 10.11).

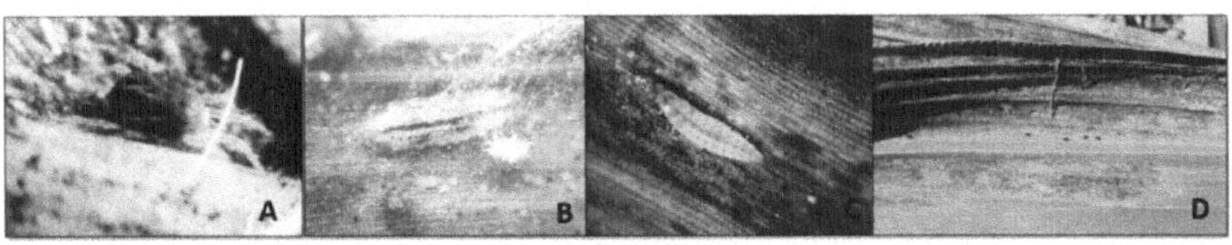

FIGURE 10.11 (A) Predatory mite spp.; (B) larva of a syrphid fly species; (C) larva of a coccinellid beetle species; and (D) adults of *Stethorus* spp. (Coleoptera: Coccinellidae) feeding on a spider mite colony.

Source: Photo Gallery of the Division of Crop Protection, SRI.

10.4.2 CHEMICAL CONTROL

Different insecticides have been recommended to manage sugarcane pests in Sri Lanka (Table 10.1). Applying small amounts of selective pesticides according to the pest forecasting schedules also reduced pest damage

(Kumarasinghe, 1999). Due to the abundance of biocontrol agents in the sugarcane ecosystem and difficulties in the foliar application of chemicals after the development of the canopy use of synthetic chemicals has become minimal among sugarcane growers in Sri Lanka except for sugarcane growers in Ampara District. They heavily depend on insecticides to control these pests. However, insecticides have never been identified as a worthy solution for sugarcane pest management due to the difficulty and high application cost, the concealed habit of most sugarcane pests, and the unpredictability of infestation and harmful effects on the environment (Wanasinghe et al., 2019).

There are few incidences when synthetic pesticides/chemical control have been important for pest management during severe pest outbreaks and in new introductions of the pests. Managing initial outbreaks of SWA using thiomethoxam was practiced until biological control agents were successfully established in infested areas. Thiomethoxam is also used in two-week intervals to manage the WLD vector population in sugarcane nurseries where vector populations are high and secondary transmission required. This is not practiced in commercial plantations considering the heavy usage of synthetic chemicals and health issues due to frequent use. Instead, it recommended to practice border treatment of Fipronil in fields with low and moderate disease vector stress. Here, soil in the furrows is treated with Fipronil at the rate of 18 kg/ha (this will reduce the termite damage to the setts, shoots, and shoot borer damage to the new crop too). After the first application, treat the 2 m width border of the field from the outer boundary line at monthly intervals for up to six months. Pesticide takes a higher place in managing live wood termites in sugarcane plantations since there are few other effective strategies. When planting systemic granular insecticide applied along the furrows, seed drenching is conducted with pesticide solution or seeds dipped in the systemic pesticidal solution to protect new plantings. Soil application of granular insecticide is the main strategy to manage after tillering. Thiamethoxam 20% + Chlorantraniliprole 20% WG (4 g/16 L dilution) and Imidacloprid 70% WG0.16 are the currently used chemicals for dipping and drenching while Fipronil (30 kg/ha) is using for soil application (Table 10.1).

The exotic invasive pest, fall armyworm (FAW), *Spodoptera frugiperda* (Lepidoptera: Noctuidae), was suddenly reported during the latter part of 2018 in commercial sugarcane growing areas in Sri Lanka. The initial population of FAW was reduced by using Chlorantraniliprole 220 g/1SC, Spinetoram 25%WG, and Emamectin Benzoate 5% SG. Other than the

TABLE 10.1 The Chemical Recommendation for Sugarcane Pests by Sugarcane Research Institute, Sri Lanka

Common Name	Compound	Toxicity	Chemical Class	Treatment Period	Target Pest	Strategy	Concentration Range
Imidacloprid	Imidacloprid	Class II	Mimic nicotine	At planting	Sugarcane termite	Soil drench and seed dipping.	0.16 g/1 L water
Phenyl pyrazole	Fipronil (PRN 1.0)	Class II	Neonicotinoids	At infestation	Sugarcane termite	Soil drench	PRN 0.75, 30 Kg/ha
	Fipronil 0.3%(W/W) GR	Class II	Neonicotinoids	At infestation	*Chilo sacchariphagus*	Soil application	PRN 1.25 and 0.75 (Liquid form)
	Fipronil	Class II	Neonicotinoids	3–6 month stage	Internode borer	Soil application	18 kg/ha
–	Fipronil 0.3%(W/W) GR	Class II	Neonicotinoids	At infestation	Pink borer	Near the root system of the soil	30 kg/ha
Thiamethoxam	Thiamethoxam 25% W/W	Class III	Neonicotinoids	At infestation	Sugarcane woolly aphid (SWA)	Foliar application	5 g/16 L
	Thiamethoxam 70%	Class III	Neonicotinoids	At infestation	Termite	Soil drench	1 ml/L
	Thiamethoxam 70%	Class III	Neonicotinoids	At infestation	Sugarcane termites	Soil drench	1,216 g/ha (1 g/L)
Chlorpyrifos	–	Class II	Organophosphate	Every 2 and 3-week interval	Sugarcane termites	Soil drench	1.07 Kg/ha
Carbosulfan	Carbosulfan 10%	Class II	Carbamate class	Once at a planting	Sugarcane termites	Granules	7.3, 9.85 Kg/ha
Furadan	–	Class I	Carbamate class	Every 2- and 3-week interval	Sugarcane termites	Granules	2.85 Kg/ha

Source: Progress Reports of Sri.

direct application of chemicals to the insects, indirect management of the host was conducted to manage the WLD vector management program, which is an effective management strategy. Vectors prefer WLD-infested sugarcane clumps compared to healthy plants for feeding and breeding (Chanchala et al., 2021). Therefore, using herbicides in rouging out programs is important in eradicating infested clumps to manage vector populations.

10.5 CULTURAL CONTROL

Several cultural control strategies have to be acquainted with and put into practice to manage economically important pest species successfully by manipulating the cropping system and/or crop management strategies.

10.5.1 WEED MANAGEMENT

Weed management is a frequently used strategy in sugarcane pest management. *Eleusine indica* (Belathana) is one of the most preferred breeding hosts for adult PB moths (Figure 10.12), and early larval instars survive on this host. Weed management is recommended and practiced in the early growth stage of the sugarcane crop to minimize PB infestation. About nine local cane field grasses are recorded as collateral hosts of *Sesamia inferans, Eragrostis diplachnoides, Panicum repens, Cyperus compressus*, etc. (Rajendran et al., 1976). In WLD vector management, avoiding *Saccharum spontaneum* (wild cane) and *Sorghum bicolor* around sugarcane fields and nursery areas are practiced, as these plants act as alternative hosts to the vector, thereby transmitting WLD to new plantings (Chanchala et al., 2015).

FIGURE 10.12 *Eleusine indica* plant infested with SB larvae (a); SB larva feeding inside *Eleusine indica* stem (b); and highly SB-infested field due to poor weed management (c).

Source: Photo Gallery of the Division of Crop Protection, SRI.

10.5.2 HABITAT MANIPULATION

Several habitat manipulation strategies have been introduced to manage economically important pests successfully. To reduce the lifespan of adult moth borers, INB and SB, introduce flowering plant species that adult moths do not prefer for feeding, such as *Scorpiadulcis, Cyanthillium cinereum, Tridax procumbens, Cyprus killingle, Sida acuta, Celosia aseatica, and Borreria ocymoides* species. *Cleome viscose* (Cleomaceae, *Wal aba*) acts as a repellent to the *D. menoni* (Chanchal et al., 2015). Most plant species belonging to the order Brassicales have a characteristic odor that eventually repels insects from their presence, and *Brassica juncea* (L.) Czern and Coss (Brassicaceae) are also an oil crop with the above characteristics.

Intercropping mustard with sugarcane minimizes the secondary transmission of the WLD with no effect on the brix and weight of the sugarcane crop (Figure 10.13). Intercropping sugarcane crops with groundnut significantly reduced termite infection in commercial cultivations in dry areas (Wanasinghe et al., 2014). The vector's secondary transmission of the WLD is reduced by increasing the spacing between two planting rows. Vector populations have been reduced with 1.5 m and 2 m spacing compared to the 1.37 m standard spacing for sugarcane cultivation. Earthing up the soil 45 days after planting and trash mulching at the base of the plant in ratoon crops are also practiced to avoid initial exposure of young sugarcane plant bases to borer larvae infestation. Removing and destroying the dead hearts along with larvae (5 to 7 cm depth from the soil) at the initial stage of the attack, collecting and destroying the egg masses and moths are done physically to manage the borer population. De-trashing of canes reduces borer infestation as it reduces protection to borer larvae in their search for boring locations and increases exposure to sunlight and predatory populations. Also, the removal of water shoots once a month from the 5th month onwards is recommended as borer larvae prefer water shoots instead of mature cane for easy penetration.

FIGURE 10.13 Mustard is incorporated in sugarcane fields.

Source: Photo Gallery of the Division of Crop Protection, SRI.

10.5.3 IRRIGATION MANAGEMENT

In Sri Lanka, irrigation management significantly reduces the population of dry seasonal pests such as live wood termites and spider mites. Proper irrigation schedule in the irrigated sector significantly reduces termite damage to the crop throughout the cropping season. The sugarcane spider mite is also a serious pest during the dry season. Sprinkler irrigation is recommended for nurseries to overcome the situation, effectively washing away the mite population and mimicking natural rainfall.

10.5.4 SEX PHEROMONES

Using sex pheromones to trap adult males of sugarcane moth borer is also an effective mass-trapping strategy (Figure 10.14). Placing pheromone traps at 5–6/ha (Delta trap) at spindle level for monitoring the population and at 20/ha from the 4th month to the 11th month of the crop for mass trapping of male moths of *C. sacchariphagus* (Z,13-Octadecenyl acetate and Z,13-Octadecenol; 7:1 Blend ratio) is recommended (Wanasinghe et al., 2023). The pheromone septa and sticky sheet must be changed fortnightly, and the Delta trap needs to be changed once every 2–3 months (If more than 3.0 months are caught per trap per week, it indicates high infestation).

FIGURE 10.14 Pheromone traps to catch moth borer, male adults.
Source: Photo Gallery of the Division of Crop Protection, SRI

10.6 VARIETAL RESISTANT/TOLERANT

Each commercially cultivating varieties are evaluated for the economically important pest spp. and confirmed their resistance/tolerant to major pests, *viz.*, variety SL 89 309, SL 92 4918 for sugarcane INB, SL 83 06, SLC 2009 01 for SWA and SL 83 06, SL 92 5588, SLT 4921 for WLD vector. Introducing resistance varieties has a great impact on pest populations in susceptible areas.

10.6.1 RECENT ADVANCES

Precision agricultural strategies, i.e., machine learning techniques and electronic and computer-based technologies, are in the picture for remote sensing and forecasting the upcoming pest issues and specified treatments. Biotechnological approaches are also progressing in accurately identifying pests and developing resistant varieties that are resistant and tolerant to available pest species. Incorporating volatile attractants into sex pheromones is another approach to the increasing problem of sugarcane moth borers in the country.

10.7 CONCLUSION

The sugarcane industry in Sri Lanka has been affected by several serious pest problems throughout the decades. They have been successfully addressed in an integrated manner using biological control agents, cultural control, resistant varieties, and other cultural management strategies. There are some advances to be added to the IPM programs with new machine learning techniques, biotechnological approaches, and volatile attractants.

ACKNOWLEDGMENTS

The authors are grateful to Dr. M. S. Perera, Director and CEO, and Dr. A. L. C. De Silva, Deputy Director (Acting) of the Sugarcane Research Institute, for the support given. Ms. V. K. A. S. M. Wanasinghe, Ms. A. N. W. S. Thushari, and all staff members of the Division of Crop Protection of SRI are appreciated for their assistance.

KEYWORDS

- **integrated pest management**
- **pest**
- **shoot borers**
- **Sri Lanka**
- **sugarcane**
- **sugarcane spider mite**
- **sugarcane termites**
- **vector**

REFERENCES

Ariyarathana, H. A. C. K., Everard, J. M. D. T., & Karunanayake, E. H. (2007). Diseased sugarcane in Sri Lanka is infected with sugarcane grassy shoot and/or sugarcane white leaf phytoplasma. *Australasian Plant Disease Notes, 2*(1), 123. doi:10.1071/dn07050.

Chanchala, K. M. G., Dayasena, Y. A. P. K., Wanasinghe, V. K. A. S. M., Hemachandra, K. S., Nugaliyadde, L., Witharama, W. R. G., & Nikpay, A. (2019). Viruliferous nature of the Sugarcane White Leaf Disease Vector *Deltocephalus menoni* (Hemiptera: Cicadellidae). In *Proceedings of the Seventh Symposium on Plantation Crop Research- "Towards Achieving Sustainable Development Goals in the Plantation Sector"* (pp. 153–162). Rubber Research Institute of Sri Lanka.

Chanchala, K. M. G., Dayasena, Y. A. P. K., Wanasinghe, V. K. A. S. M., Hemachandra, K. S., Nugaliyadde, L., Witharama, W. R. G., & Nikpay, A. (2022). Disease transmission proficiency of *Deltocephalus menoni* (Hemiptera: Cicadellidae); vector of sugarcane white leaf disease. *Agrica, 11*(1), 54–60.

Chanchala, K. M. G., Wanasinghe, V. K. A. S. M., & Hemachandra, K. S. (2015). Survival of *Deltocephalus menoni* (Homoptera: Cicadellidae), the vector of sugarcane white leaf disease in Sri Lanka on alternative host plants. *Sugarcane Journal, Sri Lanka, 2*, 36–43.

Chanchala, K. M. G., Wanasinghe, V. K. A. S. M., Ariyawansa, B. D. S. K., & Hemachandra, K. S. (2014). Relationship between the incidences of Sugarcane White Leaf Disease and the Population Dynamics of Its Vector, *Deltocephalus menoni* (Homoptera: Cicadellidae), in Sri Lanka. In A. P. Keerthipala (Ed.), *Proceedings of the 5th Symposium on Plantation Crop Research – "Towards a Green Plantation Economy"* (pp. 143–149). Sugarcane Research Institute, Uda Walawe, Sri Lanka.

Chanchala, K. M. G., Wanasinghe, V. K. A. S. M., Kulasekara, B. R., Hemachandra, K. S., Nugaliyadde, L., & Witharama, W. R. G. (2021). Tri-trophic relationship among Sugarcane White Leaf Disease Phytoplasma (WLDP), *Deltocephalus menoni* (Homoptera: Cicadellidae) and sugarcane plant in secondary transmission of the disease. *Journal of Agricultural Sciences – Sri Lanka, 16*(3), 443–451. http://doi.org/10.4038/jas.v16i03.9470.

Keerthipala, A. P. (2016). Development of Sugar Industry in Sri Lanka. *Sugar Tech, 18*(6), 612–626. doi:10.1007/s12355-016-0485-3.

Kumarasinghe, N. C. (2003). Insect fauna associated with sugarcane plantations in Sri Lanka. *Journal of Environmental Biology, 24*(4), 359–368.

Kumarasinghe, N. C. (2007). Identification of a new pest of sugarcane, Ceratovacuna lanigera Zehntner in Sri Lanka. *Sugar Tech, 9*, 220–223.

Kumarasinghe, N. C., & Jepson, P. C. (2003). Antixenotic effect of sugarcane leaves on feeding and oviposition by Pyrilla perpusilla Walker. *Sugar Tech, 5*, 11–19.

Rajendra, A. (1976). Studies on Sesamia inferens: the shoot-borer pest of sugar-cane in Sri Lanka. *Journal of the National Science Council of Sri Lanka, 4*(2), 99–108.

Rajendra, A. (1979). Insect pests attacking sugarcane in Sri Lanka and the parasites and predators recorded on them. *Ceylon Journal of Science, Biological Sciences, 13*(1/2), 29–35.

Seneviratne, J. A. U. T. (2008). An investigation of the secondary transmission of sugarcane white leaf disease in Sri Lanka (Doctoral dissertation, PhD Thesis).

Seneviratne, J. A. U. T., & Kumarasinghe, N. C. (2002). Biological control of the sugarcane plant hopper by the moth *Epiricania melanoleuca* (Fletcher) in Sri Lanka. *Sugar Tech, 4*(1), 26–32.

Song, X. P., Mo, F., Verma, K. K., Wei, J., Zhang, X., Yang, L., & Li, Y. R. (2019). Effect of sugarcane smut (*Ustilago scitaminea* Syd.) on ultrastructure and biochemical indices of sugarcane. *Biomedical Journal of Scientific & Technical Research, 17*(1), 12546–12551.

Song, X. P., Mo, F., Yan, M., Zhang, X., Zhang, B., Huang, X., Huang, D., Pan, Y., Verma, K. K., & Li, Y. R. (2022). Effect of smut infection on the photosynthetic physiological characteristics and related defense enzymes of sugarcane. *Life, 12*, 1201. https://doi.org/10.3390/life12081201.

Song, X. P., Verma, K. K., Tian, D. D., Zhang, X. Q., Liang, Y. J., Huang, X., Li, C. N., & Li, Y. R. (2021). Exploration of silicon functions to integrate with biotic stress tolerance and crop improvement. *Biological Research, 54*, 19. https://doi.org/10.1186/s40659-021-00344-4.

Verma, K. K., Song, X. P., Rajput, V. D., Solomon, S., Li, Y. R., & Rao, G. P. (Eds.). (2023). *Agro-industrial Perspectives on Sugarcane Production under Environmental Stress*. Springer, Singapore. https://doi.org/10.1007/978-981-19-3955-6_1.

Wanasinghe, V. K. A. S. M., & Chanchala, K. M. G. (2020). *Pests of Sugarcane in Sri Lanka*. Publication of Sugarcane Research Institute.

Wanasinghe, V. K. A. S. M., Chanchala, K. M. G., & Kumarasinghe, N. C. (2014). An Assessment of Major Pests of Sugarcane in Sri Lanka. *Journal of Sugarcane Sri Lanka, 1*, 21–28.

Wanasinghe, V. K. A. S. M., Chanchala, K. M. G., Gunawardena, M. K. P. C., & Wijesuriya, A. (2019). Control of Sugarcane Leaf Hopper, Pyrilla perpusilla by *Fulgoraecia (Epiricania) melanoleuca* Fletcher (Lepidoptera: Epipyropidae) in the seed cane nursery at Kantale, Sri Lanka. *Proceedings of International Conference Sugarcon 2019 at ICAR-Indian Institute of Sugarcane Research, Lucknow, and Uttar Pradesh, India from February 16-19, 2019*, pp. 119.

Wanasinghe, V. K. A. S. M., Chanchala, K. M. G., Karunathilake, A. P., Nikpay, A., & Nugaliyadde, L. (2019). First report of Fall Army Worm; *Spodoptera frugiperda* (Lepidoptera: Noctuidae) in sugarcane in Sri Lanka. *Proceedings of the international sugar and sugarcane conference*, pp. 39.

Wanasinghe, V. K. A. S. M., Chanchala, K. M. G., Nikpay, A., & Goebel, F. R. (2022). Biological control of sugarcane pests in Sri Lanka: Current status and future prospects. *Agrica, 11*, 83–90.

Wanasinghe, V. K. A. S. M., Chanchala, K. M. G., Weragoda, W. G. M. S., & Nugaliyadde, L. (2019). Evaluation of the Necessity of Insecticides and the Potential in Establishing the Larval Parasitoid: *Cotesia flavipes* (Hymenoptera: Braconidae) for Management of Sugarcane Shoot Borers in Ampara District, Sri Lanka. In V. H. L. Rodrigo, B. W. Wijesuriya, D. G. Edirisinghe, & N. M. C. Nayanakantha (Eds.), *Proceedings of the Seventh Symposium on Plantation Crop Research – "Towards Achieving Sustainable Development Goals in the Plantation Sector"* (pp. 145–152). Rubber Research Institute of Sri Lanka.

Wanasinghe, V. K. A. S. M., Kumarasinghe, N. C., & Chanchala, K. M. G. (2012). Natural Enemies of Sugarcane Woolly Aphid (*Ceratovacuna lanigera*): A survey in Passara, Sri Lanka. *Proceeding of the Fourth Symposium on Plantation Crop Research*, pp. 163–170.

Wanasinghe, V. K. A. S. M., Sithanantham, S., Chanchala, K. M. G., & Prabakaran, M. (2023). Effect of pheromone blend sources and loadings on sugarcane internode borer moth catches in Sri Lanka. *Indian Journal of Entomology. 85*, 533–535. https://doi.org/10.55446/IJE.2022.985.

CHAPTER 11

Sugarcane Management for Higher Crop and Water Productivity

SATPAL S. PAALLI, RAJNI SHARMA, and V. ANJALY

Department of Agronomy, Punjab Agricultural University, Ludhiana, Punjab, India

ABSTRACT

The sugarcane growth period differs greatly from field crops due to its longer duration. It faces year-long exposure to weather vagaries ranging from the phases of coldest, hottest, driest, and high humidity. It becomes challenging for farmers to sense the crop's need for water, assess water stress, and apply the right amount of water at the right time with precision. Owing to the knowledge gap in irrigation water management, over or under-irrigation is common, which results in lower yield and water productivity as well as higher costs. A huge amount of water can be saved with the application of agronomic interventions like laser land leveling, mulching, planting time, nutrition management, method of irrigation and its scheduling, and the Internet of Things (IoT). The future of irrigation for developing countries is IoT, where real-time data of physical parameters of soil and water is converted into digital form (data acquisition). It is stored in a cloud server that automatically guides the water pump and solenoid valves to initiate and stop the irrigation process. It is not beyond imagination that in the future, farmers will use this technology like the current generation is using mobile.

11.1 INTRODUCTION

Water is the most limiting factor in crop production, affecting the productivity and profitability of field crops. Declining irrigation water availability

Sugarcane Cultivation and Management: Challenges and Opportunities. Krishan K. Verma, Xiu-Peng Song, Munna Singh, Renato de Mello Prado, Jian-Ming Wu, and Yang-Rui Li (Eds.)

and increasing food demand have become the major concerns of present-day agriculture and necessitate the quick adoption of modern scientific technologies for efficient water management (Verma et al., 2020a, 2021a, 2022a, 2023). Climate-smart agricultural practices and technologies that prevent the deterioration of limited resources like water, increase crop yield and income, and reduce environmental footprints are the need of the hour. In the Indo-Gangetic plains of India, due to continuous mono-cropping of rice-wheat system, overuse of electric pumps, and inadequate recharge from erratic rainfall, the groundwater table has been declining at an alarming rate of 0.4-–0.9 m/annum (Brar et al., 2012). An increase in temperature due to global warming further intensifies the irrigation water demand. Recent studies predicted that in arid and semi-arid regions of Asia, there would be at least a 10% increase in irrigation water demand with a 1°C rise in temperature (Sivakumar & Stefanski, 2011). By 2050, water availability is expected to decline. In contrast, global agricultural water demand is estimated to increase by approximately 19%, and thus, water scarcity is becoming a major determinant of food security than land scarcity in the coming years (Brown & Funk, 2008; Verma et al., 2022b, c).

Sugarcane (*Saccharum officinarum* L.), a perennial plant of the *Poaceae* family, is a water-intensive crop that generally takes 12 (*Eksali*) to 18 (*Adsali*) months to mature, depending upon the variety and sowing time. The active crop season is restricted from April to September. It extends from March to October in other cane-growing areas in India due to varying soil and ambient temperatures. Being long duration, the irrigation water requirement of sugarcane is very high, especially during the hot summer months of May-June. Depending upon the yield potential, crop duration, and prevailing climatic conditions, the average crop water requirement of sugarcane is 1,800–2,200 mm. The crop requires warm weather with good soil moisture status to ensure uniform germination and emergence during the initial months. In general, the first irrigation should be given when 20–25% of plants have germinated or about 20 days after sowing (DAS); after that, irrigations are offered at 10–15 days intervals during summer, 25–30 days intervals during winter, and if there is a drought then the crop should be irrigated during the rainy season also as and when needed. The crop needs maximum water at the tillering stage (550 mm), ripening stage (650 mm), and stem elongation or grand growth phase (1,000 mm) (Table 11.1). Water requirement ranges from 1,400 to 3,500 mm in various cane-growing states of India, being the lowest in Bihar and the highest in Maharashtra. Thus, the water applied through irrigation varies from 600 to 2,000 mm in various tracts (Prasad, 2021). On average, 1 ton of cane needs about 60–70 tons of water, and thin varieties need 150 and 200 cm of water, respectively, in addition to 75 cm of rainfall (Holsambre, 1997).

Ratooning is the pivotal component of the sugarcane production system since it increases cane productivity, economizes the cost of cultivation, and improves the profit margin. Unless the ratoons have been carefully managed and improved agro-techniques are adopted, attaining a reasonably good ratoon yield is impossible. Ratoons have a lesser capacity to withstand moisture-stress conditions because of their shallow root system. Therefore, ratoon crops need to be irrigated more frequently than plant crops. The water required to produce 1 kg of plant and ratoon crops is 88 kg and 118 kg, respectively (NABARD, 2018).

TABLE 11.1 Crop Water Requirement of Sugarcane at Different Growth Stages (TNAU, 2023)

Growth Phase	Duration of Phase (Days)	Water Requirement (mm)
Germination	0–45	300
Tillering phase	45–120	550
Grand growth phase	120–270	1,000
Ripening phase	270–360	650

11.2 AGRONOMIC INTERVENTIONS FOR EFFICIENT MANAGEMENT OF IRRIGATION WATER IN SUGARCANE

11.2.1 LASER LAND LEVELING

Precise laser land leveling is the practice of smoothing the land surface using large horsepower tractors and soil movers equipped with global positioning systems (GPS) and laser-guided instrumentation to create a constant slope of 0 to 0.2%. The traditional land leveling method is labor-intensive, time-consuming, and less cost-effective. It involves field survey, staking and designing the field, calculation of cuts and fills, and using a scraper to make the land even. In the Indo-Gangetic Plains of India, about 10–25% of irrigation water is lost due to poor water management and uneven/undulating fields (Kahlown et al., 2000). Precise leveling of the field is important to enhance water use efficiency. Uneven fields not only require higher irrigation water but also have a major impact on seed germination, crop stand, and yield of crops through their nutrient-water interaction and salt-soil moisture distribution pattern. A difference of only one cm in elevation of the field requires 1,00,000 liters per hectare more water to cover the elevated spot. Thus, precise leveling should be done with a laser land leveler to ensure even irrigation water distribution and to prevent water loss through percolation. Laser land leveling reduces the time required for irrigation and enhances water productivity and other non-water farm inputs. It allows for ideal control of water distribution, reduces nutrient loss through runoff control, and thus improves seed germination, input use efficiency, and crop yields. The cost may vary depending on the topography, field shape, and equipment used. This method has raised the potential of surface irrigation efficiency, reducing water use by 20–30% and increasing crop yield by 10–20%. Jat et al. (2006) observed that water application efficiency decreases with increased standard deviation of land undulation (Figure 11.1).

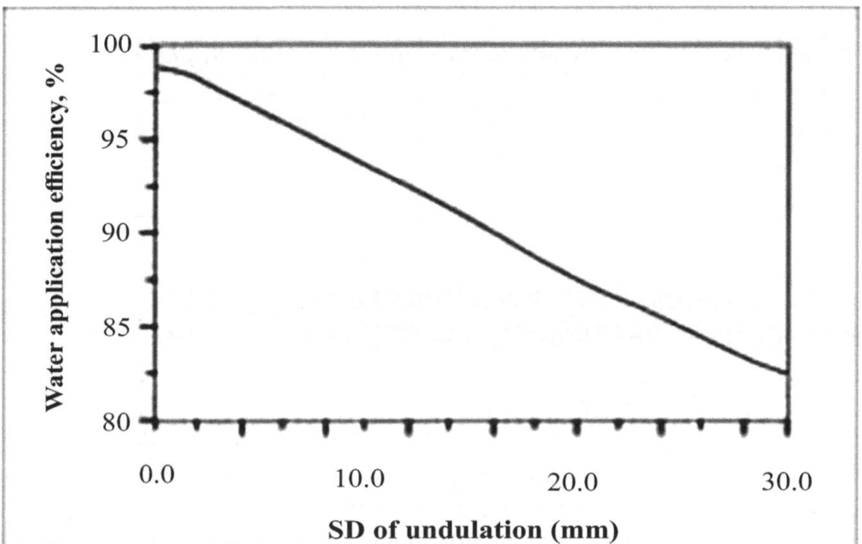

FIGURE 11.1 Simulated effect of land leveling quality indicated by the standard deviation (SD) of soil surface elevation) on water application efficiency.

Source: Reprinted from Jat et al., 2006.

Tomar et al. (2020) studied the impact of laser land leveling technology on yield, water productivity, and profitability in the alluvial soil of north Madhya Pradesh and observed that the average crop productivity in sugarcane has improved by 16.07% over traditionally leveled fields (Table 11.2).

TABLE 11.2 Sugarcane Yield and Total Duration, Applied Water Depth, and Water Use Efficiencies as Affected by Laser Land Leveling and Traditional Leveling Techniques (Tomar et al., 2020)

Treatments	Total Duration (hr ha^{-1})	Water Depth/ Irrigation (mm)	Volume of Water Applied (m^{-3})	Yield (t ha^{-1})	WUE (kg m^{-3})
Laser land leveling	112.4	80	10,400	88.6	8.50
Traditional leveling	127.8	95	12,600	76.5	6.07
Control (unlevelled)	142.5	110	15,300	64.8	4.23
C D (p = 0.05)	8.6	12.8	190	7.2	0.15

Spring sugarcane grown with rice, potato, and wheat under precision land leveling consumed 35.1% less irrigation water than late-sown sugarcane grown in rotation with rice and wheat under traditional land leveling. The precision land leveled system also recorded higher land use efficiency, energy productivity, and lower greenhouse gas emissions than those under the conventional method (Naresh et al., 2021).

11.2.2 MULCHING

Mulching is an effective tool to save water by conserving soil moisture. It is a technique of covering the soil surface with organic or plastic mulch to provide favorable conditions for plant growth and to optimize water use simultaneously. A semi-permanent protective cover is formed above the soil surface using different materials (including vegetative residues, biological geotextiles, gravel, crushed stones, dust, paper, etc.). This method helps modify the crop-growing environment by establishing a linkage between soil and agrometeorology. It not only reduces surface evaporation but also controls soil erosion, regulates soil temperature, reduces nutrient leaching, and improves the water-holding capacity of the soil (Kader et al., 2019). Mulching protects the soil from the impact of raindrops, reduces overland flow generation rates by increasing roughness, improves infiltration rate, water intake, and storage, and increases earthworm activity. The soil organic matter content is also enhanced, particularly with the use of vegetative mulches. The regulation of surface soil temperature ensures optimal seed germination and crop root development (Prosdocimi et al., 2016). Covering the soil surface with trash mulch is also an efficient method of moisture conservation in sugarcane. Jordan et al. (2010) revealed that mulch application increased soil organic matter content, and physical properties like bulk density, porosity, aggregate stability, and available water capacity. The roughness, raindrop interception, and infiltration rate were also enhanced through mulching at higher rates.

The sugarcane production system generates huge quantities of biomass every year. Nearly 10–12% of the cane yield is trash, and it contains 0.42% N, 0.15% P, and 0.57% K besides 26, 2045, 236, and 17 ppm of Zn, Fe, Mn, and Cu, respectively (Shrivastava et al., 1992). These crop residues can act as renewable organic resources with high potential for organic recycling and moisture conservation. Besides these, trash mulching is also useful in controlling weed growth and early shoot borer incidence to a substantial extent. Trash can be spread up to a thickness of 10 cm in the inter-row spaces of sugarcane once the germination has been completed. Tayade et al. (2016) reported that *in situ* trash mulching helped to increase the soil moisture content from 0.70 to 5.92% and buffered soil temperature at 25.1–27.2°C in the upper 5 cm layer of soil. This resulted in numerically higher cane weight, cane height, number of millable canes (NMC), and cane yield. The use of a crop residue layer over the soil surface significantly reduced the water use and irrigation requirement and increased the WUE in sugarcane (Olivier & Singels, 2015). Sugarcane grown with a thick mulch cover had slightly higher (5%) economic water productivity of blue water usage than those produced with a light mulch cover under subsurface drip irrigation (Adetoro et al., 2020). Ram et al. (2006) concluded that cane trash mulch application @ 7 t ha^{-1} helped improve the crop growth, herb, and essential oil yield of menthol mint.

11.2.3 METHOD OF PLANTING AND ROW SPACING

The method of planting adopted in sugarcane is determined by agroecological and socio-economic conditions, planting material used, and the nature of soil manipulation. The flat method is one of the most extensive methods followed in sub-tropical north India. The setts have been placed either head to head or bud to bud in shallow furrows of 10–15 cm depth at 90 and 75 cm distance in autumn and spring, respectively. In comparison, the ridge furrow method is common in the tropical south. Among various methods, the trench planting method can save more irrigation water. Trenches of 25 cm depth and 30 cm width are formed and incorporated with organic manures. These are covered with thin soil after the placement of cane setts, followed by irrigation. This method ensures better bud germination due to higher soil moisture, provides better anchorage to plants, and results in higher plant crop and ratoon yields (Prasad, 2021). However several problems like more seed requirements, high labor demand, and difficulty in the adoption of intercultural operations are confronted in single trench planting. Thus, paired row trench planting can be adopted for better crop stand and water saving. In this method, two rows of sugarcane can be planted in 20–25 cm deep and 30 cm wide trenches, separated at a distance of 90 cm (Sanghera et al., 2018). The ring pit method is another water-saving planting method developed by IISR, Lucknow. This method saves irrigation water by 25–30% and increases irrigation WUE by 30–40% in sugarcane since more than 40% of the surface area remains dry with the application of irrigation water in pits only (Shukla et al., 2020). Nearly 20 healthy setts from heat-treated canes are arranged horizontally in each circular pit of 75–90 cm diameter and 45 cm depth. About 25–35 millable canes are generated in each pit, giving a substantially higher cane yield than the flat method. The spring-planted sugarcane raised in furrows of FIRB (furrow irrigated raised bed) of standing wheat generates higher output (Singh et al., 2012).

The most important variable contributing almost 40% to cane yield is the number of millable canes, which is altered by manipulating the crop micro-environment. The closely planted canes tiller profusely without pause and produce one or two thin stalks at harvest. Conversely, wide-spaced sugarcane grows slowly and has abundant tillers over a long period, giving a false impression of higher yield. Thus, large yield increases are obtained at closer spacing (Irvine & Benda, 1980). Dhanapal et al. (2019) observed that bud chip settling planted at 150 x 45 cm spacing was superior to bud chip settling planted at 150 x 60 cm spacing regarding cane yield and irrigation water use efficiency (IWUE).

On the contrary, Manimaran et al. (2009) reported that the adoption of wider row spacing (120 cm) with cross-planting gave higher cane and sugar yield than narrow row spacing (80 cm) under the same planting method. The wider row spacing in sugarcane ensures better light interception and space availability, thus providing scope for successful intercropping in its early growth phases and for mechanization during intercultural operations and harvesting. The cane yield, WUE, apparent water productivity (AWP), and total water productivity (TWP) as influenced by planting methods and mulching have been given in Table 11.3. It could be concluded that paired row trench planting (30:120 cm) along with mulch @ 6 t ha^{-1} is the best strategy to attain higher crop and water productivity over the conventional flat (75 cm apart) planting with and without mulch. The adoption of paired row trench planting reduced the wettable area to 40% and saved irrigation water (Singh et al., 2022).

TABLE 11.3 Cane Yield, Apparent Water Productivity (AWP), Total Water Productivity (TWP), and Water Use Efficiency (WUE) as Influenced by Planting Methods and Mulching (Singh et al., 2022)

Planting Method	Cane Yield ($t\ ha^{-1}$)	AWP ($kg\ m^{-3}$)	TWP ($kg\ m^{-3}$)	WUE ($kg\ m^{-3}$)
Conventional flat with mulch @ 6 t ha^{-1}.	96.4	14.0	10.2	8.4
Conventional flat without mulch.	90.2	13.0	9.5	8.2
Paired row trench planting with mulch @ 6 t ha^{-1}.	103.0	31.2	16.6	9.6
Paired row trench planting without mulch.	94.7	28.5	15.2	9.1
LSD ($p = 0.05$).	2.7	0.7	0.4	–

11.2.4 TIME OF PLANTING

The planting and ratooning time influences the productivity of the sugarcane crop cycle. The sugarcane planting time window is wider in the tropics. In contrast, the planting time is restricted to certain months in subtropics, owing to the low temperature from mid-November to February. The optimum time for sugarcane planting in subtropical India is February-March. Thermal regimes remain congenial for cane bud sprouting during this period. However, the most critical tillering stage coincides with the hot, desiccating, harsh summer months of April to July. There will be a drastic reduction in the yield unless the crop has been irrigated at this stage. The farmers in North India also opt for autumn planting of cane in September-October to ensure an optimum soil and atmospheric temperature during the germination and tillering phases. It is superior to spring planting in cane yield and juice quality. However, it is not popular among farmers as it keeps the field engaged for over a year.

Generally, the longer the crop duration, the higher the evapotranspiration. Delayed planting reduces the length of the pre-monsoon desiccating period, thus cutting down the crop water requirement. According to Yadava (1991), eight irrigations were required for the February-planted crop to maintain 75% available soil moisture (ASM). In contrast, five irrigations were sufficient to maintain the same moisture level in the April-planted crop.

11.2.5 NUTRITION MANAGEMENT

Judicious use of nutrients is necessary for sustainable sugarcane cultivation, and underapplication of nutrients may result in loss of yield and quality. Since it is a high biomass-yielding crop, large amounts of nutrients are required to achieve optimum productivity. For every ton of cane produced, the crop removes about 0.56–1.20 kg of N, 0.38–0.82 kg of P_2O_5, 1.00–2.50 kg of K_2O, 0.25–0.60 kg of Ca, 0.20–0.35 kg of Mg, 0.02–0.20 kg Na and 2.0–2.7 kg of SO_4 in addition to micronutrients (Zende, 1990). Canes have a relatively higher demand for potassium (K), and the application of potassic fertilizers, especially in K-deficient soils, improves plant performance and input use efficiency (Bhatt et al., 2021). Potassium is essential for the uptake of water, stomatal leaf opening, maintenance of cell turgidity, translocation of proteins and carbohydrates, accumulation of sucrose, and formation of proline during moisture-stress conditions (Hunsigi, 2011). Water deficit conditions may hasten flowering, reduce nutrient uptake and utilization, and increase cane fiber and bagasse percentage. Spray application of K can be recommended either alone or in combination with urea at very low concentrations to cope with the moisture stress conditions and to attain higher yields. Foliar application of urea with potash (2.5% each) during the formative phase was found to be beneficial under moisture-limiting situations. There was an increase in cane yield by 18.6% over control (Mathew et al., 2004; Gopalasundaram et al., 2012). Rajanna & Patil (2003) reported that applying N and K through drip fertigation from 30 to 240 DAP at six-day intervals gave 24.34% higher yield and saved 46.52% water compared to surface irrigation with the recommended fertilizer dose. The setts can be soaked in saturated lime solution (80 kg in 400 liters of water) for 2 hours before planting to improve cane and juice quality under such situations (Prasad, 2021).

Silicon (Si) fertilization significantly mitigates the negative impacts of drought stress on sugarcane plant growth. Applying Si improved plant water status and photosynthetic activity and activated the antioxidant machinery (catalase, peroxidase, and superoxide dismutase) in sugarcane (Verma et al., 2019, 2020c, 2021c, 2022c, d). The role of Si in plants is not just through the formation of a physical barrier in cell walls, lumens, and intercellular voids, but it also alters physiological activities, particularly in plants subjected to stress conditions (Sacala, 2009; Verma et al., 2022d, 2023). Supplying Si through the nutrient solution to pre-sprouted sugarcane seedlings reduced the damage caused by water deficit in the first 30 days after transplanting, with a 30% level of soil water retention. This led to the increased quantum efficiency of photosystem II, leaf water potential, chlorophyll content, and water content, decreased cell electrolyte leakage, and free proline content, resulting in higher growth and biomass production (Teixeira et al., 2020). Press mud is a compressed sugar industry waste product, rich in organic carbon, macro, and micronutrients. Irrigation water saving of up to 4% was achieved in sugarcane by applying press mud alone (Shukla et al., 2020). Its application to the soil helps add nutrients to the soil, sustaining microbial growth and enzymatic activities, and enhancing the productivity of ratoon crop (Yang et al., 2013; Verma et al., 2021b).

11.2.6 METHOD OF IRRIGATION AND ITS SCHEDULING

Surface irrigation is the widely adopted irrigation method in sugarcane as it is cultivated as a row crop planted widely and due to the convenience in its adoption. Furrow irrigation is advocated over flood irrigation due to the even distribution of water in the field and efficient water use. Skip furrow irrigation could result in 36.5% water savings and increase WUE by 64%. In this method, irrigation is given in alternate rows instead of applying water to all the rows and inter-row spaces (Srivastava & Johari, 1979). Wijayawardhana et al. (2017) proved that alternate-row furrow irrigation saved irrigation water by 35 to 45% and increased crop water use efficiency by 46% compared with every furrow irrigation. The quality of the cane was also superior through the adoption of the former method.

Drip irrigation can be adopted in water-scarce areas. It is the method of applying water at the base of the plant at a rate nearly equal to the plant's consumptive use rate, thereby minimizing water losses through percolation, runoff, and evaporation from the soil. The water applied from the point source spreads in all directions in the soil outward from the source. This method is also known as trickle irrigation, drip irrigation, sip irrigation, localized irrigation, daily flow irrigation, and micro-irrigation. Drip irrigation systems are suitable for uneven topography with varying soil textures and in areas where water is scarce or expensive. Precise and timely application of nutrients, herbicides, insecticides, and fungicides is also possible through irrigation water. The water application efficiency is much higher in drip irrigation (more than 90%) compared to surface irrigation methods (about 60%) and overhead sprinkler irrigation (about 75%) (Thomas, 2022). Significant influence of irrigation methods (flood, furrow, and drip) on cane stalk, cane top, cane trash, and biological yield at harvest (Table 11.4) was observed by Singh & Singh (2020). Drip irrigation revealed its superiority over flood irrigation and furrow irrigation. An increased cane yield of 22.30% was observed under drip irrigation compared to flood irrigation. Similar results were also observed in the case of cane top, cane trash, and biological yield of sugarcane due to flourishing growth, higher biomass accumulation, and favorable yield parameters. Narayanamoorthy (2010) also concluded that the cane yield of sugarcane cultivated under the drip irrigation method is much higher than under surface irrigation.

TABLE 11.4 Effect of Irrigation Methods on Sugarcane Productivity (Singh & Singh, 2020)

Irrigation Method	Cane Yield (t ha^{-1})	Cane Tops Yield (t ha^{-1})	Cane Trash Yield (t ha^{-1})	Biological Yield (t ha^{-1})	Harvest Index (%)
Flood irrigation	137.2	21.1	13.1	171.4	80.0
Furrow irrigation	143.5	22.0	13.7	179.2	80.0
Drip irrigation	167.8	24.8	15.2	207.9	80.7
SEm+	3.23	0.46	3.37	3.37	0.40
LSD (p = 0.05)	11.19	1.58	11.67	11.67	NS

Pawar et al. (2013) assessed the performance of sugarcane under drip fertigation in the clayey soil of Rahuri and found that drip irrigation saved water to the extent of 57% and increased yield to the extent of 17.6% over conventional irrigation. Considerable saving of irrigation water was noticed in sugarcane under both sub-surface drip (30.73%) and surface drip irrigation (22.70%) over surface irrigation (Mahesh et al., 2016). Based on an investigation carried out by Gouri et al. (2014) during 2012–2013 and 2013–2014 to determine the effect of surface and subsurface drip fertigation on yield and water use efficiency of sugarcane, it was found that higher water use efficiency was registered by subsurface (1.24 and 1.33) and surface (1.21 and 1.28) irrigations as compared to furrow method of irrigation (0.68 and 0.77), respectively. Table 11.5 indicates the influence of different irrigation methods on the yield and water use efficiency of sugarcane. It could be found that drip irrigation at 2 days resulted in 108% higher irrigation water use efficiency (IWUE) in comparison to furrow irrigation, in which the IWUE was 787 kg ha^{-1}cm^{-1}. It could also be concluded that the adoption of drip irrigation not only saved 42.6% of irrigation water but increased sugarcane yield by 13.3–19.7% in different day intervals (Singandhupe et al., 2008). Based on a study conducted by Mahesh et al. (2013) to determine the performance of sugarcane under subsurface drip irrigation, it was found that subsurface drip irrigation helped to save water by 38.84% and increased water use efficiency by up to 161.40 kg ha^{-1} mm^{-1}. Based on a similar study, Kombali et al. (2015) pointed out that subsurface drip irrigation results in a 39% higher cane yield over surface irrigation due to considerable irrigation water saving.

Irrigation scheduling determines the correct frequency and duration of watering and is an essential tool for good water management. Irrigation must be given to sugarcane when the soil does not cope freely with the plant water requirements to ensure better crop growth. Among various approaches, irrigation at critical growth stages of sugarcane is appropriate in the subtropical northern parts of India. Any water shortage during the tillering phase may

TABLE 11.5 Cane Yield, Sugar Yield, and Irrigation Water Use Efficiency (IWUE) as Influenced By Irrigation Methods (Singandhupe et al., 2008)

Irrigation Method	Cane Yield ($t\ ha^{-1}$)	Sugar Yield ($t\ ha^{-1}$)	IWUE ($kg\ ha^{-1}\ cm^{-1}$)
Furrow method at 75 mm CPE	131.4	15.0	787
Drip at 2 days	157.3	19.0	1,644
Drip at 3 days	152.7	18.2	1,593
Drip at 4 days	149.0	17.6	1,551
SE	2.7	0.8	26

negatively impact cane growth, development, and quality. Recently, irrigation based on irrigation water: cumulative pan evaporation (IW: CPE) has resulted in higher cane yield with minimum water use (Prasad, 2021). The IW: CPE ratio approach is easier to adopt since the depth for each irrigation has already been determined to be 75 mm for sugarcane crops grown in subtropical India (Srivastava et al., 2022). Singh et al. (2007) experimented with optimizing the soil moisture regime in sugarcane in subtropical India. They concluded that at a lower IW: CPE (0.25) ratio (means fewer irrigations), increased water extraction by crop roots from deeper layers under a drier regime due to limited moisture availability in upper layers. The soil moisture depletion was higher under silty clay loam than sandy loam soil, at 0.75 and 1.0 IW:CPE. This was due to better water use in silty clay loam soil than in sandy loam soil. In another study, subsurface drip irrigation at 100% pan evaporation (E_{pan}) recorded significantly higher cane yield and yield parameters viz., millable canes, cane length, number of internodes per cane, cane girth, single cane weight, and cane yield (Kombali et al., 2015). It can be interpreted from Table 11.6 that the number of millable canes (NMC), cane yield, and sugar yield increases successively with the rise in the frequency of irrigation from IW:CPE of 0.5 to 1.0, while the reverse trend was seen for

TABLE 11.6 Effect of Irrigation Schedules on the Number of Millable Canes (NMC), Cane Yield, and Water Productivity of Sugarcane (Singh & Brar, 2015)

Irrigation Schedule (IW: CPE)	NMC ($000\ ha^{-1}$)	Cane Yield ($t\ ha^{-1}$)	AWP ($kg\ m^{-3}$)	TWP ($kg\ m^{-3}$)
0.50	90.6	64.5	1.193	0.713
0.75	103.6	73.7	0.912	0.637
1.00	109.0	78.7	0.747	0.569
CD (p = 0.05)	9.8	2.9	0.045	0.029

water productivity, because of an increase in the volume of irrigation water applied (Singh & Brar, 2015).

Significantly higher cane yield (219 t ha^{-1}) and juice quality parameters with the least amount of reducing sugars (2.82%) were recorded with subsurface drip irrigation at 100% E_{pan}, which is statistically similar to drip irrigation at 75% E_{pan} but significantly higher than furrow irrigation (148 t ha^{-1}) with a higher amount of reducing sugars (2.99%) (Table 11.7). This increase in yield was mainly due to efficient water and nutrient absorption and their accumulation by the crop and maintenance of an excellent soil-water-air relationship with increased oxygen concentration in the root zone. The early development of canes with even maturity under drip irrigation resulted in good-quality juice with the least amount of reducing sugars (Kombali et al., 2016).

TABLE 11.7 Cane Yield and Juice Quality of Sugarcane as Affected by Subsurface Drip Fertigation (Kombali et al., 2016)

Treatment	Cane Yield (t ha^{-1})	Juice Extraction (%)	Brix (%)	Pol (%)	Purity (%)	CCS (%)	Reducing Sugars (%)	Sugar Yield (t ha^{-1})
75% Epan	205	64.52	20.77	19.73	93.73	14.10	2.87	28.80
100% Epan	219	65.71	21.15	20.09	95.18	14.36	2.82	31.48
Furrow irrigation	148	62.73	19.99	18.99	92.96	13.57	2.99	20.03
SEm+	5.3	0.70	0.30	0.33	0.61	0.10	0.22	0.72
LSD (p = 0.05)	20.7	2.21	0.90	0.98	2.00	0.30	0.06	2.83

11.2.7 INTERNET OF THINGS (IOT)

Smart agriculture systems are designed to promote agricultural development and solve issues in the agricultural system. The Internet of Things (IoT) is used to develop smart farming in all aspects, with a proposed model consisting of sensors used to supervise soil moisture content. In this process, data on the physical parameters of soil and water is converted into digital form (data acquisition), then stored and transferred to the desired location to determine the right time for scheduling irrigation. Barapatere & Patel (2019) conducted research on real-time sensing of soil moisture data in sugarcane, then uploaded the data to a cloud server, which further guided the water pump and solenoid valves to initiate the irrigation process as required by the crop, using CROPWAT 8.0 software (Figure 11.2). CROPWAT 8.0 is a decision support tool developed by the Land and Water Development Division of FAO for Windows that calculates the crop water requirements and irrigation needs based on soil, atmospheric, and crop information.

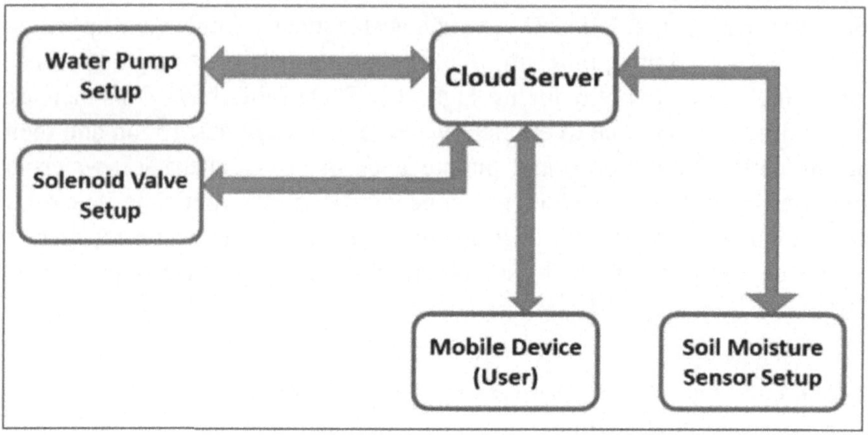

FIGURE 11.2 IoT-based smart irrigation system.

Source: Reprinted from Barapatere & Patel, 2019. https://creativecommons.org/licenses/by/4.0/

Ragavi et al. (2019) reported that smart IoT devices used in the field helped identify the water level, utilized fertilizers and pesticides, and improved sugarcane productivity due to better usage of resources. A cybernetic closed-loop solution was piloted on a sugarcane farm in northeastern Australia to improve irrigation management by integrating the WiSA automated irrigation system with the IrrigWeb irrigation decision support tool. The simulation results indicated that irrigation scheduling could be effectively managed through a closed-loop solution that incorporates irrigation decision support tools with practical constraints (Wang et al., 2020). Hincapié et al. (2019) used sensors to monitor the soil matric potential and managed irrigation scheduling in sugarcane. The data received from sensors indicated that irrigation is to be started at–85 kPa. Viedman et al. (2018) used wireless sensor networks (WSN) based on IoT with middleware SAMIoT and concluded that irrigation scheduling is possible by capturing sensor data, processing, and calculating the matric potential.

11.3 CONCLUSION AND FUTURE PERSPECTIVES

Using agronomic technologies to efficiently manage water resources can significantly increase sugarcane yield and water productivity. Now, the challenge is to apply this to farmers' field conditions. Particularly for developing countries, IoT is the future where we can imagine that every farmer adopts it like the mobile used by the current generation.

KEYWORDS

- **crop management**
- **internet of things**
- **irrigation scheduling**
- **laser land leveling**
- **productivity**
- **row spacing**
- **sugarcane**
- **water-use**

REFERENCES

Adetoro, A. A., Abraham, S., Paraskevopoulos, A. L., Owusu-Sekyere, E., Jordaan, H., & Orimoloye, I. R. (2020). Alleviating water shortages by decreasing water footprint in sugarcane production: The impacts of different soil mulching and irrigation systems in South Africa. *Groundwater for Sustainable Development, 11*, 100464.

Barapatre, P., & Patel, J. N. (2019). Development of Internet of Things (IoT) based smart irrigation system for sugarcane crop. *International Journal of Innovative Technology and Exploring Engineering, 8*(9S), 650–654.

Bhatt, R., Singh, J., Laing, A. M., Meena, R. S., Alsanie, W. F., Gaber, A., & Hossain, A. (2021). Potassium and water-deficient conditions influence the growth, yield and quality of ratoon sugarcane (*Saccharum officinarum* L.) in a semi-arid agroecosystem. *Agronomy, 11*(11), 2257.

Brar, S. K., Mahal, S. S., Brar, A. S., Vashist, K. K., Sharma, N., & Buttar, G. S. (2012). Transplanting time and seedling age affect water productivity, rice yield and quality in north-west India. *Agricultural Water Management, 115*, 217–222.

Brown, M. E., & Funk, C. C. (2008). Food security under climate change. *Science, 319*, 580–581.

Dhanapal, R., Tayade, A. S., Bhaskaran, A., & Geetha, P. (2019). Efficient water management in sugarcane with composted coir pith and sugarcane trash under tropical Indian conditions. *Sugar Tech., 21*, 256–264.

Gopalasundaram, P., Bhaskaran, A., & Rakkiyappan, P. (2012). Integrated nutrient management in sugarcane. *Sugar Tech., 14*, 3–20.

Gouri, V., Devi, T. C., Kumari, M., Bharatalakshmi, M., Rao, K. P., & Murthy, K. R. (2014). Effect of surface and subsurface drip fertigation on yield and quality of sugarcane. *Indian Journal of Sugarcane Technology, 29*(01), 35–37.

Hincapie Gómez, E., Sánchez Benítez, J., & Carbonell González, J. A. (2019). IoT Network Applied to Agriculture: Monitoring Stations for Irrigation Management in Soils Cultivated with Sugarcane. In J. Corrales, P. Angelov, & J. Iglesias (Eds.), *Advances in Information and Communication Technologies for Adapting Agriculture to Climate Change II. AACC 2018* (pp. 293–301). Springer. https://doi.org/10.1007/978-3-030-04447-3_17.

Holsambre, D. G. (1997). Irrigation management techniques in sugarcane. *Farming Systems Journal, IWRS, 17*(3), 1–10.

Hunsigi, G. (2011). Potassium management strategies to realize high yield and quality of sugarcane. *Karnataka Journal of Agricultural Sciences, 24*(1), 45–47.

Irvine, J. E., & Benda, G. T. A. (1980). Sugarcane spacing. II. Effects of spacing on the plant (No. DOE/ET/20002-T1). Department of Agriculture, New Orleans, LA (USA).

Jat, M. L., Chandna, P., Gupta, R., Sharma, S. K., & Gill, M. A. (2006). Laser land leveling: A precursor technology for resource conservation. *Rice-Wheat Consortium Tech. Bull. Ser., 7*, 48. http://www.knowledgebank.irri.org/csisa/images/FactsheetsAndReferences/Techbulletins/lasertechbull.pdf

Jordán, A., Zavala, L. M., & Gil, J. (2010). Effects of mulching on soil physical properties and runoff under semi-arid conditions in southern Spain. *Catena, 81*(1), 77–85.

Kader, M. A., Singha, A., Begum, M. A., Jewel, A., Khan, F. H., & Khan, N. I. (2019). Mulching as a water-saving technique in dryland agriculture. *Bulletin of the National Research Centre, 43*(1), 1–6.

Kahlown, M. A., Raoof, A., & Hanif, M. (2000). Rice yield as affected by plant densities. *Mona Exp. Project Bhalwal, Report No. 238.*

Kombali, G., Sheshadri, T., Nagaraju, M. T., Basavaraja, P. K., & Mallikarjuna, G. B. (2015). Performance of sugarcane under varied levels of irrigation and nutrients through subsurface drip fertigation. *Mysore Journal of Agricultural Sciences, 50*, 290–293.

Kombali, G., Sheshadri, T., Nagaraju, Thimmegowda M. N., Basavaraja, P. K., & Mallikarjuna, G. B. (2016). Influence of subsurface drip fertigation on yield and quality of sugarcane. *Annals of Biology, 32*, 169–173.

Mahesh, R., Asoka, R. N., & Archana, H. A. (2016). Performance of surface and subsurface drip fertigation on yield and water use efficiency of sugarcane. In *Proceedings of the 2nd World Irrigation Forum* (pp. 6–8). Chiang Mai, Thailand.

Mahesh, R., Krishnasamy, S., Gurusamy, A., & Mahendran, P. P. (2013). Performance of subsurface drip fertigation on yield attributes, yield, water saving and water use efficiency of sugarcane (*Saccharum officinarum* L.). *Madras Agricultural Journal, 100*(4–6), 455–459.

Manimaran, S., Kalyanasundaram, D., Ramesh, S., & Sivakumar, K. (2009). Maximizing sugarcane yield through efficient planting methods and nutrient management practices. *Sugar Tech., 11*, 395–397.

Mathew, T. M., Kurian, B. G., Sosamma, J., Kuriakose, M., Sreekumar, K., & Jayakumar, G. (2004). Effect of time and mode of application of potassium on the growth, yield and quality of sugarcane ratoon. *Sugar Tech., 6*, 81–83.

NABARD. (2018). [Sustainable Sugarcane Initiative](https://www.nabard.org/demo/auth/writereaddata/File/11%20_ Sustainable_S sugarcane_Initiative). Accessed 30 Nov 2021.

Narayanamoorthy, A. (2010). Can drip method of irrigation be used to achieve the macro objectives of conservation agriculture? *Indian Journal of Agricultural Economics, 65*, 902–2016-67945.

Naresh, R. K., Bhatt, R., Chandra, M. S., Laing, A. M., Gaber, A., Sayed, S., & Hossain, A. (2021). Soil organic carbon and system environmental footprint in sugarcane-based cropping systems are improved by precision land leveling. *Agronomy, 11*(10), 1964.

Olivier, F. C., & Singels, A. (2015). Increasing water use efficiency of irrigated sugarcane production in South Africa through better agronomic practices. *Field Crops Research, 176*, 87–98.

Pawar, D. D., Dingre, S. K., & Surve, U. S. (2013). Growth, yield and water use in sugarcane (*Saccharum officinarum*) under drip fertigation. *Indian Journal of Agronomy, 58*, 396–401.

Prasad, R. (2002). *Textbook of Field Crops Production-Commercial Crops*. Indian Council of Agricultural Research, New Delhi, pp. 610.

Prosdocimi, M., Tarolli, P., & Cerdà, A. (2016). Mulching practices for reducing soil water erosion: A review. *Earth-Science Reviews, 161*, 191–203.

Ragavi, V., Subburaj, J., Keerthana, P., & Soundaryaveni, C. (2019). Smart agriculture to increase farmers' profitability using Internet of Things. *Indian Journal of Science and Technology, 12*(8), 1–6.

Rajanna, M. P., & Patil, V. C. (2003). Effect of fertigation on yield and quality of sugarcane. *Indian Sugar, 52*(12), 1007–1011.

Ram, D., Ram, M., & Singh, R. (2006). Optimization of water and nitrogen application to menthol mint (*Mentha arvensis* L.) through sugarcane trash mulch in a sandy loam soil of semi-arid subtropical climate. *Bioresource Technology, 97*(7), 886–893.

Sacala, E. (2009). Role of silicon in plant resistance to water stress. *Journal of Elementology, 14*(3), 619–630.

Sanghera, G. S., Singh, R. P., Singh, O., & Tyagi, V. (2018). Initiatives for sustainable sugarcane farming in Punjab. *Journal of Plant Science Research, 34*(2), 137–152.

Shrivastava, A. K., Ghosh, A. K., & Agnihotri, V. P. (1992). *Sugarcane Ratoons*. Oxford and IBH, New Delhi.

Shukla, S. K., Nagargade, M., Pathak, A. D., & Blessey, M. (2020). Water footprint in sugarcane. *Technical Bulletin ICAR-Indian Institute of Sugarcane Research, Lucknow*, pp. 1–36.

Singandhupe, R. B., Bankar, M. C., Anand, P. S. B., & Patil, N. G. (2008). Management of drip irrigated sugarcane in western India. *Archives of Agronomy and Soil Science, 54*(6), 629–649.

Singh, H., & Singh, R. K. (2020). Irrigation and fertigation scheduling effect on phenology and yield of spring sugarcane (*Sachharum officinarum*). *Indian Journal of Agricultural Sciences, 90*(6), 1054–1057.

Singh, K., & Brar, A. S. (2015). Effect of planting methods and irrigation schedules on cane yield, quality, economics and water productivity of spring sugarcane (*Saccharum officinarum*) in South Western Punjab. *Indian Journal of Agronomy, 60*(4), 601–605.

Singh, K., Gill, M. S., Singh, A., Singh, D., Uppal, S. K., & Singh, J. (2012). Sugarcane planting in standing wheat using furrow irrigated raised bed (FIRB) method. *Sugar Tech., 14*, 351–356.

Singh, K., Pal, R., Chalotra, N., & Brar, A. S. (2022). Water productivity of sugarcane influenced by planting techniques, mulching and irrigation scheduling in Indo-Gangetic plains of India. *Sugar Tech., 24*(2), 408–418.

Singh, P. N., Shukla, S. K., & Bhatnagar, V. K. (2007). Optimizing soil moisture regime to increase water use efficiency of sugarcane (*Saccharum spp.* hybrid complex) in subtropical India. *Agricultural Water Management, 90*, 95–100.

Sivakumar, M. V. K., & Stefanski, R. (2011). Climate Change in South Asia. In Lal, R., Sivakumar, M. V. K., Faiz, S. M. A., Rahman, A. H. M. M., & Islam, K. R. (Eds.), *Climate change and food security in South Asia* (pp. 13–28). Springer Science+Business Media B.V.

Srivastava, S. C., & Johari, D. P. (1979). *The irrigated sugarcane in India*. AICRP (Sugarcane), Indian Institute of Sugarcane Research, Lucknow, pp. 36.

Srivastava, T. K., Hasan, S. S., & Verma, R. R. (2022). Mobile 'app' for precise irrigation scheduling in sugarcane cultivation under Indian subtropics. *Sugar Tech., 24*(3), 822–831.

Tayade, A. S., Geetha, P., Dhanapal, R., & Hari, K. (2016). Effect of in situ trash management on sugarcane under wide row planting system. *Journal of Sugarcane Research, 6*(1), 35–41.

Teixeira, G. C. M., de Mello Prado, R., Rocha, A. M. S., dos Santos, L. C. N., dos Santos Sarah, M. M., Gratão, P. L., & Fernandes, C. (2020). Silicon in pre-sprouted sugarcane seedlings mitigates the effects of water deficit after transplanting. *Journal of Soil Science and Plant Nutrition, 20*, 849–859.

Thomas, C. G. (2022). *Irrigation and Water Management*. Ane Books Pvt. Ltd., Chennai.

TNAU. (2023). Expert system for sugarcane. Accessed: http://www.agritech.tnau.ac.in/expert_system/sugar/irrigationmanagement.html.

Tomar, S. S., Singh, Y. P., Naresh, R. K., Dhaliwal, S. S., Gurjar, R. S., Yadav, R., Sharma, D., & Tomar, S. (2020). Impacts of laser land leveling technology on yield, water productivity, soil health and profitability under arable cropping in alluvial soil of north Madhya Pradesh. *Journal of Pharmacognosy and Phytochemistry, 9*(4), 1889–1898.

Verma, K. K., Li, D. M., Singh, M., Rajput, V. D., Malviya, M. K., Minkina, T., Singh, R. K., Singh, P., Song, X. P., & Li, Y. R. (2020b). Interactive role of silicon and plant-rhizobacteria mitigating abiotic stresses: A new approach for sustainable agriculture and climate change. *Plants, 9*, 1055. DOI: 10.3390/plants9091055.

Verma, K. K., Liu, X. H., Wu, K. C., Singh, R. K., Song, Q. Q., Malviya, M. K., Song, X. P., Singh, P., Verma, C. L., & Li, Y. R. (2019). The impact of silicon on photosynthetic and biochemical responses of sugarcane under different soil moisture levels. *Silicon, 12*, 1355–1367.

Verma, K. K., Song, X. P., Budeguer, F., Nikpay, A., Enrique, R., Singh, M., Zhang, B., Wu, J. M., & Li, Y. R. (2022c). Genetic engineering: an efficient approach to mitigating biotic and abiotic stresses in sugarcane cultivation. *Plant Signaling & Behavior, 17*(1), e2108253.

Verma, K. K., Song, X. P., Li, D. M., Singh, M., Wu, J. M., Singh, R. K., Sharma, A., Zhang, B. Q., Li, Y. R. (2022d). Silicon and soil microorganisms improves rhizospheric soil health with bacterial community, plant growth, performance and yield. *Plant Signaling & Behavior, 17*(1), e2104004.

Verma, K. K., Song, X. P., Lin, B., Guo, D. J., Singh, M., Rajput, V. D., Singh, R. K., Singh, P., Sharma, A., Malviya, M. K., Chen, G. L., & Li, Y. R. (2021c). Silicon induced drought tolerance in crop plants: physiological adaptation strategies. *Silicon, 14*, 2473–2487. DOI: 10.1007/s12633-021-01071-x.

Verma, K. K., Song, X. P., Rajput, V. D., Boldyreva, V., Zhang, B. Q., Minkina, T., & Li, Y. R. (2022b). Morpho-physiological, biochemical and ultrastructural modifications on sugarcane to prolonged water deficit. Verma, K. K., et al. (eds.), *Agro-industrial perspectives on sugarcane production under environmental stress*. https://doi.org/10.1007/978-981-19-3955-6_11.

Verma, K. K., Song, X. P., Rajput, V. D., Solomon, S., Li, Y. R., & Rao, G. P. (Eds.) (2023). *Agro-industrial Perspectives on Sugarcane Production under Environmental Stress*. Springer, Singapore. https://doi.org/10.1007/978-981-19-3955-6_1.

Verma, K. K., Song, X. P., Tian, D. D., Singh, M., Verma, C. L., Rajput, V. D., Singh, R. K., Sharma, A., Singh, P., Malviya, M. K., & Li, Y. R. (2021a). Investigation of defensive role of silicon during drought stress induced by irrigation capacity in sugarcane: physiological and biochemical characteristics. *ACS Omega, 6*, 19811–19821.

Verma, K. K., Song, X. P., Verma, C. L., Chen, Z. L., Rajput, V. D., Wu, K. C., Liao, F., Chen, G. L., & Li, Y. R. (2021b). Functional relationship between photosynthetic leaf gas exchange in response to silicon application and water stress mitigation in sugarcane. *Biology Research, 54*, 15. https://doi.org/10.1186/s40659-021-00338-2.

Verma, K. K., Song, X. P., Yadav, G., Degu, H. D., Parvaiz, A., Singh, M., Huang, H. R., Mustafa, G., Xu, L., & Li, Y. R. (2022a). Impact of agroclimatic variables on proteogenomics in sugarcane (*Saccharum spp.*) plant productivity. *ACS Omega, 7*, 22997–23008.

Verma, K. K., Song, X. P., Zeng, Y., Li, D. M., Guo, D. J., Rajput, V. D., Chen, G. L., Barakhov, A., Minkina, T. M., & Li, Y. R. (2020c). Characteristics and correlation of leaf stomata and its relationship with photosynthesis on *Saccharum spp.* under different irrigation and silicon application. *ACS Omega, 5*, 24145−24153.

Verma, K. K., Wu, K. C., Verma, C. L., Li, D. M., Malviya, M. K., Singh, R. K., Singh, P., Chen, G. L., Song, X. P., & Li, Y. R. (2020a). Developing mathematical model for diurnal dynamics on photosynthetic responses in sugarcane responsive to different irrigation and silicon application. *PeerJ, 8*, e10154. DOI: 10.7717/peerj.10154.

Viedman, J. L. R., Benítez, J. S., Gonzáles, J. A. C., & Hincapié, E. (2018). SAMIoT – Middleware based on IoT for irrigation planning on large-scale crops. Accessed: *The Third International Conference on Universal Accessibility in the Internet of Things and Smart Environments.* Accessed: https://www.researchgate.net/profile/Hugo-Ordonez-2/publication/325063572_SAMIoT__Middleware_based_on_IoT_for_Irrigation_Planning_on_Large-scale_Crops/links/5af449bda6fdcc0c030ade94/SAMIoT-Middleware-based-on-IoT-for-Irrigation-Planning-on-Large-scale-Crops.pdf.

Wang, E., Attard, S., Linton, A., McGlinchey, M., Xiang, W., Philippa, B., & Everingham, Y. (2020). Development of a closed-loop irrigation system for sugarcane farms using the Internet of Things. *Computers and Electronics in Agriculture, 172*, 105376.

Wijayawardhana, L. M. J. R., Witharama, W. R. G., & De Silva, A. L. C. (2017). Evaluation of alternate-row furrow irrigation technique on the growth and performance of sugarcane (*Saccharum officinarum L.*) grown in reddish brown earth soil in Sri Lanka. *Sugarcane Sri Lanka, Vol. 3,* pp. 1–9.

Yadava, R. L. (1991). *Sugarcane Production Technology: Constraints and Potentialities.* Oxford & IBH Publishing Company.

Yang, S. D., Liu, J. X., Wu, J., Tan, H. W., & Li, Y. R. (2013). Effects of vinasse and press mud application on the biological properties of soils and productivity of sugarcane. *Sugar Tech, 15*, 152−158.

Zende, G. K. (1990). Soil fertility management for higher sugar and sugarcane production. In: *Recent Advances in Sugarcane*, (Ed. P.N. Rao), Vuyyuru, Andhra Pradesh: KCP Ltd., pp. 99–200.

CHAPTER 12

Economic Importance and Yield Potential of Sugarcane in Pakistan

SHAHID AFGHAN,[1] MUHAMMAD EHSAN KHAN,[1] KRISHAN K. VERMA,[2] and AMIN NIKPAY[3]

[1]*Sugarcane Research and Development Board, Ayub Agricultural Research Institute, Faisalabad, Punjab, Pakistan*

[2]*Sugarcane Research Institute, Guangxi Academy of Agricultural Sciences/Key Laboratory of Sugarcane Biotechnology and Genetic Improvement (Guangxi), Ministry of Agriculture and Rural Affairs/ Guangxi Key Laboratory of Sugarcane Genetic Improvement, Nanning, Guangxi, China*

[3]*Department of Plant Protection, Sugarcane and By-Products Development Company, Salman Farsi Agro-Industry, Ahwaz, Iran*

ABSTRACT

Sugarcane is mainly cultivated in tropical and subtropical regions; nearly 85% of sugar is consumed worldwide. In the last five years, the area under sugarcane has greatly increased worldwide. In Pakistan, the production and yield of sugarcane have greatly improved compared to other crops. It is the 2nd largest economically important crop after cotton. It is a high-value cash crop significant for Pakistan's sugar and sugar-related industries. It contributes about 0.6% to GDP and 2.9% of total value added in agriculture. It creates huge revenue for the Government and is used as a source of energy/power. The climate of Pakistan suits sugarcane production, especially in Punjab and Sindh provinces. Different climatic factors like sunlight, temperature, germination, tillering, growth, humidity, dew, frost, hailstorms, windstorms, sunburn, and drought greatly affect sugarcane production. Pakistan is a

Sugarcane Cultivation and Management: Challenges and Opportunities. Krishan K. Verma, Xiu-Peng Song, Munna Singh, Renato de Mello Prado, Jian-Ming Wu, and Yang-Rui Li (Eds.)

leading cane-growing country and ranks 5th in area, 6th in cane sugar production, and 9th largest sugar producer globally. This chapter describes the economic importance, climate, and yield potential of sugarcane in Pakistan.

12.1 INTRODUCTION

Sugarcane is a globally significant crop, providing nearly 85% of the sugar consumed worldwide. In recent years, the planting of sugarcane for the production of biofuels has been expanded rapidly as energy canes (Verma et al., 2020a, b; Deng et al., 2022; Xu et al., 2022). The cultivation of sugarcane is one of the most important activities worldwide due to its dietary, environmental, social, and economic implications and potential productive diversification with co-products and byproducts (Huang et al., 2021; Oliveira et al., 2022). Sugarcane is a worldwide crop cultivated in more than 105 countries. From its very origin, in earlier times to its present-day production, sugarcane has played its role in improving the socio-economic conditions of human society (Qureshi, 2004; Verma et al., 2021a, b, 2023).

Previous trends have shown a tremendous increase in the area of sugarcane production worldwide. But for 2019–2020, global sugar production is estimated to decline by 3% to 1,745 million tonnes while sugarcane was grown on 26.5 million ha. This diminution is attributed to the 5 million tonnes drop in India's production, resulting from lower area and expected yields. However, Brazil and India are essentially tied as the top sugar-producing countries, while Pakistan occupies the fifth place (Annual Report of Pakistan, 2020–2021).

It is mostly used as a food crop to produce raw and refined sugar, gur, and shakkar. Sugarcane improvement has traditionally focused on sucrose-yield traits. In the future, energy canes with higher yields of fermentable sugars and fiber (bagasse) for biofuel and electricity applications will be developed (Chu & Kong, 1971; Verma et al., 2022a, b). Sugarcane has gained importance for its dietary value and industrial utilization of several products. Its products and by-products have revolutionized native and international trade, and the crop production trends have played a dominant role in altering countries' economic and fiscal positions.

Its juice is used for making white sugar, brown sugar (khand), and jaggery (gur) (Qureshi & Afghan, 2005). The sugarcane bagasse, a byproduct, is nowadays valued by producers from the sugar sector since it is presented as the main feedstock source for bioenergy and biofuel production. In Pakistan, almost 3.50 million metric tonnes of bagasse are consumed, with an average recovery rate of 30%. Bagasse is conventionally recycled as confined energy in sugar factories, i.e., boiler steam is required to generate and drive the prime movers (Dotaniya, 2014). Bagasse is used to produce compressed fiberboard, paper, plastics, and furfural. Due to the high silicon content of sugarcane, bagasse can be used to produce silica, a valued material for industry and health products. Molasses is used in distilleries to manufacture ethyl alcohol, butyl alcohol, citric acid, etc. Sugarcane filter cake (press mud) has good potential as an organic fertilizer (Prasannamedha et al., 2020).

12.2 IMPORTANCE OF SUGARCANE

12.2.1 DIETARY VALUE

Sugar is used as a sweetener in a number of kitchen dishes of varying tastes, beverages, and pharmaceutical products. It is an important constituent of the human diet, having a pleasant taste and a high caloric value of 387 per 100 grams (Dotaniya, 2014). Some fruits and vegetables also have some forms of sugar (Table 12.1), but it is not derived as a commercial product (Fahim, 2014). The trend of using Low Caloric Sweeteners (LCS) has now developed in some quarters. Among the low-energy sweeteners, saccharin is the oldest sweetener. Sugarcane juice contains zero fat, cholesterol, fiber, and protein. This is the healthiest and most nutritious drink one can think of consuming. The latest research showed that sugar-sweetened beverages (SSBs) and LCS beverages were linked to an increased risk of developing type 2 diabetes. Sugar percentage in different fruits has been presented in Table 12.1. Some forms of sugar are produced from palm trees, sugar maple, sweet sorghum, maize, sugar beet, and sugarcane (Qureshi, 2004) (Table 12.2).

12.2.2 GROWER'S PROSPERITY

Sugarcane is a high-value cash crop that has significance for sugar and sugar-related industries in Pakistan. It contributes about 0.6% to GDP and 2.9% of total value added to agriculture, bringing prosperity to the growers' community (GOP, 2019). In areas of cane concentration, healthy socio-economic change is witnessed, which has improved the living standard of growers. The dilemma is that farmers normally do not receive a fair market price for their sugar crops. General problems faced by almost every farmer regarding sugarcane production are lack of irrigation water, non-availability of improved varieties of sugarcane, high cost of inputs for land preparation,

TABLE 12.1 Sugar Contents in Different Fruits

Fruits	Sugar (%)
Mangoes	12.7
Strawberries	4.1
Pomegranates	11.6
Guavas	4.7
Papaya	5
Watermelon	5.3
Grapefruit	5.9
Figs	13.8
Peaches	7.1
Oranges	8
Apricots	7.8
Plums	8.4
Pears	8.4
Pineapple	8.4
Cherries	10.9
Apples	8.8
Grapes	13.8
Bananas	10.4

Source: Fuhr (2016).

TABLE 12.2 Plants Utilized for Commercial Sugar Production

Palm Trees	9% sucrose on the weight of palm juice.
Sugar Maple	2% sugar maple sap.
Sweet Sorghum	9% sucrose (brix 19%, pol 15%).
Maize	2–4% glucose.
Sugar Beet	12–18% sucrose.
Sugarcane	10–13% sucrose.

Source: Sahari et al. (2014).

diseases, insect pests, weeds, and marketing problems (Kakade, 1985). The adversities faced by sugarcane farmers have been unequivocally portrayed by the multitude of protests organized by sugarcane farmers. Given the failure of the governments to resolve these issues, it is apparent that a different course of action needs to be taken to address the fundamental problem. We need to look beyond our conventional strategies.

In developing countries, sugar factories delay the timely payments of sugarcane farmers. Because of millions of sugarcane farmers, governments must swiftly resolve the most pivotal issue before it is further retrograde. In return, governments mainly have two options. First, one is to pay back the dues to the farmers from the state treasury or provide subsidies to sugar factories. Either way, the government must shift the monetary burden to the taxpayers (Qureshi & Afghan, 2005). In the short run, given the obstinate demand of sugarcane farmers to receive their payments and the incapability of sugar factories to make the payments, this seems to be a plausible option.

12.2.3 SUGAR INDUSTRY OF PAKISTAN

A total of 89 sugar mills are operational (Annual Report of Pakistan, 2019). Sugarcane was cultivated in an area of 1.16 million ha during 2020–2021 (Annual Report of PSMA, 2021). On a regional basis, Punjab's share in the total sugarcane area of Pakistan was 65.7%, followed by Sindh at 26.8% and KPK at 7.5%. The sugarcane yield increased from the last five in 2020–2021. The highest yield was recorded at 69.55 tonnes/ha, while the maximum share in harvest was Punjab at 59.5 tons (Table 12.3).

TABLE 12.3 Provincial Distribution of Sugarcane Production in Pakistan in 2020

Province	Area (000 ha)	Cultivated Area (%)	Production (000 ton)	Provincial Production (%)	Yield (t/ha)
Punjab	643.4	4.71	43346.6	65.7	67.37
Sindh	286.0	9.95	17233.8	26.8	60.24
KPK	109.4	6.31	5753.9	7.5	52.62
Pakistan	1038.9	5.37	66334.4	100.0	63.85

Source: Annual Report of Pakistan Sugar Mills Association (2020).

The area under sugarcane in Pakistan has increased from 9,60,000 ha in 2001–2002 to 1,038,879 ha in 2019–2020, with an annual growth rate of 1.0% (GOP, 2020). The total increase in area, production, and per hectare yield during this period is around 26.7%, 73.1%, and 27.5% (Annual Report of Pakistan Sugar Mills Association, 2020–2021).

12.2.4 LABOR AND WORKFORCE

Sugarcane is a source of employment for millions of people in the farm and sugar industries. For example, in Pakistan, between 15,000 and 25,000 families are involved in the sugarcane zone of one factory. Almost 9,80,000 farmers were engaged in sugarcane cultivation in Pakistan, of which 7,07,000 farmers worked in Punjab, 2,00,000 in Sindh, and 81,000 in KPK, respectively. On average, 1,200 employees work in a sugar mill with a workforce of 106.8 thousand individuals in the production and processing departments. Overall, around 4 million employees are engaged directly or indirectly in the sugar business in Pakistan (Malik, 2018). Due to the lack of mechanical sugarcane farming (planting and harvesting) in Pakistan, innumerable workers are deployed in cane harvesting, transport, loading, and unloading. The huge involvement of technicians in farm machinery with the transport system of tractors, trolleys, and trucks depicts the quantum of business in various sectors (Qureshi & Afghan, 2005).

12.2.5 SOURCE OF FOOD FODDER AND FERTILIZER

Sugarcane is used as a source of food and fodder. The press mud and fly ash products of the sugar industry are highly valued organic fertilizers. Filter Cake is used as fertilizer and in brick kilns (Nasir et al., 1994), also used with distillery effluent and nitrogen-fixing bacteria. Some processors/factories have acquired certificates of halal and Organic Products like Sugar/Gurr/Shakar. Production of organic Gur and Shakar is about 1,000 tons having a business of around 30 crores (Annual Report of Shakarjang Sugar Mills, 2018).

12.2.6 GOVERNMENT REVENUES

In Pakistan, the revenue is derived from GST (general sales tax) on the sale of sugar, which is 17% of the sugar price. A substantial amount is recovered as a sugarcane Cess Fund at Rs. 2.00 per maund of sugarcane delivered to sugar mills. Land revenue and water rates, at Rs. 400 per acre of cane, amount to billions of rupees per annum and are the direct source of income for the government. The excise duty on molasses and alcohol is also a big source of revenue for the government as well (Usman, 2016).

12.2.7 SOURCE OF ENERGY/POWER

Molasses, the main by-product of the sugar industry, is utilized to produce alcohol and rum and as feed for livestock. Ethanol has now gained high importance as an energy product (Sugarcane Handbook, 2017). Sugarcane is used as feedstock for ethanol production and as biofuel in vehicles. Using high-pressure steam boilers and efficient energy use in the sugar factories has given new vistas of co-generation from bagasse and trash to export surplus electric power to the national grid system. Production of biogas from spent wash/vinasse fermentation is also a source of cheaper energy in distilleries (Sugarcane Handbook, 2017).

12.2.8 INDUSTRIAL UTILIZATION

Sugarcane occupies a prominent place for high biomass-fiber production and molasses. As a raw material, sugarcane has attained worldwide importance for dozens of industrial derivatives (Verma et al., 2020c, d, 2021c, d). The main by-products of cane are molasses used for alcohol, bagasse utilized for energy production, co-generation, manufacture of particleboard, and filter press cake used as organic fertilizer (Pakistan Agricultural Research Council, 2018). In the present-day market economy, the profits derived from the manufacture of sugar are decreasing, and more attention is paid to the manufacture of by-products. Today, the focus is to adopt sugar production technology to yield energy-based co-products, ethanol, and electricity. Cane acreage fluctuates depending on market forces, socioeconomic conditions, and crop competition (Sugarcane Handbook, 2017).

12.3 CANE YIELDS ON THE GLOBAL LEVEL

On the global level, sugarcane is grown in 105 countries in equatorial, tropical, and sub-tropical regions. The cane area, production, and yields of principal cane-growing countries are shown in Table 12.4. The world data show that in 2017, sugarcane was grown in an area of 25.98 million hectares with a cane yield of 70.9 tch. Concerning countries, Brazil occupies the leading position by growing cane in an area of 10.18 million hectares, followed by India, China, Thailand, and Pakistan, growing cane in an area of 4.39, 1.38, 1.37, and 1.22 million hectares, respectively (Legendre & Martin, 1977).

During the middle of the 20th century, the average yield was hardly 42.5 tons of cane per hectare, which has increased to 70.9 tons. During the 1950–1953 period, principal cane-growing countries, Brazil, India, China, and Thailand, had cane yields of 38.7, 32.1, 35.2, and 17.5 t/ha, respectively, much below the world average. During five decades of progressive development, Brazil, China, and Thailand have improved their yields to 74.5, 76.1, and 75.2 t/ha, respectively, which exceeds the world average of 70.9 t/ha.

Still, there are countries with an average yield of 121.0 tonnes per hectare (Guatemala) and 112.7 tonnes per hectare (Egypt), as shown in Table 12.4. Pakistan grasps an important position in cane area and production in the world and ranks in the top 5th position, but with respect to cane yield (69.55 t/ha), it ranks much below the principal cane-growing countries (Muhammad et al., 2019). Comparing other countries, Pakistan took a late start to meet yield gaps. Even India has touched 70 t/ha, while Pakistan has yet to travel a lot to reach this yield level.

TABLE 12.4 Sugarcane Area, Production, and Yields of Top 20 Cane Growing Countries During 2017

Country	Area (000 ha)	Production (M ton)	Yield (t/ha)
World	25976.94	–	70.89
Brazil	10184.34	758.55	74.48
India	4389.00	306.07	69.74
China	1377.11	104.79	76.10
Thailand	1368.27	102.95	75.24
Pakistan	1216.89	73.40	60.31
Mexico	722.00	56.95	73.78
Australia	457.47	36.56	80.62
Philippines	439.47	29.29	66.94
Indonesia	430.11	21.21	49.32
Colombia	397.39	34.64	87.16
Cuba	387.70	16.07	47.45
Argentina	378.82	19.16	50.54
USA	365.88	30.15	82.41
Vietnam	281.15	18.36	65.29
Guatemala	278.97	33.76	121.01
South Africa	264.50	17.39	65.74
Burma	163.25	10.37	63.52
Bolivia	151.99	8.05	52.96
Cameroon	136.29	1.29	9.45
Egypt	135.40	15.26	112.70

Source: Annual Report of Pakistan Sugar Mills Association (2017).

The highest cane yield of 121.01 t/ha is reported in Guatemala, followed by 112.70 tons in Egypt. Among the top 20 cane-growing countries, Colombia (87.16), the USA (82.41), and Australia (80.82) have yielded over 80 tons of cane per hectare. In contrast, China, Thailand, Brazil, and Mexico have exceeded 70 tons in yield (Legendre & Martin, 1988). Pakistan has just 60 tons against about 70 tons yield of India. Pakistan, though, stands in fifth position in cane area and production in the world but has to travel a lot in the compatible field of yields per hectare.

12.4 EFFECT OF LATITUDE

The data in Table 12.5 indicates that the world's largest cane area is in S. America (44.5%), closely followed by Asia (40.3%). The continents of N.C. America, Africa, and Oceania grow cane in small fractions of 8.85%, 8.91%, and 1.82% of the area, respectively. The maximum cane area (55%) falls in countries situated in latitude ranges of 10°–20°N or S. The countries in latitude ranges of 0–10°, 20–25°, and 25–35° grow 10.5%, 16.5%, and 19.5% of the total world cane average, respectively. The continent of Asia has a relatively higher proportion of cane grown in 25–35° latitudes, S. America in 10–20°, while cane area in N. America and Oceania mostly fall in latitude ranges of 20–25°.

TABLE 12.5 Cane Area (000 ha) and Cane Yield (Tha1) of Various Continents Located in Different Latitude Ranges During (2017)

Continent	**Latitude Ranges – Cane Area and Cane Yield**					**Share in Area (%)**
	0–10	**10–20′**	**20–25′**	**25–35′**	**Total**	
Asia Area	691.8	3496.4	2421.4	3869.3	10478.9	40.3
Yield	61.1	66.3	66.5	62.5	64.7	
Africa Area	523.5	400.9	138.9	468.3	1531.6	5.9
Yield	34.5	52.6	42.7	82.8	54.7	
N.C. America Area	765.0	14.8	1162.0	365.9	2307.7	8.9
Yield	89.2	59.7	62.9	82.4	74.7	
S. America Area	733.0	10336	118.0	385.4	11572.7	44.5
Yield	82.7	74.2	56.0	50.7	74.6	
Europe Area	–	–	–	0.08	0.08	0.00
Yield				27.0	27.0	
Ocenea Area	7.1	35.2	435.5	–	477.8	1.8
Yield	28.7	42.3	80.6		73.4	
Total World Area Yield	2720.4	14283.6	4275.8	5088.9	25976.9	100
	70.0	71.7	65.9	64.9	70.9	
% Share in Area	10.5	55.0	16.5	19.6	100	–

Source: FAOSTAT (2017).

The latitudes have a great impact on cane yields. The highest cane yields are observed in latitude ranges of 0°–20° (70.0–71.7 tch), followed by 20°–25° (65.9 tch) and 25°–35° (64.9 tch). It indicates that cane yields are gradually declining as we move away from the Equator. With respect to overall global cane yields, North and South America have the highest yield of 74.7 tons and are closely followed by Oceania (73.4 tch), while the yields of Asia (64.7 tch), Africa (54.7 tch), and Europe (27.0 tch) are considerably low. The latitude range of 10°–25° is considered the most favorable for cane production.

12.5 CORRELATION OF LATITUDE WITH CANE YIELDS AND JUICE QUALITY

12.5.1 CANE YIELD

On the global level, the effect of latitude on cane yields is well-marked. The yield level towards the equator is high, and as the cane growing is shifted away from the equator, yields show a gradual decline (Mendelsohn, 2014). The survey of the global area indicates some very interesting climatic features of countries lying in favorable climatic zones but having very low yields (Mendelsohn, 2014; Verma et al., 2020e, f, 2023). In present-day agriculture, the word 'favorable climate' has been substituted by 'favorable environment' (Kurukulasuriya et al., 2006; Hussain et al., 2018). High cane and sugar yields are obtained under such dry environments where satisfactory soil water balance can be achieved by irrigation water under optimum soil and crop management practices (Marin et al., 2019). The sugarcane varieties have been developed, which owing to their great adaptability, can be cultivated under a wide range of climate and soil conditions (Qureshi et al., 2010; Verma et al., 2020f, 2021, e; Oliveira et al., 2022). Sugarcane cultivation is therefore spread in tropical, sub-tropical, and temperate regions between latitude ranges of 0 and 37° N and S. World data indicate that the best cane-producing areas in Asia are situated in 25°–35° latitude (Kurukulasuriya et al., 2006). On a global level as well, after 10–20° latitude, the 25–35° latitude region produces the maximum cane in the world.

12.5.2 CANE QUALITY

Lower latitudes ensure longer days with increased sunshine duration during the growth phases of the plant. As such, high sucrose contents in cane are noted at around 18° S and 18 N. Sugar contents drop rapidly from these latitudes towards the subtropics and less rapidly towards the equator. High sugar recoveries demand a relative temperature disparity (RTD) value around 14–16°C, low daily mean temperature of 10–12°C, and low relative humidity (<50%) during ripening (Naoko et al., 2009). The wider range between the day (maximum) and night (minimum) temperature during ripening results in higher sucrose contents (Das, 1941). High altitudes of 1,000 meters or more within equatorial climates and low altitudes of subtropics also produce good-quality cane (Blume, 1986).

On the other hand, tropical countries or regions with high average temperatures (25–27°C) and high precipitation of more than 1,500 mm produce more biomass of cane but are of low sugar content. According to Blume (1986), lower sucrose contents in equatorial tropics than in much drier environments of sub-tropical regions are due to higher temperatures, more cloudiness, high precipitation, and consequently less light insulation. In tropical areas where the weather is distinctly wet and dry, moisture status inside the cane is a dominating factor in the synthesis and translocation of sugars. Under such conditions, control of irrigation and fertilizer is exercised to hasten or delay maturity (Humbert, 1968). In modern agriculture, high sucrose content is achieved irrespective of latitude in regions of low precipitation in both tropics and sub-tropics if optimum soil and crop water balance have been achieved by human control. In light of the preceding discussions, climatic features have a distinct effect on plant growth and maturity phases.

- The high cane yield per hectare results from a large biomass production caused by moderately elevated temperatures, increased precipitation, and a longer growing season. However, these do not favor high sucrose recovery in cane.
- High sucrose contents in cane are a function of relatively low temperature and low precipitation during the maturity phase, which otherwise depress growth (Naoko et al., 2009).

12.6 CLIMATIC FACTORS WITH EFFECTS

The change in weather factors like sunlight, temperature, rainfall, humidity, and solar radiation affect the different phases of growth, maturity, and ripening phases of the plant (Hussain et al., 2018; Verma et al., 2020f, 2022c).

12.6.1 SUNLIGHT

Sunlight is used as a photoperiod to regulate various growth processes in the vegetative phase and for flowering. The sugarcane yield is governed by the following:

- The amount of sunlight exposure to the crop over its lifetime, during various phases of growth.
- Ability to intercept and use light energy to produce carbohydrates, the main source of sugars.

For sugarcane, the desirable locations for light availability are within 30° N or °S, and more so on tropical latitude ranges. Crops with high yield and great potential need longer periods with greater light. For the best result, 18 months are preferred over 12 months (Hussain et al., 2018). Tillering is badly affected in long hazy seasons. For good tillering, excessive light is required if plants grow too close to light is curtailed, causing tiller mortality.

12.6.2 TEMPERATURE

Sugarcane crops might be exposed to scorching heat with a maximum mean temperature of 40–42°C, reaching the highest maximum of 48–50C in Pakistan, India, and other countries. A little deviation was observed in maximum and minimum temperatures during summer and winter in tropical and equatorial climates, except in mountain heights or regions under the influence of cold or hot sea currents.

12.6.3 GERMINATION

Germination estimates revealed that in Pakistan the optimum temperature range was 28–33°C (Biswass, 1988). Germination is accelerated in the temperature range of 26 to 33°C, is sensitive at 22°C, and is decreased at 18°C and below. The 32–38°C is reported to be the optimum temperature for germination. In Pakistan, September and March are considered the ideal conditions for good germination. When the temperature is above 25°C, maximum germination of almost 75% was obtained within 20–25 days of cane planting. The temperature is somewhat lower in February-March, and germination takes four or five weeks. Rains may prolong the germination period (Afghan & Jamil, 2013).

12.6.4 TILLERING

In sugarcane, tillering is noticeably correlated with temperature. Tillering gradually increases with increasing temperature until the maximum reaches around 30°C.

12.6.5 GROWTH

Temperature is considered the chief growth monitoring factor. The 26–27°C is the ideal temperature for optimum growth; growth is checked at 10°C and 27°C is optimum for both growth and nutrient absorption (Humbert, 1968). The critical temperature for cane growth is 18–20°C (Bacchi, 1977; Biswas, 1988), and below 12°C, growth ceases. The leaves become pink if it is less than 5°C (Parthasarathy, 1972; Biswas, 1988). Canopy development is governed by the prevalence of moderate temperature between 21°C and 38°C with a relative humidity of 50% (Kakade, 1985). At a temperature of 35°C, the growth rate decreases due to increased photorespiration (Chu & Kong, 1971). Cane thrives at temperatures as high as 45°C in Pakistan. The lower leaves die, and the uppers show a yellow-green appearance. The 27°C is the optimum temperature for growth and nutrient absorption (Burr, 1957). A sevenfold decrease in P uptake was noticed as the temperature fell from 22.1 to 16.7°C. A drop in root temperature from 23°C to 19C cuts P intake to one-third and reduces N intake to about one-half (Burr, 1957).

12.6.6 MATURITY AND RIPENING

Large differences between day and night temperatures favor the process. A temperature of 10 to 12°C is considered the minimum temperature for ripening (Parthasarathy, 1972). However, the ripening process is accelerated with the drop in humidity in the environment and leaf moisture in the plant. Mild drought conditions may be created by withholding irrigation to hasten the ripening process. For sugar content in cane, this process needs further clarification. The most important ripening agents are low temperature and moderate water deficit associated with nitrogen (Alexander, 1973). Some authors propose that in addition to air temperature and soil moisture, variables such as photoperiod and solar radiation must also be considered (Legendre & Martin, 1977). Cloudy days are limiting factors for sucrose accumulation in plants. Solar radiation is directly related to sugar ripening. Factors such as air temperature, precipitation, and cloud-free bright shining days increase the time period available for photosynthesis (Cardoza & Sentelhas, 1973). A temperature of 20–25°C during the day and 10–14°C at night, associated with bright sunny days with low humidity, are ideal for sucrose synthesis and leaf accumulation.

12.6.7 HUMIDITY

The growth of cane crops depends on atmospheric humidity and proper soil moisture. High humidity for a long duration may inhibit evapotranspiration and affect development. High humidity causes infection of some viral and fungal diseases. Evapotranspiration rate increases with high temperature and low humidity to balance the water stability in the plant, thus hindering growth and may also increase the sensitivity of sugarcane to some sap-sucking pests such as mites. Atmospheric humidity is associated with coastal environments as well as with the rainfall pattern of the area. A positive correlation has been found between the rate of cane elongation and rainfall. Rainfall distribution is more important than total rainfall, and light showers are more beneficial than heavy rains. The water needs of plants influence cane growth through moisture absorption by leaves, raising atmospheric humidity, and keeping the leaves surfaces clean for optimum transpiration and respiration. Moderate rainfall of 750–1,000 mm supplemented by sufficient and timely irrigation is considered best for healthy crop growth. As already discussed, the ripening period demands moderate moisture stress.

12.6.8 DEW

Dew deposits on leaf surfaces help in the foliar absorption of moisture. It also delays the rise in leaf temperature and, thus, reduces the evapotranspiration rate. Dew deposits are estimated to be 0.25 to 0.40 mm per night, with a total of 25 to 30 mm per annum (Biswas, 1988). Thus, dew deposits help to mitigate the severity of water stress to a certain degree in moisture-stress areas.

12.8 SUGARCANE AGRO-CLIMATIC ZONES WITH YIELD AND PRODUCTION POTENTIAL IN PAKISTAN

Among cane-producing countries, Pakistan, with cane sugar production of 70 million tons during 2016–2017, ranked fifth in the world (Afghan et al., 2016). Influenced by climatic conditions and cane/sugar marketing trends, cane production varies yearly. The sugar industry has developed a daily cane crushing capacity of 5,67,920 tons for 125 days of crushing duration. The Pakistan sugar industry achieved a sugar recovery level of 10.47% during 2018–2019. With further propagation of quality cane varieties, it is not far from achieving an average sugar recovery of 11% (Fahim, 2014). As for cane production, the average yield so far is just 64.30 tons per hectare; nevertheless, some districts of Punjab and Sindh have achieved an average cane yield of 82 tons per hectare (GOP, 2019). Improved production technology and inputs need to be mobilized to enhance the national yield to 70 tons per hectare and then plan for 75 tons. However, sugarcane cultivation is confined to part of coastal areas and plains, and plains of river Indus and adjoining rivers in Sindh, Punjab, and Khyber Pakhtunkhwa provinces (Shakoor et al., 2011). With better inputs and management practices, average cane yields in Pakistan can be increased to 75 tonnes per hectare. In the present scenario, if the interest of millers prevails with somewhat more investment in cane yield maximization campaigns, 70 tonnes per hectare should be the national goal for Pakistan, and this would need extraordinary efforts.

12.8.1 PUNJAB

Punjab has a very hot summer and a very cold winter. The hot season starts in April and continues until August. The highest temperature (48–50°C) occurs in extreme periods from May to July, with a mean maximum range of 37–42°C and a mean minimum range of 23–28°C for the corresponding period. During winter, a mean minimum range of 4–6°C is associated with occasional mild frost. The climate is more extreme in the southwest regions in Multan and Khanpur zones. Two-thirds of the rainfall is received in the monsoon season (mid to late June). The rainfall pattern of upper Punjab is higher and declines gradually towards central and southern Punjab (Annual Program of Research Work, 2018). D. G. Khan is also situated in the dry Northwest zone of Pakistan. Atmospheric humidity is low, almost 33–40% during summer and 55–65% during winter due to low precipitation.

The cane area is concentrated in Faisalabad, Bahawalpur, Dera Ghazi Khan, and Sargodha Divisions. Cane yields in Punjab have gradually increased to 64 t/h. However, within the province, the average yield of 75 and 71 tons per hectare is obtained in Bahawalpur and D.G. Khan Divisions, respectively. The average yield of Rajanpur District is reported to be 82 t/h, while the yield of R. Y. Khan is 77 t/h. The highest yield of 82.09 tonnes per hectare was obtained in Rajanpur District, followed by 77.75 tonnes per hectare recorded in R. Y. Khan District. The yield of districts located in Bahawalpur, D. G. Khan, and part of Faisalabad divisions is between 60 and 70 tonnes per hectare. In the rest of the districts, yields vary between 50 and 60 tonnes per hectare.

12.8.2 SINDH

Sindh is divided into two zones, and the lower Sindh constitutes Hyderabad and Mirpur Khas divisions bordered by the Arabian Sea coast. Compared with other regions, daily mean temperatures are relatively higher in winter and lower in summer, and the monthly mean shows lesser variability in maximum and minimum temperatures. During May and June, the mean maximum is higher (40–42°C), but the mean minimum (26–28°C) is favorable for the growth and maturity phases.

In Badin and Thatta, the district temperature is relatively milder, with cloudy weather from May to July. Coastal winds increase the humidity level as annual rainfall in Hyderabad (178 mm) and Badin (222 mm) districts is very low. In summer, the temperature increases during noon but the humidity level is favorable for crop growth. Low relative humidity (46–66%) and mild mean minimum temperatures (8–16°C) free from frost during ripening are favorable for good sugar recoveries (Qureshi et al., 2010).

The upper Sindh, including the Sukkur and Larkana divisions, is much away from the influence of coastal climate. The summer and winter months experience extreme weather with extremely low rainfall (88.2 mm). During summer, the relative temperature is higher than in lower Sindh, and crops are subject to moisture stress. Better soil, crop, and water management practices mitigate stress conditions in the Sukkur division. The yield of the Sukkur division is the highest, around 75 t/ha, in Sindh, followed by 67 t/ha in Benazir Abad. Larkana faces poor soils and more waterlogging; hence yields of Larkana are much lower. Climatically, the lower Sindh, including the Bhanbore division, is close to the coast and should be more conducive to cane production than the central and upper Sindh areas. Infected due to higher cane and sugar yields, the cane fields of Thatta and Badin districts, with part of Hyderabad and Mirpurkhas, have been declared the Sugarland of Pakistan (Qureshi et al., 2010).

The highest yield of 82.7 tonnes per hectare was recorded in the Ghotki district; Larkana and Nawabshah showed cane yields of around 69 tonnes per hectare. Eight districts showed yields between 60 and 68 tonnes per hectare, while yields of the rest of the districts ranged between 40 and 60 tonnes per hectare. It may be noted with great concern that the yields of districts Thatta and Badin in lower Sindh's coastal area have dropped to 59.3 and 46.5 tons, respectively (Qureshi et al., 2010). This is attributed to waterlogging with a large area under rice crops and fish farms.

12.8.3 KHYBER PAKHTUNKHWA

The province faces extreme climate during summer and winter, with mean maximum and minimum temperature ranges of 36–41C and 4–5°C, respectively. D. I. Khan is considered to have hot weather in summer with 271 mm of rainfall annually, and the Peshawar division is considered to have cold weather in winter with 404 mm of rainfall. Both zones have huge differences in rainfall patterns. This region has very low humidity and a higher RTD factor, as there is no favorable effect on cane growth. The mean minimum temperature is between 4°C and 5°C, and frosts are a common occurrence in the area. Sugarcane is concentrated in Dera Ismail Khan, Peshawar, and Mardan Divisions. The cane yield of the province is hardly around 51 t/ha. However, Dera Ismail Khan has attained a good position in cane yields (61.27 t/ha). D. I. Khan District is leading in cane yield with 62 t/ha. All of the remaining districts have yields lower than 51 t/ha.

12.9 CANE AREA AND PRODUCTION

Pakistan is a leading cane-growing country and stands in 5th position in cane area and production worldwide. It has a history of development, starting from a small cane area and cottage industry to a vast area under cane and a huge industrial setup, making the country self-sufficient in white sugar production. During the initial stage of the creation of Pakistan in 1947–1950, the cane was grown on an area of 2,01,900 hectares, and here too, most of the cane was planted in Punjab (1,49,300 ha), a little in KPK (44,800 ha) and quite negligible in Sindh (7,830 ha). One sugar factory in KPK and one very small factory in Punjab. Cane was mostly grown for the cottage industry to produce Gur, Shakkar, and Khand, and only 2% of total cane production was utilized for white sugar production. During the past seven decades, gradual expansion in the sugar industry was associated with spontaneous expansion in the area under cane, reaching 1.039 million hectares during 2019–2020 (Annual Report of Pakistan Sugar Mills Association, 2020). The decade-wise development in the area with cane yields from 1947–1950 to 2019–2020 period in the cane-growing provinces of Pakistan is reproduced in Table 12.6.

Table 12.7 indicates that in the initial periods of Pakistan's existence (1950–1960), cane yields were around 26–33 tons per hectare. For the following 40 years (1960–1990), cane yields oscillated between 33 and 42 tons per hectare. Nevertheless, after 1990, cane yield started to increase with a gradual rise to 69.55 tonnes per hectare during 2020–2021. The Sindh province even reached 65.56 tons per hectare during 2020–2021, while Punjab reached 73.36 tons per hectare during 2020–2021. However, yields in KPK are still static, around 52.38 tonnes/ha. Smallholders, those who own 2–5 acres, are characterized by the continuous growth of sugarcane with and without ratooning. Many farmers have moved to growing rice and other crops to meet the food security needs of their families. Medium farmers, those who own 5–10 acres of land, have a little more flexibility to grow rice in addition to sugarcane. These farmers can purchase sugarcane inputs from private sources. Their primary concern is the timely availability of extension services from the public sector (Babu, 2015). Increasing the share of sugarcane in these farmers' crop portfolios is the main goal of sugar mills.

TABLE 12.6 Growth in Area and Cane Yields in Cane-Growing Provinces of Pakistan from 1947–1950 to 2019–2020

Year	Punjab		Sindh		KPK		Pakistan	
	Area (ha)	Yield (t/ha)	Area (ha)	Yield (t/ha)	Area (ha)	Yield (t/ha)	Area (ha)	Yield (t/ha)
1947–1950	170.0	37.7	8.10	34.1	41.30	28.40	201.9	33.47
1959–1960	286.6	27.09	20.20	41.70	65.20	30.40	396.6	26.80
1969–1970	454.9	43.3	83.40	40.9	81.70	39.7	620.00	42.50
1979–1980	501.4	38.7	129.90	35.90	87.10	39.20	718.50	38.3
1989–1990	501.0	37.30	250.7	49.30	102.10	43.40	854.3	41.5
1999–1900	672.1	40.3	230.6	51.27	108.3	46.3	1009.8	45.9
2009–2010	607.4	51.6	233.9	57.7	100.8	44.7	942.9	52.36
2016–2017	777.82	63.78	320.51	63.05	118.57	47.50	12.16.9	62.00
2017–2018	859.13	64.10	333.30	61.85	148.53	51.2	1340.9	62.11
2018–2019	710.61	63.19	279.50	59.72	110.10	48.8	1101.1	60.97
2019–2020	643.430	67.37	286.090	62.93	109.36	52.62	1038.9	63.85
2020–2021	776.98	73.36	279.69	65.56	107.44	52.38	1164.12	69.55

Source: Annual Report of Pakistan Sugar Mills Association (2021).

Fast expansion in the sugar industry created competition in cane procurement that persuaded millers to initiate development activities in mill zones. Several sugar mills invested in seed propagation of new cane varieties, fertilizers, and plant protection measures with technical guidance to growers.

TABLE 12.7 Sugarcane Area and Yields of Various Districts in Punjab, Sindh, and Khyber Pakhtonkha, During 2017–2018

District	Cane Area (000 hectare)	Cane Yield (ton/ha)	District	Cane Area	Cane Yield
		Punjab			
Punjab	859.3	64.10	–	–	–
Sargodha	65.96	57.18	Sheikhupura	2.02	59.30
Khushab	8.90	56.91	Nankana sb.	13.76	56.26
Mianwal	3.24	50.45	Lahore	0.40	50.17
Bhakkar	29.95	65.21	Kasur	30.76	51.37
Faisalabad	112.50	57.00	Multan	6.47	52.85
T. T. Singh	38.44	62.35	Lodhran	5.26	69.27
Jhang	55.04	59.30	Khanewal	6.88	63.36
Chiniot	43.71	55.43	Vehari	17.81	57.46
Gujrat	2.43	48.61	Muzaffargarh	61.11	68.44
M. Bahauddin	23.88	52.48	Layyah	18.21	65.58
Sialkot	1.21	35.05	D. G. Khan	10.12	69.36
Narowal	1.62	31.91	Rajanpur	32.37	82.09
Gujranwala	202	39.06	Bahawalpur	30.35	66.41
Hafizabad	6.47	48.42	R. Y. Khan	193.03	77.75
Okara	14.57	49.90	Bahawalnagar	13.76	58.11
Sahiwal	4.86	54.05	–	–	–
Pakpattan	2.02	53.95	–	–	–
		Sindh			
Sindh	327.29	62.96	–	–	–
Thatta	36.37	59.30	Nawabshah	35.71	69.19
Badin	42.60	46.45	Nawshehra Feroz	22.26	63.26
Hyderabad	6.39	57.33	Sanghar	17.34	67.21
T. M. Khan	18.88	64.14	Sukkar	7.86	66.22
T. Allahyar	20.45	45.47	Khairpur	22.76	62.27
Matiari	15.84	66.22	Ghotki	53.40	82.71
Jamshoro	0.58	44.48	Larkana	0.71	69.68
Dadu	4.50	53.37	Kambar	0.19	54.36
Mirpur Khas	18.90	55.37	Shikarpur	0.47	45.46
Tharparkar	0.28	60.29	Jaccobabad	0.15	40.52
Umarkot	1.64	63.60	–	–	–
		Khyber Pakhtonkha			
Khyber Pakhtonkha	148.53	51.23	–	–	–
Bannu	0.43	39.74	Naushehra	2.04	50.77
Charsada	32.20	43.48	Peshawar	10.71	51.94
D. I. Khan	63.08	62.00	Swabi	2.21	38.75
Malakand	4.88	38.17	Tank	1.09	39.96
Mardan	30.39	42.48	Others	1.50	25.38

Source: Muhammad et al. (2019).

12.10 SUGARCANE INDUSTRY

Initially, the cane was grown to meet the 'Gur,' 'Shakkar,' and 'Khandsari' demands of the local market and household needs (Qureshi & Afghan, 2005). For white sugar, there was a small sugar factory of 300 TCD at Rahwali (Gujranwala) and one sugar mill of 3500 TCD at Mardan (KPK). With the creation of Pakistan in 1947, only two sugar mills (1 in Sindh and 1 in Punjab) were operational. After every 10 years, we find a group of new sugar mills scattered all around the cane-growing regions. There are 89 sugar mills in operation, including 45 in Punjab, 38 in Sindh, and 6 in KPK.

In the initial period of industrial development, sugar mills installed crushing capacities were hardly around 1,500–4,000 tons of cane daily. Still, daily cane crushing capacities range from 12,000 to more or less 24,000 tons. The magnitude of the sugar industry expansion may be realized from the fact that during 1947–1950 the sugar mills had a crushing capacity of just 2,800 tons of cane daily, while the daily cane crushing capacity of the present-day sugar industry is around 5,90,000 tons, for 120 days of crushing duration. Pakistan's sugar industry has passed through different phases of development. In its initial stage, almost all the cane was crushed in a local crusher for 'Gur,' 'Shakkar,' and 'Khandsari' production. Only 2% of total cane production was utilized for white sugar manufacture in two small factories. Gradually, the sugar industry expanded to the extent that 95% of Sindh, 67% of Punjab, and 67% of KPK cane production were crushed in Sugar Mills. In KPK, there is more trend of 'Gur' usage, and two sugar mills have been closed due to the non-availability of cane for sugar mills.

12.11 CANE CRUSHING DURATION

The crushing duration of cane pertains to the period for which the required quantum of cane is available in the area to meet the crushing requirements of a sugar factory. In the earlier period of industrial development, while the crushing capacities of sugar factories were around 1,500–4,000 TCD, sugar mills operated for 180–240 days a year, even at 3.5–5% recoveries in summer months. Mills used to start sometime in October and continued crushing till June. Now we find that sugar mills with usual crushing capacities of 8,000–12,000 TCD still have several factories with 18,000–24,000 TCD. These days cane is sufficient to complete crushing in 120–135 days. The objective is to complete the crushing of cane during the period of its peak maturity. Nevertheless, the Cane Act allows crushing from 1st November to 15th March for 160 days. While finalizing the crushing duration and especially the start of the crushing season, sugar recovery is of prime consideration.

12.12 SUGAR RECOVERIES

Sugar contents in cane play a leading role in regulating the economic viability of a sugar mill; in usual terms, it is indicated by 'sugar recovery.' It is the amount of sugar recovered per quintal/100 kg of cane crushed in a sugar factory. It is affected by a number of factors in the environment, field, and factory (Aslam et al., 2019). The matter has been discussed at length in the text; however, important ones are listed below.

12.12.1 FACTORS AFFECTING SUGAR RECOVERIES

12.12.1.1 CANE VARIETIES

Cane varieties have a dominant role in improving sugar mill recovery. High sugar-yielding varieties to be grown are preferred to early maturing varieties with high sugar content.

12.12.1.2 IMPROVING AGRONOMIC CONDITIONS

Improving agronomic conditions has a significant role in enhancing sugar mill recovery. The following are some agronomic conditions that improve the sugar recovery of cane varieties:

- Nutrient rich soil (good fertile and salt-free land).
- Sweet water zone.
- Planting for a longer growth period is preferred.
- Judicious use of balanced fertilizer, with 'K' application not to be ignored.
- Timely application of fertilizer; delayed application should be avoided, and 'N' contents in plant tissues should be exhausted by maturity time.
- Judicious use of irrigation water, with excess application to be avoided.
- Restrict irrigation before crop harvest.
- Measures to save the crop from lodging

12.12.1.3 NATURAL CALAMITIES

Measures to be taken to save the crop from drought and frost.

12.12.1.4 CROP HARVEST STRATEGIES

Crop harvest schedule according to crop maturity and maturity period of cane varieties. Cane harvest according to the crushing capacity of sugar mills and milling operations to be regular and consistent. Cane stalks are to be free of trash/extraneous matter, cane stalking must be avoided.

12.12.1.5 SUGAR FACTORY MILLING AND PROCESSING

All possible measures will be taken to reduce sugar recovery losses during various milling and processing operations. During 65 years of development, the Pakistan sugar industry expanded from 2 small factories in 1947 to 89 sugar mills in 2016–2017. The first 35–40 years phase is not very appreciable for sugar recoveries attained in the factories. Due to its relatively favorable climate and somewhat better sugar varieties, the sugar recoveries in Sindh province have been observed at a little over 9%. However, in Punjab, the major cane-producing area, the sugar recoveries have been oscillating between 8 and 8.5% up to the year 2000 (Afghan & Hussnain, 2000). The same is true for the KPK province (Table 12.8). The sugar industry realized the importance of sugarcane varieties by the end of the past century, and several industrial units deployed resources to develop high-sugar-content varieties in their respective zones.

The sugar industry now has a good number of quality varieties including BL4, Thatta10, CP77-400, Nia 2004, NIA 2011, HS12, SPF234, CPF 237, CPF246, CPF 249, CPF 250, CPF 252, CPF 253, SG 676, Th 2109, and Th 326. These varieties have a sugar recovery range of 9.5 to 12.5%. As a result, after the year 2010, the Pakistan sugar industry has shown a progressive rise in its sugar recovery, reaching an average level of 10.47% during 2018–2019 (Table 12.8).

12.13 CONCLUSION

Sugarcane is the 2nd most economically important crop in Pakistan after cotton. The results showed that the passage of time, area, production, yield, and sugar recovery of sugarcane increased greatly. However, in Pakistan, sugarcane is mostly used as multiple sources of food, fodder, and fertilizer, as well as in many beverages such as sweeteners. A huge number of people are engaged in the sugar industry of Pakistan, thus creating human resources and a workforce for the country. The sugar industry of Pakistan earns huge

TABLE 12.8 Sugar Recovery Trends in Sugar Mills of Sindh, Punjab, and KPK During the 1947–2020

Years	Sugar Recovery Percent Cane			
	Sindh	Punjab	KPK	Pakistan
1947–1950	–	7.23	7.08	7.16
1950–1955	–	7.73	7.43	7.58
1955–1960	–	7.54	8.08	7.81
1960–1965	7.97	7.91	8.03	7.97
1965–1970	8.85	8.41	8.62	8.63
1970–1975	8.88	8.67	7.68	8.41
1975–1980	9.24	8.42	8.01	8.56
1980–1985	9.10	8.28	8.39	8.59
1985–1990	9.09	8.23	8.85	8.72
1990–1995	9.35	8.23	8.59	8.72
1995–1900	9.55	7.93	8.02	8.50
2000–2005	9.35	8.61	8.10	8.69
2005–2010	9.49	8.78	8.12	8.80
2010–2015	10.10	9.75	9.20	9.68
2015–2016	9.94	10.65	9.44	10.16
2016–2017	9.77	10.16	9.43	9.87
2017–2018	9.79	10.55	9.52	10.02
2018–2019	10.31	10.79	10.41	10.47
2019–2020	9.71	10.21	10.30	9.89
2020–2021	9.43	10.02	9.86	9.61

Source: Annual Report of Pakistan Sugar Mills Association (2021).

revenue for the government. As for climatic conditions, Pakistan's climate suits its cultivation, especially in Punjab, Sindh, and KPK provinces. In Punjab, Rajanpur and Rahim Yar Khan are the most suitable areas for sugarcane cultivation, while Sindh's Ghotki and Thatta districts provide the best environment for sugarcane cultivation. In KPK, D. I. Khan and Mardan districts have the highest production. The sugar industry is gradually improving its status concerning its sugar recovery level. While checking the sugarcane varieties being grown and crushed in the factory, there appears to be a direct relationship between the cane varieties and the sugar recovery of the mills. The sugar mills having a large area under high sugar varieties would depict much better recoveries than the mills crushing a large percentage of low-quality cane. It is a matter of programming to harvest and crush cane according to its maturity, and varietal combination matters much. The sugar industry of Pakistan needs a series of new varieties in the pipeline; for that matter, there should be a close liaison between the Research Institutes and the sugar industry for the betterment of the sugarcane crop.

KEYWORDS

- **cane recovery**
- **cane varieties**
- **economic significance**
- **Pakistan**
- **sugarcane**
- **sugarcane agro-climatic zones**
- **yield potential**

REFERENCES

Afghan, S., & Hussnain, Z. (2000). Clonal evolution program at Shakarganj Sugar Research Institute, Jhang. *Pakistan Sugar Journal, 25*(6), 76–97.

Afghan, S., & Jamil, M. (2013). Climate Change Impact on Sugar Industry of Pakistan – An Overview. *Annual Convention Pakistan Society of Sugar Technologists, 24*, 7–14.

Afghan, S., Hussnain, Z., Ali, K., & Shahzad, A. (2013). Performance of promising sugarcane clones developed from imported fuzz (true seed) in Pakistan. *Proceedings of the International Society of Sugar Cane Technology, 28*, 10.

Afghan, S., Shahzad, A., Comstock, J. C., Zhao, D., & Ali, A. (2016). Registration of 'CPSG-3481' sugarcane. *Journal of Plant Registrations, 10*, 124–129.

Agarwal, R. A. (1959). Insect resistance in Coimbatore canes: A review. *Indian Sugar, 8*, 1–10.

Ahmed, M. N., & Schmitz, M. (2011). Economic assessment of the impact of climate change on the agriculture of Pakistan. *Business and Economics Horizons, 4*, 1–12.

Alexander, A. G. (1973). *Sugarcane Physiology*. Elsevier Science Publishers. pp. 399–485.

Ali, S., Liu, Y., Ishaq, M., Shah, T., Ilyas, A., & Din, I. U. (2017). Climate change and its impact on the yield of major food crops: Evidence from Pakistan. *Foods, 6*(6), 39.

Alvarez, J., Deren, C. W., & Glaz, B. (2003). Sugarcane selection for sucrose and tonnage using economic criteria. *Proceedings of the Sugar Cane International Conference*, Nov.-Dec. 6–10, 2003.

Annual Program of Research Work. (2017). Sugarcane Research Institute, Ayub Agricultural Research Institute, Faisalabad.

Annual Program of Research Work. (2018). Sugarcane Research Institute, Ayub Agricultural Research Institute, Faisalabad.

Annual Report of Pakistan Agriculture Research Council 2017-18, Federal Government, Islamabad, Pakistan.

Annual Report of Pakistan Meteorological Department (2019). Islamabad, Pakistan.

Annual Report of Pakistan Sugar Mills Association (2018–2021). Islamabad.

Annual Report of Shakarganj Sugar Mills (2018). Jhang, Pakistan.

Aslam, C., Waqas, M., Ahmad, R., Khaliq, A., & Ahmad, R. (2019). Improving the productivity and sugar recovery of cane by potash nutrition under different planting methods. *Pakistan Journal of Agricultural Sciences, 55*(3), 557–566.

Babu, S. C. (2015). Private Sector Extension with Input Supply and Output Aggregation: Case of Sugarcane Production System. In Yuan Zhou & Suresh Chandra Babu (Eds.), *Public Policy and Global Development, Knowledge Driven Development*. Academic Press, pp. 73–90.

Bacchi, O. S. (1977). Minimum threshold temperature for sugarcane growth. *Proceedings of the International Society of Sugarcane Technology*, pp. 16.

Bhutta, E., Ilyas, M., & Usman, M. (2019). The need for transforming agriculture produce markets: evidence from Punjab, Pakistan. *Pakistan Journal of Agricultural Sciences, 56*(3), 767–773.

Biswass, B. C. (1988). Agroclimatology of the sugarcane crop (Technical Note N0- 193). Secretariat of the World Meteorological Organization, Geneva, Switzerland.

Blume, H. (1986). Environment and cane sugar yield. *Proceedings of the International Society of Sugarcane Technology, 19*, 277–288.

Burr, G. O. (1957). The sugarcane Plant. *The Annual Review of Plant Physiology, 8*, 257–308.

Cardoza, N. P., & Sentelhas, P. E. (1973). Climatic effect on sugarcane ripening under the influence of cultivars and crop age. *Scientia Agricola, 70*(6), 449–456.

Chinnadurai, C. (2017). Potential Health Benefits of Sugarcane. In *Sugarcane Biotechnology: Challenges and Prospects*, pp. 1–12.

Chu, C. C., & Kong, L. (1971). Photo-respiration of Sugarcane. *Taiwan Sugar Experimental Station Annual Report*, 1–14.

Corcoran, S. G., Hashemi, M., Sadeghpour, A., Jahanzad, E., Keshavarz, R. A., Liu, X., & Herbert, S. J. (2020). Understanding intercropping to improve agricultural resiliency and environmental sustainability. In Donald L. Sparks (Ed.), *Advances in Agronomy*, 162, 199–256. Academic Press.

Das, U. K. (1941). Cane breeding in Coimbatore. *Hawaiian Plant Record*, 45, 97–120.

Deng, Z. N., Huang, H. R., Huang, C. M., Verma, K. K., Luo, H. B., Jiang, S. L., Li, Y. J., Cao, H. Q., Wu, K. C., & Xu, L. (2022). Short-term limited water irrigation influences the photosynthetic pigments, enzymatic and non-enzymatic activities in *Saccharum spontaneum* L. at vegetative stage. *Sugar Tech. 25, 110–118*. https://doi.org/10.1007/s12355-022-01175-7.

Dotaniya, M. L. (2014). *Role of bagasse and press mud in phosphorus dynamics* (1st ed.). Lap Lambert Academic Publisher, Germany.

Drewnowski, A., & France, B. (2007). Liquid calories, sugar, and body weight. *American Journal of Clinical Nutrition*, 85, 651–661.

FAO. (2020). *World Food and Agriculture – Statistical Yearbook 2017*. Rome. https://doi.org/10.4060/cb1329en.

FAO. (2020). *World Food and Agriculture – Statistical Yearbook 2020*. Rome. https://doi.org/10.4060/cb1329en.

FAOSTAT 2018. (Food and Agriculture Organization Statistics). Global Information and Early Warning System.

Food and Agriculture Organization of the United Nations (FAO). (2013). *The State of Food Insecurity in the World: The Multiple Dimensions of Food Security*. FAO: Rome, Italy, pp. 76.

Food and Agriculture Organization of the United Nations (FAO). (2014). *The State of Food Insecurity in the World: Strengthening the Enabling Environment for Food Security and Nutrition*. FAO: Rome, Italy.

Fuhr. (2016). High or Low? The Sugar in Your Favorite Fruits. https://www.popsugar.com/fitness/Sugar-Content-Fruit-20134844.

GOP. (2017). *Pakistan Economic Survey (2016–17)*. Ministry of Food Agriculture and Livestock, Federal Bureau of Statistics, Government of Pakistan, Islamabad, Pakistan, 27–34.

GOP. (2019). *Pakistan Economic Survey (2019–20)*. Ministry of Food Agriculture and Livestock, Federal Bureau of Statistics, Government of Pakistan, Islamabad, Pakistan, 27–34.

GOP. (2020). *Pakistan Economic Survey (2020–21)*. Ministry of Food Agriculture and Livestock, Federal Bureau of Statistics, Government of Pakistan, Islamabad, Pakistan, 27–34.

Huang, H. R., Gao, Y. J., Malviya, M. K., Verma, K. K., Solanki, M. K., Huang, Y. X., Li, X., Deng, Y., Yan, J., Tang, S. Y., Wang, L. W., & Xu, L. (2021). Genetic diversity analysis of sugarcane (*Saccharum spp.* Hybrids) among high-sucrose clones of GT series and commonly used parents by using microsatellite markers in Guangxi, China. *Sugar Tech. 24, 397*–407. https://doi.org/10.1007/s12355-021-01033-y.

Humbert, R. P. (1968). *The Growing of Sugarcane*. Elsevier Publishing Co. Amsterdam, London, New York.

Hussain, S., Khaliq, A., Mehmood, U., Qadir, T., Saqib, M., Iqbal, M. A., & Hussain, S. (2018). Sugarcane production under changing climate: effects of environmental vulnerabilities on sugarcane diseases, insects and weeds. *Climate Change and Agriculture*, doi: http://dx.doi.org/10.5772/intechopen.81131, pp. 1–17.

Iqbal, M. A., & Iqbal, A. (2014). Sugarcane production, economics and industry in Pakistan. *American Eurasian Journal of Agricultural & Environmental Sciences*, 14(12), 1470–1477.

Kakade, J. R. (1985). *Agricultural Climatology*. Metropolitan Book Co. New Delhi.

Khan, A. Q., Muhammad, I., & Fahad, I. (2016). An assessment of main problems faced by farming community in sugarcane production of district Peshawar. *International Journal of Agricultural Extension and Rural Development*, 3(1), 149–155.

Kurukulasuriya, P., Mendelsohn, R., Hassan, R., Benin, J., Deressa, T., Diop, M., Eid, H. M., Fosu, K. Y., Gbetibouo, G., & Jain, S. (2006). Will African agriculture survive climate change? *World Bank Economic Review*, 20, 367–388.

Legendre, B. L., & Martin, F. A. (1977). Ripening studies with Glyphosine in Louisiana Sugarcane. *Proceedings of the American Society of Sugar Cane Technologists*, 6, 62–64.

Luo, J., Pan, Y. B., Xu, L., Grisham, M. P., Zhang, H., Zhang, H., & Que, Y. (2015). Rational regional distribution of sugarcane cultivars in China. *Scientific Reports*, 5, 15721.

Malik, K. B. (2018). *Agricultural and Industrial Aspects of Sugarcane Production*. Al Madina Printers, Lahore, Pakistan.

Marin, F. R., Edreira, J. I. R., Andrade, J., & Grassini, P. (2019). On-farm sugarcane yield and yield components as influenced by number of harvests. *Field Crops Research*, 240, 134–142.

Mendelsohn, R. (2014). The impact of climate change on agriculture in Asia. *Journal of Integrative Agriculture*, 13, 660–665.

Ministry of National Food Security and Research. (2018). *Agricultural Statistics of Pakistan*. Economic Wing, MNFS&R, Islamabad.

Muhammad, I. T., Mohammad, I. J., Iftekhar, N., Naeem, A., & Abid, M. (2019). A face for enhancing cane & sugar yield in Pakistan. *Global Scientific Journal*, 7(3), 670–686.

Naoko, U., Haruto, S., Naohiro, & Ryu, O. (2009). Effects of the Temperature Lowered in the Daytime and Night-time on Sugar Accumulation in Sugarcane. *Plant Production Science*, 12(4), 420–427.

Nasir, N. M., Afghan, S., & Qureshi, S. A. (1994). Utilization of Bio-compost Produced from Filter Cake and Stillage at Shakarganj Sugar Research Institute, Jhang. *Pak. Sugar J.*, 8, 21–26.

Oliveira, M. A., Verma, K. K., Bhatt, R., & Oliveira, T. B. A. (2022). Impact of green and organic fertilizers on soil fertility and sugarcane productivity. In Verma KK., et al. (eds.), *Agro-industrial Perspectives on Sugarcane Production under Environmental Stress*. https://doi.org/10.1007/978-981-19-3955-6_11.

Pakistan Agricultural Research Council. *Sugarcane Crop in Pakistan*. PARC, Islamabad. Retrieved 31-10-2018, from http://edu.par.com.pk/wiki/sugarcane/.

Pakistan Sugar Mills Association (PSMA). (2019). *Annual report, 2018-19*. Pakistan Sugar Mills Association, Islamabad. http://www.psmacentre.com/documents/PSMA%20Annual%20Report%202019.pdf.

Pakistan Sugar Mills Association (PSMA). (2020). *Annual report, 2019-20*. Pakistan Sugar Mills Association, Islamabad. http://www.psmacentre.com/documents/PSMA%20Annual%20Report%202020.pdf.

Parthasarathy, S. V. (1972). *Sugarcane in India*. K. C. P. Publishers, Madras, India.

Prasannamedha, Kumar, G., Mehala, P. S., Sharumitha, R., & Surendhar, T. J. D. (2020). Enhanced adsorptive removal of sulfamethoxazole from water using biochar derived from hydrothermal carbonization of sugarcane bagasse. *Journal of Hazardous Materials*, *407*, 124825.

Qureshi, A., Sarwar, P. G., McCornick, A., Sarwar, & Sharma, B. R. (2010). Challenges and prospects of sustainable groundwater management in the Indus Basin, Pakistan. *Water Resources Management*, 24(8), 1551–1569.

Qureshi, M. A., & Afghan, S. (2005). Sugarcane cultivation in Pakistan. *Pakistan Sugar Book*. Pak. Soc. Sugar Technol., Karachi, Pakistan.

Qureshi, S. (2004). Significance of sugar industry in National Economy. *Economic & Social Review*, 2, 17–21.

Raza, H. A., Amir, R. M., Saghir, A., & Tahir, M. (2020). Sugarcane production and protection constraints faced by the growers of Punjab, Pakistan with special focus on the role of agricultural extension worker in related mitigation. *Pak. J. Agricul. Sci.*, 57(6), 1681–1688.

Raza, H. A., Amir, R. M., Wudil, A. H., Usman, S., Shoaib, M., Ejaz, R., & Khan, H. A. (2018). Economic analysis of Jaggery (Gur) production in Tehsil Shakargar. *J. Global Innovation Agric. Social Sci.*, 6(2), 69–73.

Sahari, J., Sapuan, S. M., Zainudin, E. S., & Maleque, M. A. (2014). Physico-chemical and thermal properties of starch derived from sugar palm tree (*Arenga pinnata*).

Shakoor, U., Saboor, A., Ali, I., & Mohsin, A. Q. (2011). Impact of climate change on agriculture: empirical evidence from arid region. *Pak. J. Agri. Sci.*, 48(4), 327–333.

Sugarcane Handbook. (2017). Shakarganj Sugar Mills Limited, Jhang, Pakistan.

Usman, M. (2016). Contribution of Agriculture Sector in the GDP Growth Rate of Pakistan. *J. Glob. Econ.*, 4, 184–187.

Verma, K. K., Anas, M., Chen, Z., Rajput, V. D., Malviya, M. K., Verma, C. L., Singh, R. K., Singh, P., Song, X. P., & Li, Y. R. (2020d). Silicon supply improves leaf gas exchange, antioxidant defense system and growth in sugarcane responsive to water limitation. *Plants*, 9, 1032, DOI:10.3390/plants9081032.

Verma, K. K., Li, D. M., Singh, M., Rajput, V. D., Malviya, M. K., Minkina, T., Singh, R. K., Singh, P., Song, X. P., & Li, Y. R. (2020b). Interactive role of silicon and plant-rhizobacteria mitigating abiotic stresses: A new approach for sustainable agriculture and climate change. *Plants*, 9, 1055, DOI: 10.3390/plants9091055.

Verma, K. K., Liu, X. H., Wu, K. C., Singh, R. K., Song, Q. Q., Malviya, M. K., Song, X. P., Singh, P., Verma, C. L., & Li, Y. R. (2020f). The impact of silicon on photosynthetic and biochemical responses of sugarcane under different soil moisture levels. *Silicon*, 12, 1355–1367.

Verma, K. K., Singh, P., Song, X. P., Malviya, M. K., Singh, R. K., Chen, G. L., Solomon, S., & Li, Y. R. (2020e). Mitigating climate change for sugarcane improvement: role of silicon in alleviating abiotic stresses. *Sugar Tech.*, 22(5), 741–749, DOI: 10.1007/s12355-020-00831-0.

Verma, K. K., Song, X. P., Budeguer, F., Nikpay, A., Enrique, R., Singh, M., Zhang, B., Wu, J. M., & Li, Y. R. (2022c). Genetic engineering: an efficient approach to mitigating biotic and abiotic stresses in sugarcane cultivation. *Plant Signaling & Behavior*, 17(1), e2108253.

Verma, K. K., Song, X. P., Li, D. M., Singh, M., Wu, J. M., Singh, R. K., Sharma, A., Zhang, B. Q., & Li, Y. R. (2022d). Silicon and soil microorganisms improve rhizospheric soil health with bacterial community, plant growth, performance, and yield. *Plant Signaling & Behavior*, 17(1), e2104004.

Verma, K. K., Song, X. P., Lin, B., Guo, D. J., Singh, M., Rajput, V. D., Singh, R. K., Singh, P., Sharma, A., Malviya, M. K., Chen, G. L., & Li, Y. R. (2021d). Silicon-induced drought tolerance in crop plants: physiological adaptation strategies. *Silicon*, *14*, 2473–2487. DOI: 10.1007/s12633-021-01071-x.

Verma, K. K., Song, X. P., Rajput, V. D., Boldyreva, V., Zhang, B. Q., Minkina, T., & Li, Y. R. (2022b). Morpho-physiological, biochemical, and ultrastructural modifications on sugarcane to prolonged water deficit. In Verma, K. K., et al. (Eds.), *Agro-industrial perspectives on sugarcane production under environmental stress*. https://doi.org/10.1007/978-981-19-3955-6_11.

Verma, K. K., Song, X. P., Rajput, V. D., Solomon, S., Li, Y. R., & Rao, G. P. (Eds.) (2023). *Agro-industrial Perspectives on Sugarcane Production under Environmental Stress*. Springer, Singapore. https://doi.org/10.1007/978-981-19-3955-6_1.

Verma, K. K., Song, X. P., Tian, D. D., Guo, D. J., Chen, Z. L., Zhong, C. S., Nikpay, A., Singh, M., Rajput, V. D., Singh, R. K., Minkina, T., & Li, Y. R. (2021a). Influence of silicon on biocontrol strategies to manage biotic stress for crop protection, performance, and improvement. *Plants*, 10, 2163. https://doi.org/10.3390/plants10102163.

Verma, K. K., Song, X. P., Tian, D. D., Singh, M., Verma, C. L., Rajput, V. D., Singh, R. K., Sharma, A., Singh, P., Malviya, M. K., & Li, Y. R. (2021b). Investigation of defensive role of silicon during drought stress induced by irrigation capacity in sugarcane: physiological and biochemical characteristics. *ACS Omega*, 6, 19811–19821.

Verma, K. K., Song, X. P., Verma, C. L., Chen, Z. L., Rajput, V. D., Wu, K. C., Liao, F., Chen, G. L., & Li, Y. R. (2021c). Functional relationship between photosynthetic leaf gas exchange in response to silicon application and water stress mitigation in sugarcane. *Biol. Res.*, 54, 15. https://doi.org/10.1186/s40659-021-00338-2.

Verma, K. K., Song, X. P., Verma, C. L., Malviya, M. K., Guo, D. J., Rajput, V. D., Sharma, A., Wei, K. J., Chen, G. L., Solomon, S., & Li, Y. R. (2021f). Prediction of photosynthetic leaf gas exchange of sugarcane (*Saccharum spp.*) leaves in response to leaf positions to foliar spray of potassium salt of active phosphorus under limited water irrigation. *ACS Omega*, 6, 2396–2409.

Verma, K. K., Song, X. P., Yadav, G., Degu, H. D., Parvaiz, A., Singh, M., Huang, H. R., Mustafa, G., Xu, L., & Li, Y. R. (2022a). Impact of agroclimatic variables on proteogenomics in sugarcane (*Saccharum spp.*) plant productivity. *ACS Omega*, 7, 22997–23008.

Verma, K. K., Song, X. P., Zeng, Y., Guo, D. J., Sing, M., Rajput, V. D., Malviya, M. K., Wei, K. J., Sharma, A., Li, D. P., Chen, G. L., & Li, Y. R. (2021e). Foliar application of silicon boosts growth, photosynthetic leaf gas exchange, antioxidative response and resistance to limited water irrigation in sugarcane (*Saccharum officinarum* L.). *Plant Physiol. Biochem.*, 166, 582–592.

Verma, K. K., Song, X. P., Zeng, Y., Li, D. M., Guo, D. J., Rajput, V. D., Chen, G. L., Barakhov, A., Minkina, T. M., & Li, Y. R. (2020c). Characteristics and correlation of leaf stomata and its relationship with photosynthesis on *Saccharum spp.* under different irrigation and silicon application. *ACS Omega*, 5, 24145–24153.

Verma, K. K., Wu, K. C., Verma, C. L., Li, D. M., Malviya, M. K., Singh, R. K., Singh, P., Chen, G. L., Song, X. P., & Li, Y. R. (2020a). Developing mathematical model for diurnal dynamics on photosynthetic responses in sugarcane responsive to different irrigation and silicon application. *PeerJ*, 8, e10154. DOI: 10.7717/peerj.10154.

Xu, L., Deng, Z. N., Wu, K. C., Malviya, M. K., Solanki, M. K., Verma, K. K., Pang, T., Li, Y. J., Liu, X. Y., Kashyap, B. K., Dessoky, E. S., Wang, W. Z., & Huang, H. R. (2022). Transcriptome analysis reveals a gene expression pattern that contributes to sugarcane bud propagation induced by Indole-3-Butyric Acid. *Front. Plant Sci.*, 13, 852886. doi: 10.3389/fpls.2022.852886.

Yasar, A., Ali, A., Tabinda, A. B., & Tahir, A. (2015). Waste to energy analysis of Shakarganj sugar mills: biogas production from the spent wash for electricity generation. *Renewable and Sustainable Energy Reviews*, 43, 126–132.Yasar, A., Ali, A., Tabinda, A. B., & Tahir, A. (2015). Waste to energy analysis of Shakarganj sugar mills: biogas production from the spent wash for electricity generation. *Renewable and Sustainable Energy Reviews*, 43, 126–132.

Index

D

E

F

G

M

N

Q

T

U

V